Library of Congress Cataloging-in-Publication Data

In the line of duty : Army art, 1965–2014 / general editor, Sarah G. Forgey ; art photography, Pablo Jimenez-Reyes.
pages cm.
1. United States. Army—Pictorial works. 2. Soldiers—United States—Pictorial works. 3. United States—Armed Forces—Foreign countries—Pictorial works. 4. United States—History, Military—Pictorial works. 5. United States. Army—Art collections—Catalogs. 6. Soldiers in art—Catalogs. 7. Armies in art—Catalogs. 8. United States—History, Military—Pictorial works. I. Forgey, Sarah G., editor. II. Jimenez-Reyes, Pablo, photographer.
UA25.I64 2014
355.009730222—dc23

2014028850

First Printing—CMH Pub 70–123

In the Line of Duty

Army Art, 1965–2014

General Editor
Sarah G. Forgey

Layout and Design
Gene Snyder

Art Photography
Pablo Jimenez-Reyes

Center of Military History
United States Army
Washington, D.C., 2015

Table of Contents

Foreword

Those unfamiliar with the Army Art Program are often surprised that the U.S. Army, an organization focused on mastering the art and science of war, killing, and selective destruction, has such an initiative at all. In fact, soldier art and soldier art programs in the Army have a long tradition dating back to World War I. Artists, even more than photographers, are often able to capture the very essence of war, from the harsh realities of combat to the humdrum routine of soldiers' lives. The best combat artists can bring to life the immediacy of soldiers' experiences for even the casual viewer. It is that sense of seizing the moment and preserving a glimpse of the reality of military life that makes such artwork valuable to the Army and the nation.

The U.S. Army Center of Military History is proud to present *In the Line of Duty: Army Art, 1965–2014*, an updated volume on the Army Art Collection. It is long overdue. The Center published two books, *Portrait of an Army* and *Soldiers Serving the Nation*, in the 1990s, when both the Army and the Army's art program looked very different. Two electronic publications, *Army Artists Look at the War on Terrorism, 2001 to the Present,* and *Art of the American Soldier: Documenting Military History Through Artists' Eyes and in Their Own Words*, featured the work of more recent Army artists.

In the Line of Duty contains art ranging from the Vietnam War to Desert Storm and from Bosnia-Herzegovina to Iraq and Afghanistan. While there have been dramatic changes in uniforms, equipment, and weapons over those years, the essential character of U.S. Army soldiers remains remarkably the same, and it is that character that these artists capture. Soldier-artists are dedicated to documenting the history of the Army in a unique and impressionistic way. Their focus, as the reader will find, is on the heart of the Army, the men and women who serve their country around the world, rather than on technology or equipment. The individual soldier endures as the strength of the Army and that shines through in most of these works. I encourage the reader to peruse these pages to discover more about the collective experience of the American soldier. Successful art can withstand the test of time, and it remains relevant by depicting memories and customs shared by many generations of those who served their country. I believe that the art in this book will continue to appeal to soldiers and veterans for decades to come.

Washington, D.C.
1 December 2014

RICHARD W. STEWART, PH.D.
Chief of Military History

Acknowledgments

This book draws its origins from the 2011 debut of the electronic publication *Art of the American Soldier: Documenting Military History Through Artists' Eyes and in Their Own Words*, by Renée Klish. Klish, a former Army art curator, produced a book of selected art that included the artists' words and other documentation to "open a window into the creative process." During her tenure, she contacted many former artists or their families to add information to the Army Art Collection files.

In 2012, Beth MacKenzie, chief of the Center's Historical Products Branch, introduced a proposal to modify and expand *Art of the American Soldier* to highlight works produced from Vietnam to the present conflicts in Iraq and Afghanistan. Her team, consisting of Diane Sedore Arms, Cheryl L. Bratten, and Gene Snyder, began to compile images and quotes to be published in a printed edition. This new art book would supplement the two books published in the 1990s by Gordon R. Sullivan and Marylou Gjernes—*Portrait of an Army* and *Soldiers Serving the Nation.*

In May 2012, as the current Army art curator, I chose additional images to represent the artists of each period and wrote the text to supplement each section. My 2012 summer interns, Kelly Russo, Madeline Hodges, Victoria Kolton, and Jackie Streker, assisted me by researching the artists' biographical information and gathering more quotes. With the support of the collection photographer, Pablo Jimenez-Reyes, and curatorial assistant Summer Brown, all of the artwork selected for the book was digitally photographed at a high resolution.

I would also like to thank M. Sgt. Martin J. Cervantez and Sfc. Amy Brown for many conversations about art in general, the Army's collection, their own work (which appears in the last section of this book), and what it means to be an artist in the Army. Similarly, in the course of my research I have spoken to many former Army artists, all of whom have enriched my perspective on their work by generously sharing their recollections. I wish to thank all of them for their service and for sharing their talents with generations to come.

Finally, *In the Line of Duty: Army Art, 1965–2014,* is a product of successful collaboration between the Center's Histories and Museum Divisions and the talents of the many dedicated and capable individuals mentioned above.

Introduction

War is one of the oldest subjects in the history of art, yet almost every time the words *Army Art Program* are spoken, the speaker is met with astonishment and interest. Most people are not surprised that the Army has a historical art collection, but their assumption is usually that most of the art is large-scale battle scenes painted by great masters. Many are aware of the military art programs that existed during the world wars, and some Vietnam veterans recall meeting artists or viewing their works, but the vast majority is unaware that the Army still has an active artist program. Furthermore, virtually everyone—from soldiers to scholars to the average citizen—is amazed to hear that today's Army artists are themselves soldiers.

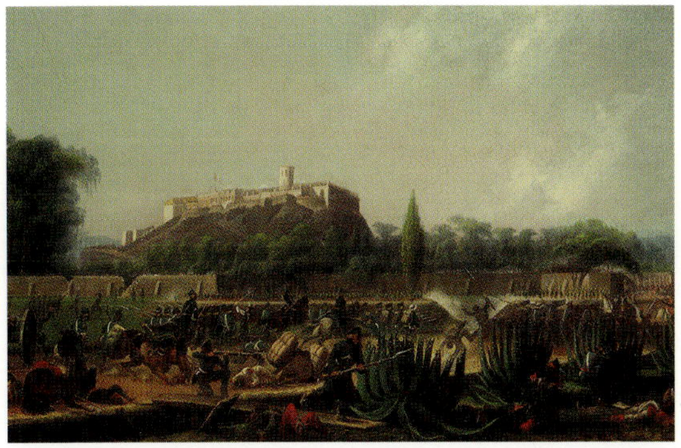

SOUTH SIDE OF THE CASTLE OF CHAPULTEPEC
Mexico
James Walker, 1848
Oil on board, 11" x 16"

To understand the Army's modern-day artist program, a brief historical context is necessary. The Army's relationship with the arts began in the early days of the Army, when an artist's talents were the only tools available to visually record battles, troop positions, fortifications, engineering projects, and other activities. Among the earliest eyewitness works in the Army Art Collection are twelve small panel paintings showing Mexican War scenes by James Walker, who observed the events and volunteered his artistic services to the Army.

While the Army unofficially used the skills of artistic soldiers and volunteer civilians throughout the nineteenth century, most Army-related art was produced by civilian artists working on private commissions, such as Winslow Homer's famous scenes of the Civil War that appeared in *Harper's Weekly*. Beginning in the Civil War, the emerging medium of photography excited the American people, and a pioneering era of war documentation began. As in the civilian art world, however, the traditional arts of painting and drawing were not abandoned entirely.

The Army Art Program officially began in World War I, when the War Department selected eight prominent illustrators to chronicle the activities of the American Expeditionary Forces.

These artists—civilians before the war—were commissioned as captains in the Army and given the broad mission to record any Army-related enterprise that they observed firsthand in the medium and style of their choice. While their compositions were subject to the same wartime censorship restrictions as photography and writing, the artists experienced a high degree of creative freedom. The artists produced several hundred works of art before the program concluded at the end of the war. The art was placed in the custody of the Smithsonian Institution, which was at that time the repository of the Army's historical property. There was no Army art program during the years between the world wars.

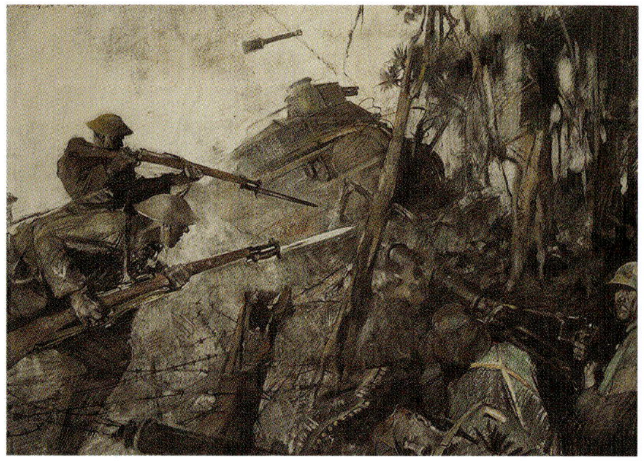

STORMING MACHINE GUN
France
George Harding, 1918
Charcoal on paper, 18¼" x 27¼"

As the United States began preparations to enter World War II, both the civilian art world and the military recognized the need for artistic documentation of the war. Frances Brennan, newly appointed as the director of the Office of War Information, underlined this urgency in a statement that "the American People need their artists now—to charge them with the grave responsibility of spelling out their anger, their grief, their greatness, and their justice."[1] Civilian magazines such as *Life* began hiring artists as war correspondents, while the War Department formed the War Art Advisory Committee, consisting of prominent civilian art experts, to select the artists. By spring of 1943, the committee had identified forty-two of the most talented American artists—both soldier and civilian—to participate. As in World War I, the artists were given a wide degree of creative freedom. Committee chairman and War Art Unit member George Biddle enthusiastically wrote to the artists that

> any subject is in order, if as artists you feel that it is part of War; battle scenes and the front line; battle landscapes; the dying and the dead; prisoners of war; field hospitals and base hospitals; wrecked habitations and bombing scenes; character sketches of our own troops, of prisoners, of the natives of the countries you visit; never official portraits; the tactical implements of war; embarkation and debarkation scenes; the nobility, courage, cowardice, cruelty, boredom of war; all this should form part of a well-rounded picture. Try to omit nothing; duplicate to your heart's content. Express, if you can, realistically or symbolically, the essence and spirit of war. You may be guided by Blake's mysticism, by Goya's cynicism and savagery, by Delacroix's romanticism, by Daumier's humanity and tenderness; or better still follow your own inevitable star.[2]

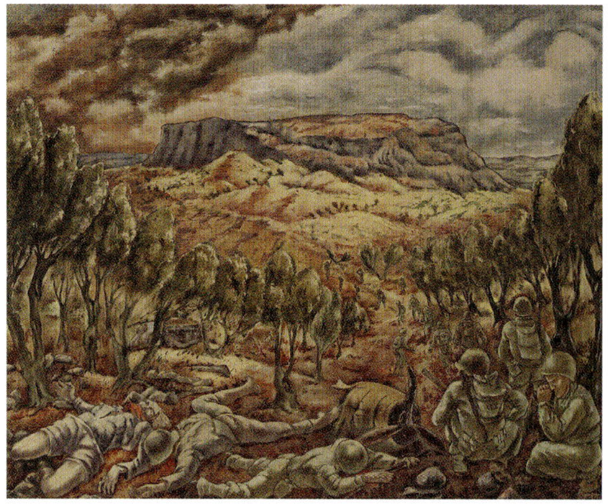

HILL 609—TUNISIA
Tunisia
George Biddle, 1943
Oil on canvas, 40" x 50"

By May 1943, the first artists of the War Art Unit arrived in their assigned theaters all over the world to begin work.

Just two months into the program, however, Congress cut funding for the War Art Unit. Faced with no other option, the Army reassigned the soldier-artists and released the civilians.

---

[1] *New York Times*, "Artists' War Work Centralized in OWI," 9 Aug 1942, p. 42.
[2] Memo, George Biddle for War Artists to Be Sent Overseas, 1 Mar 1943, Army Art Collection files, CMH.

Recognizing an opportunity, *Life* approached the War Department with an offer to employ the civilian artists. Artistic coverage of the war continued through *Life*'s art program, art programs sponsored by other private corporations such as Abbott Laboratories, sketches and cartoons produced by *Yank* magazine artists, and the perseverance of the soldier-artists, who carried on with their official duties and were eventually authorized to create art in their spare time as well. By the end of the war, the Army had acquired over two thousand works of art, which became the foundation of the Army Art Collection. *Life* donated its World War II art to the Department of Defense in 1960. This important gift reunited the soldier and civilian art into one collection that displayed various artists' perspectives on the war, as the War Art Advisory Committee had envisioned in 1943.

With the end of World War II, the Army's art program came to an end. During the Korean War, the Army briefly considered the possibility of deploying artists, but the initiative never gained momentum. Several civilian magazines continued the World War II tradition by sending artist-correspondents to Korea. Robert "Weldy" Baer, an Army civilian temporarily assigned to the Eighth Army Historical Section, sketched and painted his observations of military operations in Korea.

With the lack of artistic documentation of the Korean War in mind, the Army visualized an extensive undertaking during the Vietnam War that used the talents of both soldier and civilian artists. Administered by the Office of the Chief of Military History and the Army Arts and Crafts Program, this venture sought to achieve the same mission as the art programs during the world wars, allowing the participants a great degree of artistic freedom. The decision to involve civilian as well as soldier artists was meant to ensure that the art reflected the perspectives of both young soldiers and mature artists and that the quality of art coverage of the Vietnam War was consistent with that of World War II.

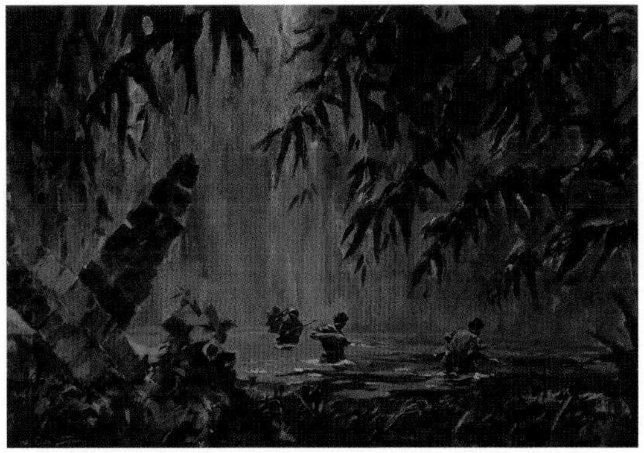

PATROL IN THE JUNGLE
Vietnam
Roger Blum, 1966
Oil on canvas, 24" x 36"

The Army Arts and Crafts Program recruited the soldier-artists and organized them into Combat Artist Teams. Team members were issued cameras, though many of the artists preferred to sketch on the spot. The supervisor of one of the artist teams reported that "none of the men seem too interested in relying totally on film for their sketches. All of them seem more determined to record on paper on the spot when time and circumstances permit."[3] By 1970, when the program expanded to include other countries, nine Combat Artist Teams had been sent to Vietnam. Each team spent sixty days there, during which time it traveled extensively to ensure maximum coverage. Artists participated in patrols and photographed, sketched, and kept notes on their experiences. At the end of their time, each team traveled to Hawaii, where the artists spent the next sixty days in a studio creating finished artwork.

---

[3] Ltr, Fitzyallen Yow, Technical Supervisor of Combat Artist Team VII, to Mrs. Nowlin, 2 Sep 1968, Army Art Collection files, CMH.

Members of the Combat Artist Teams came from a variety of backgrounds, though all shared an interest in art. Artists included enlisted men, officers, those who voluntarily served, draftees, conscientious objectors, art students, experienced illustrators, architects, cartoonists, and urban planners. The military personnel who applied to be combat artists came from a myriad of Army jobs, such as postal clerks, illustrators, platoon leaders, mechanics, medical clerks, dental technicians, and aviation mechanics. Some extended their foreign duty tours in order to participate. At the end of their service, most Army artists who traveled to Vietnam continued in art-related careers, and many stayed in touch with the other members of their team.

Civilian artists—all volunteers—journeyed to Vietnam for thirty days, during which time they were given the status of official visitors. Following their stints and drawing on their experiences, the artists returned to their own studios to create eyewitness artwork, which would be presented to the Army. The artists agreed not to use the material from their visit for any other purpose for one year.

In comparison to the World War II art program, the Army's Vietnam art program failed to attract the prominent artists of the period. The World War II program was specifically envisioned to attract the best artistic talent of the generation, and, true to that vision, many prominent artists contributed to the war, either through the Army Art Program, as creators of war posters, or as correspondents for magazines such as *Life*. Art was essential in the marketing of World War II, with top illustrators, experienced in advertising and illustration art, crafting the image of the war that the American public saw through posters. In contrast, no prominent artists designed posters during the Vietnam War; more commonly, posters were purely graphic text, illustrated with photographs, or were reprints of popular posters from earlier periods, including James Montgomery Flagg's iconic *I Want You for the U.S. Army*. Influenced by the counterculture of the 1960s and inclined against traditional authority, many artists were drawn toward the antiwar movement rather than government-sponsored art programs. The war was simply unpopular, and prominent artists were not interested in participating, either due to political objections to the war or due to the risks inherent in traveling to a volatile region.

Regardless of the lack of involvement by America's top artists, the Army's Vietnam War art quickly gained critical praise. After viewing the art submitted by the first Combat Artist Team, Lester Cooke, curator of paintings at the National Gallery of Art, stated that the Army had "hit the jackpot."[4] An exhibit of artists' compositions toured military installations, and a television program entitled *The Big Picture* chronicled Combat Artist Team IX's experience. The success of the Combat Artist Program inspired many individual unit artist programs, the most

---

[4] Quote from Memo, Dr. Lester Cooke, 10 Apr 1967, Army Art Collection files, CMH.

extensive of which was the 25th Infantry Division's venture. Unfortunately, little of the unit artists' work was turned over to the Army Art Collection, so it has gone largely unpreserved.

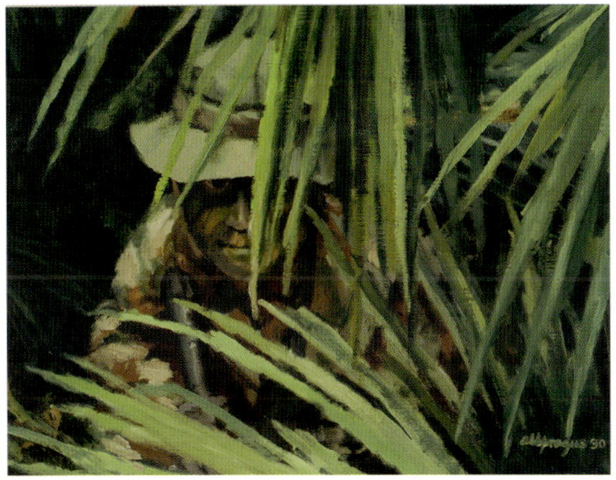

REMEMBER—IT'S A JUNGLE OUT THERE
Panama
Al Sprague, 1990
Oil on board, 18" x 24"

By early 1969, the Army had begun to review the Combat Artist Program in Vietnam with the intention of expanding artistic coverage to other geographic areas. This initiative, renamed the Army Artist Program, was envisioned to ensure full historical attention to ongoing operations, to improve troop morale, and to increase public awareness of the Army's worldwide role. Following the same model as the Vietnam program, it included a combination of soldier art teams and volunteer civilian artists. As in the art programs of the world wars and Vietnam, artists were given the broad mission to "record all aspects of military operations. No effort will be made to furnish explicit instructions as to subject matter. Artists are to be permitted freedom of expression."[5] Soldier-artist teams were assigned to various locations for forty-five days to observe and gather material, followed by sixty days in the studio. Volunteer civilian artists were issued invitational travel orders and spent thirty days in a specific country before returning to their studios to create art, which they donated to the Army. The expanded program was successful in recruiting a wider array of civilian volunteers, many of whom were interested in the program in Vietnam but were reluctant to travel to that region.

From 1970 to 1988, thirteen artist teams recorded Army activities all over the world. Selection of artists was similar to the Combat Artist Program in Vietnam. All applicants volunteered for the duty and submitted applications containing examples of their work. Competition was often fierce, with the selection committee routinely rejecting dozens of applications or holding them for future consideration. Artist teams journeyed to European countries, Japan, Korea, Panama, Thailand, and various locations within the United States. They recorded combat actions and patrols, training exercises, Corps of Engineers undertakings, refugee camps, and other Army subjects. Artist teams experienced triumphs, such as press coverage and acclaimed exhibits, as well as difficulties including extreme weather conditions, transportation breakdowns, and even theft of their supplies. While travel conditions ranged from good to bad, artist teams endeavored to keep a positive attitude and to create art matching the quality produced by past Army artists.

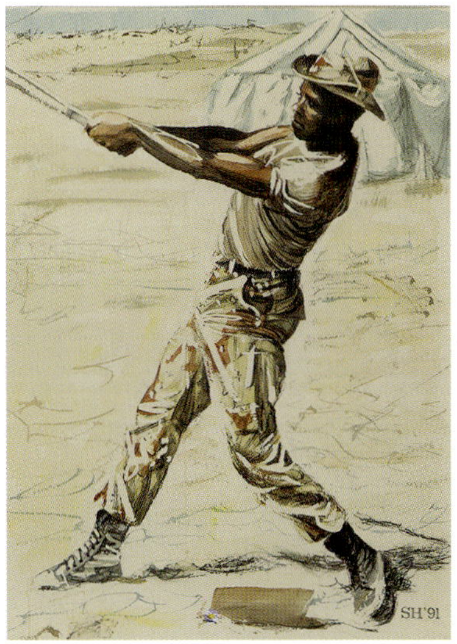

PLAYING BASEBALL
Operation DESERT SHIELD
Sieger Hartgers, 1991
Watercolor and ink on paper, 30" x 40"

Two teams of artists deployed to cover the conflict in the Persian Gulf. The first team went to Saudi Arabia for forty days from November to December 1990 in support of Operation DESERT SHIELD. The artists, Sfc. Peter G. Varisano and Sfc. Sieger Hartgers, had both been members of previous Army Artist Teams and volunteered to participate again. Their work during this period depicts not only the preparations for war, but the lighter side of the Army, including many personal images of soldiers relaxing or reflecting.

---

[5] Memo, Lt Col C. F. Moore, Chief, Plans and Programs, for Deputy Chief of Military History, 5 Jan 1972, Army Art Collection files, CMH.

The second Army Artist Team, consisting of reservists Capt. Mario H. Acevedo and Lt. Col. Frank M. Thomas, deployed in support of Operation DESERT STORM. While the artists arrived after the informal cease-fire agreement and did not see combat firsthand, they spent six weeks in Kuwait and Saudi Arabia sketching, taking notes, and talking to soldiers about their experiences. The battle scenes that the two artists produced were based on extensive research and observation of scenery, soldiers, and the aftermath of combat in the area. Both voluntarily extended their tours on active duty to complete their art.

In the summer of 1991, the U.S. Army Center of Military History sent a number of artists, most of them former participants in Army Artist Teams, all over the United States to record homecoming parades and reunification of soldiers with their families. These works became part of the Army Art Collection and remain an important part of the visual story of the war in the Persian Gulf.

In April 1991, an exhibit of Varisano's and Hartgers' DESERT SHIELD art opened at The Old Guard Museum, Fort Myer, Virginia. During the show, Chief of Staff of the Army General Gordon R. Sullivan suggested the addition of a permanent artist-in-residence position to the staff of the U.S. Army Center of Military History. The Center moved quickly to turn this vision into a reality later that same year. The exhibition, as well as a later one featuring Acevedo's and Thomas' DESERT STORM art, received high praise and solidified the continuing need for an active Army Art Program.

In July 1991, Varisano became the first Army Artist-in-Residence. Serving as the administrator of the Army Artist Program as well as the senior artist, he was responsible for selection of artist teams, arranging their deployments, assisting the curatorial staff, guiding the work of junior artists, recording Department of the Army–level activities, and traveling throughout the world to document ongoing Army actions. The creation of the Artist-in-Residence slot was not initially intended to replace the previous system of Army Artist Teams, and the two programs coexisted for several years. By 2000, however, the Army Artist Program had evolved into the single Artist-in-Residence position.

Under the Artist-in-Residence's direction in the 1990s, Army Artist Teams deployed all over the world to record the Army's activities. Like their predecessors, they were given broad parameters in which to work: to observe, to select their own subjects from within that experience, and to create documentary art in the medium and style of their choice. They created art based on Army actions in Bosnia-Herzegovina, Guantanamo Bay, Haiti, Korea, Kosovo, Panama, Rwanda, and Somalia. Artists were also assigned to cover Army training including at the Air Assault School at Fort Campbell, Kentucky; the Combat Maneuver Training Center at Hohenfels, Germany; and the Joint Readiness Training Center at Fort Polk, Louisiana, as well as the Army's role in relief efforts following disasters such as Hurricane Andrew.

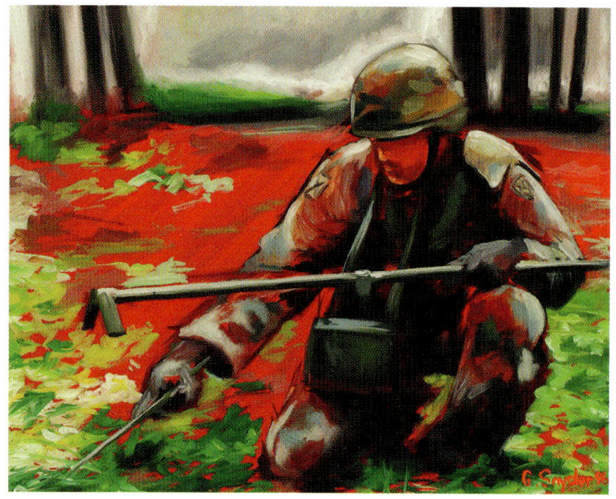

PROBING FOR LAND MINES IN BOSNIA
Bosnia-Herzegovina, Operation JOINT ENDEAVOR
Carl E. "Gene" Snyder, 1996
Oil on canvas, 16" x 20"

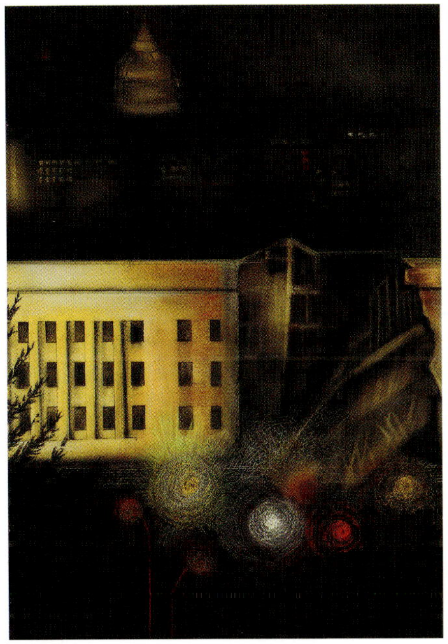

SEPTEMBER 11
Pentagon, Arlington, Virginia
Henrietta M. Snowden, 2001
Pastel on paper, 30" x 22½"

On 11 September 2001, Artist-in-Residence M. Sgt. Henrietta M. Snowden undertook the monumental task of chronicling the Pentagon just after the terrorist attack. Traveling to various vantage points to peruse the destruction and rescue efforts and reviewing numerous photographs for reference, Snowden ultimately produced a haunting pastel image of the damaged Pentagon at night, with rescue lights visible in the foreground and the dome of the Capitol rising out of a hazy background.

Since the terrorist attacks against the United States in 2001, six Artists-in-Residence have documented Army and joint activities in Afghanistan and Iraq. While this program ensures that the Army will always have one official artist, the inherent weakness of it is that the responsibility for recording the Army's contemporary history in art falls to just a single soldier. With only one official Army artist at a given time, the Army Art Collection lacks the same level of coverage of the Global War on Terrorism and Overseas Contingency Operations that it has had for the major twentieth-century wars and conflicts. This deficiency is partially rectified by donations from talented soldier-artists who are not part of the official program. M. Sgt. Timothy B. Lawn, a member of the U.S. Army Reserve who deployed to Iraq from 2004 to 2005 as a staff sergeant, spent time during his tour creating drawings and watercolor paintings for donation to the Army Art Collection. Capt. Heather C. Englehart, a member of the Louisiana Army National Guard, deployed to Iraq and Kuwait from 2003 to 2004 as a first lieutenant. Her command allowed her time to paint and draw scenes of her experiences, which she also gave to the Army Art Collection. Englehart and Lawn mark the first time since the unit artist programs of the Vietnam period that soldier-artists outside of the official art program have made a notable contribution to the documentation of Army events through art.

The great value of the Army Art Collection is that it is a visual account of routine Army life that links generations of soldiers. Though the faces, the technology, and the enemy change from war to war, some aspects of a soldier's career are timeless. While a World War II or Vietnam veteran may not recognize the technology used in Iraq or Afghanistan, all soldiers bond over the typical experiences they have shared: engaging in basic training, waiting around for transportation, performing guard duty, or finding ways to relax in the midst of chaos. These themes recur throughout the Army Art Collection, and many of the pieces in this book were chosen with that commonality in mind.

The story of the Army Art Program is about artists and their works. This book presents their art beginning with that produced during the Vietnam War up to 2014. It is divided into five chronological sections, with a brief introduction at the start of each. The text is minimal because good art tells the tale far better than words, but, whenever possible, the artist's inspiration for an individual composition will be included. The question is often asked why the Army maintains its artist program rather than relying on photography for documentation, and it is the Center's sincerest hope that the value of an artist's interpretation will be illustrated by the art on these pages and thus the query answered. The best way to describe art's worth is to echo a sentiment voiced repeatedly by Army artists since World War I: a camera records events; the artist imbues them with soul.

Vietnam War

Army art from the Vietnam War is unlike Army art from any other period. In striking contrast to the more conventional artistic approaches of the world wars, the Vietnam period is characterized by a wide stylistic diversity with many artists working in bright, vibrant colors and an expressive manner. Some artists preferred quick sketches or watercolors, with figures and landscape elements defined by only a few swift brushstrokes, such as Sp4c. Roger Blum's *Infantry Soldier* (page 8). Figures are often thin or ghost-like and are sometimes dwarfed by the landscape or war machines surrounding them. It is common to see soldiers portrayed alone or in very small groups, as in Sp5c. Burdell Moody's *Point Man* (page 36) and in civilian artist Horatio A. Hawks' *Long Range Patrol* (page 57). Stylistic variety ranges from impressionistic, such as 1st Lt. John O. Wehrle's *Landing Zone* (page 22), to hyperrealistic, such as Sp4c. Michael R. Crook's *Jonah '67* (page 29). While many artists worked in the traditional mediums of oil paints and watercolors, others embraced the still-new medium of acrylic paints, which had become commercially available in the 1950s. For the first time in the history of the Army Art Program, some artists submitted collages, three-dimensional works, and multimedia artwork.

The subject matter varies widely as well. While the artists were given the same general instructions as Army artists during World Wars I and II, there are drastically fewer pieces in this period that show the light side of soldier life. Many World War II artists painted scenes of soldiers relaxing or engaging in leisure activities. In contrast, soldiers featured in Vietnam artwork look tense, even when they are at rest. The soldiers represented in action scenes are frequently anonymous, their faces turned away from the viewer or otherwise obscured, as if to suggest that they can be replaced by anyone. Soldiers experiencing deep emotions are similarly anonymous, like the subject portrayed in Sp4c. Victory Von Reynolds' *A Soldier* (page 60). Conversely, many other portraits depict recognizable soldiers. The subjects of these compositions may look exhausted, bored, disinterested, and aged beyond their years, but all share a certain stoicism and strength.

Much of the Vietnam artwork is beautiful. Despite the "detached from reality" feeling that many Vietnam artists recall, most of the artists were able to isolate moments of beauty within the chaos of war, as in Sp5c. Burdell Moody's *Firefly Mission* (page 37). In other works, such as Sp5c. David M. Lavender's *Cheeta Gets a Bath* (page 13), the beauty of the vibrant sunset is contrasted with the harsh lines of the equipment. This disparity between beauty and violence highlights the many contradictions of the Vietnam War. In explaining such an incongruity in one of his pieces of art, Robert Knight stated that it "represents for me the contrasts that I saw while in Vietnam. I saw a beautiful country, culture and people caught up and entangled in a complicated war. I saw in the people, a great strength and resilience."[1] Contradictions are expressed in much of the Vietnam artwork, as the artists grappled with the violence, low morale of American troops, and growing uncertainty of support for the troops from the American people.

The Army art of this conflict is heavily influenced by the era in which it was painted. Experimentation in new artistic mediums and the loosening of styles echo trends in the civilian art world, the source of the training for many of the young artists. Many of the artists were draftees, and none could escape the influence of negative public opinion on the war. Despite this, there are very few blatant protest pieces within the body of Vietnam Army art. The artists are honest in their portrayal of events, conditions, and their fellow soldiers and do not shy away from criticism of the destruction and loss of life that they experienced, but, without exception, their art shows respect for and commiseration with their comrades in uniform.

[1] E-mail, Robert Knight to general editor, 2 Aug 2010, Army Art Collection files, CMH.

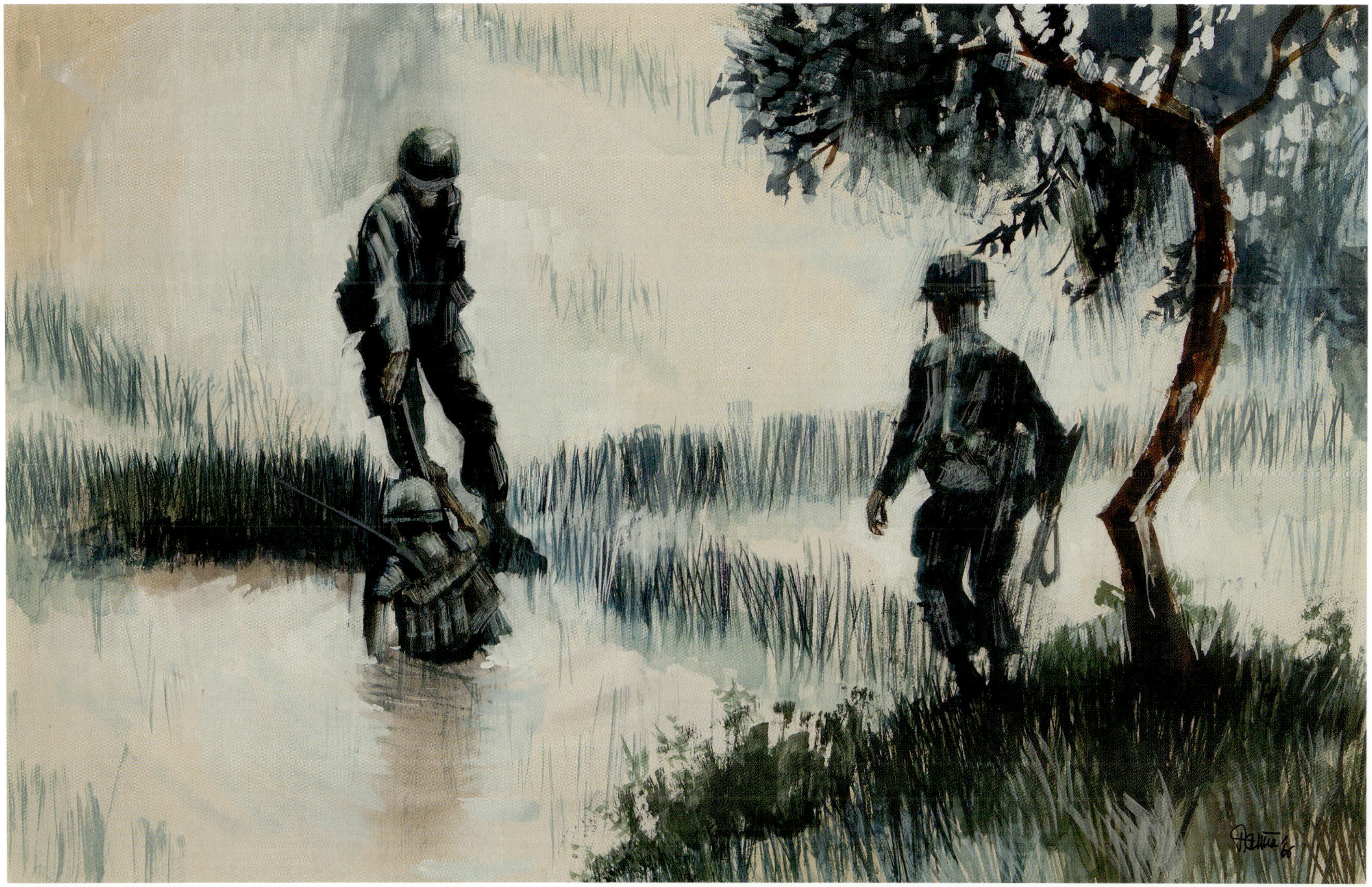

"I didn't tell my other team members that I had never had any formal art instruction (especially since I was the Team Leader and considered the 'Old Man'). I just looked over their shoulders a lot and learned how to work with all types of mediums."
—Augustine G. Acuña

THE RICE PADDIES
Vietnam
Augustine G. Acuña, 1966
Watercolor on board, 18" x 27¾"

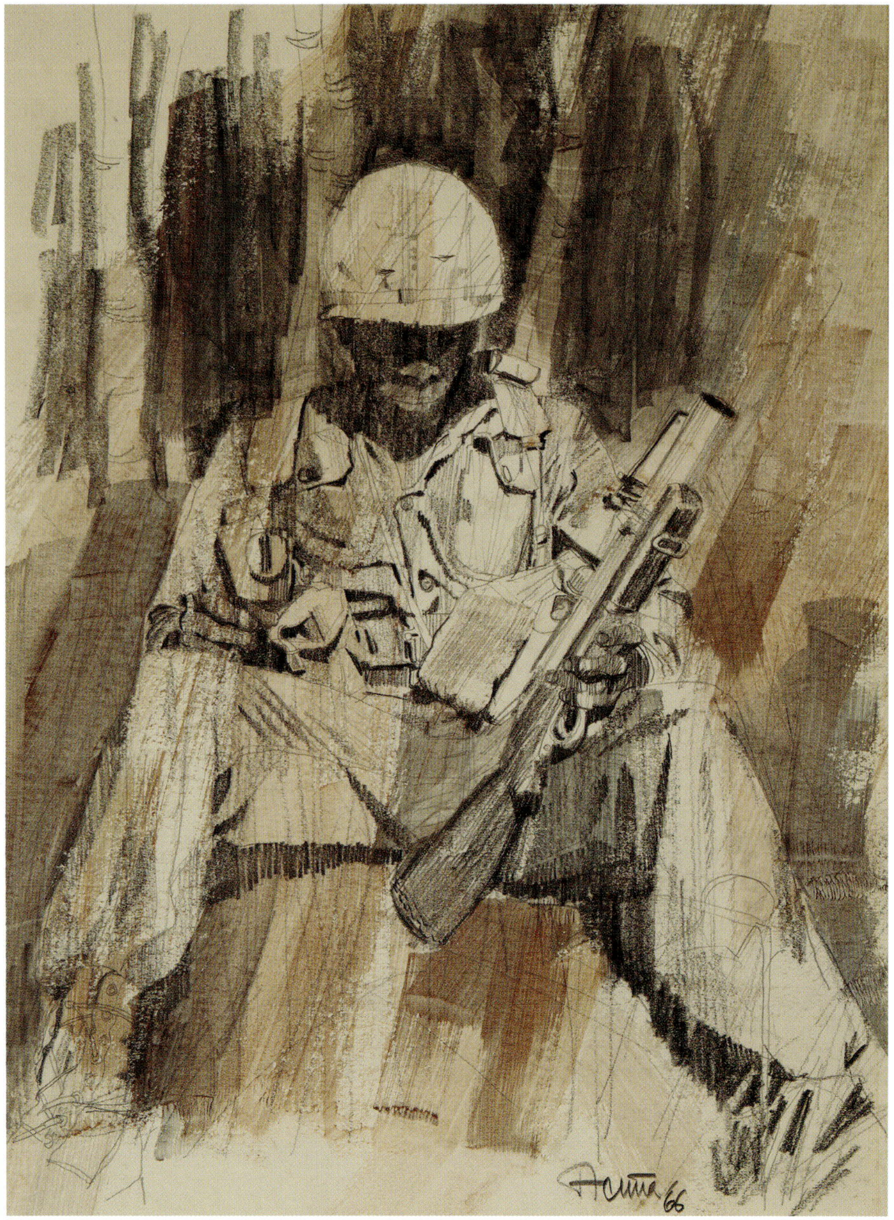

BEAT
Vietnam
Augustine G. Acuña, 1966
Pencil on illustration board, 25" x 16¾"

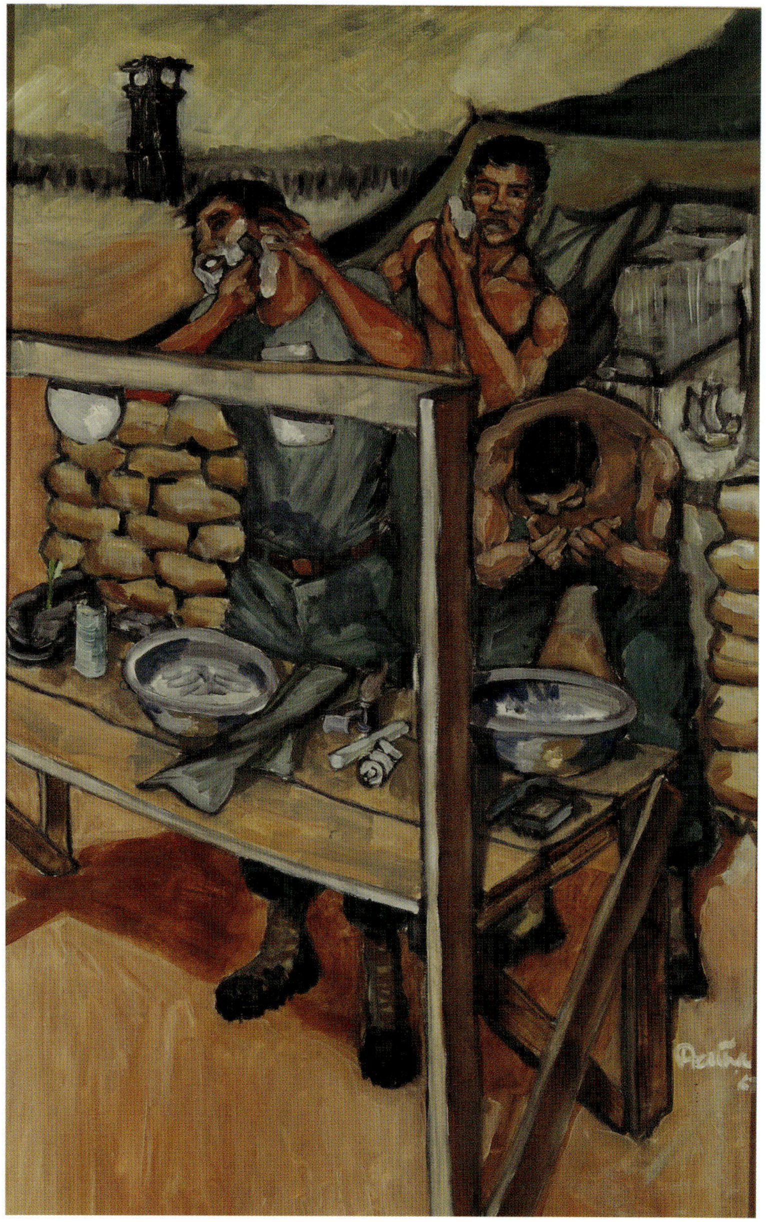

MORNING RITUAL
Vietnam
Augustine G. Acuña, 1966
Oil on canvas, 38" x 24"

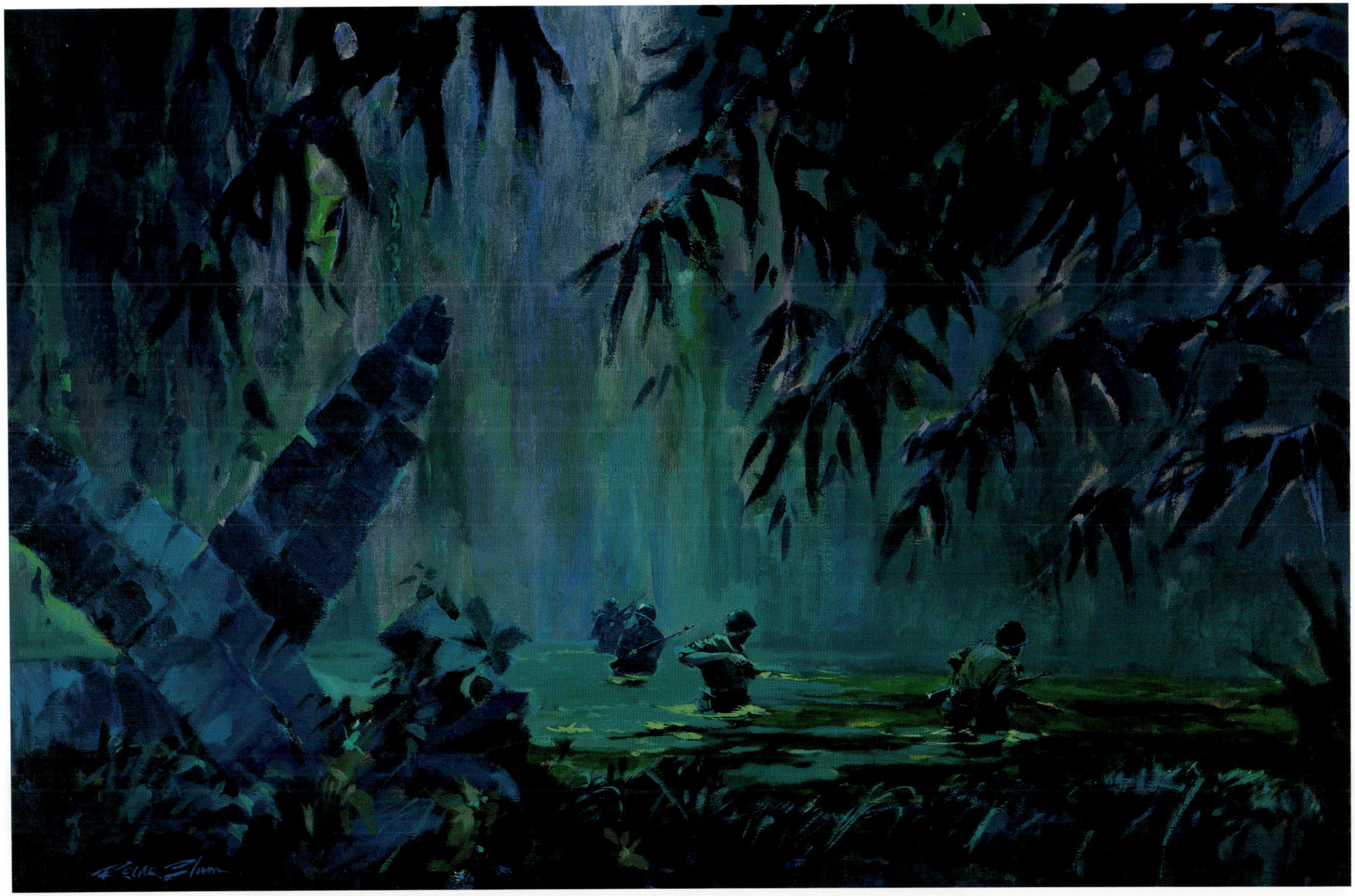

"This day required traversing numerous terrains in order to reach our mountain top destination. Before heading out, we were advised to: 'Be very observant to our surroundings, don't pick up anything on or off the trails, and be extremely quiet in conversation and movement.' These instructions were the emotional catalyst that inspired my painting."

—Roger Blum

PATROL IN THE JUNGLE
Vietnam
Roger Blum, 1966
Oil on canvas, 24" x 36"

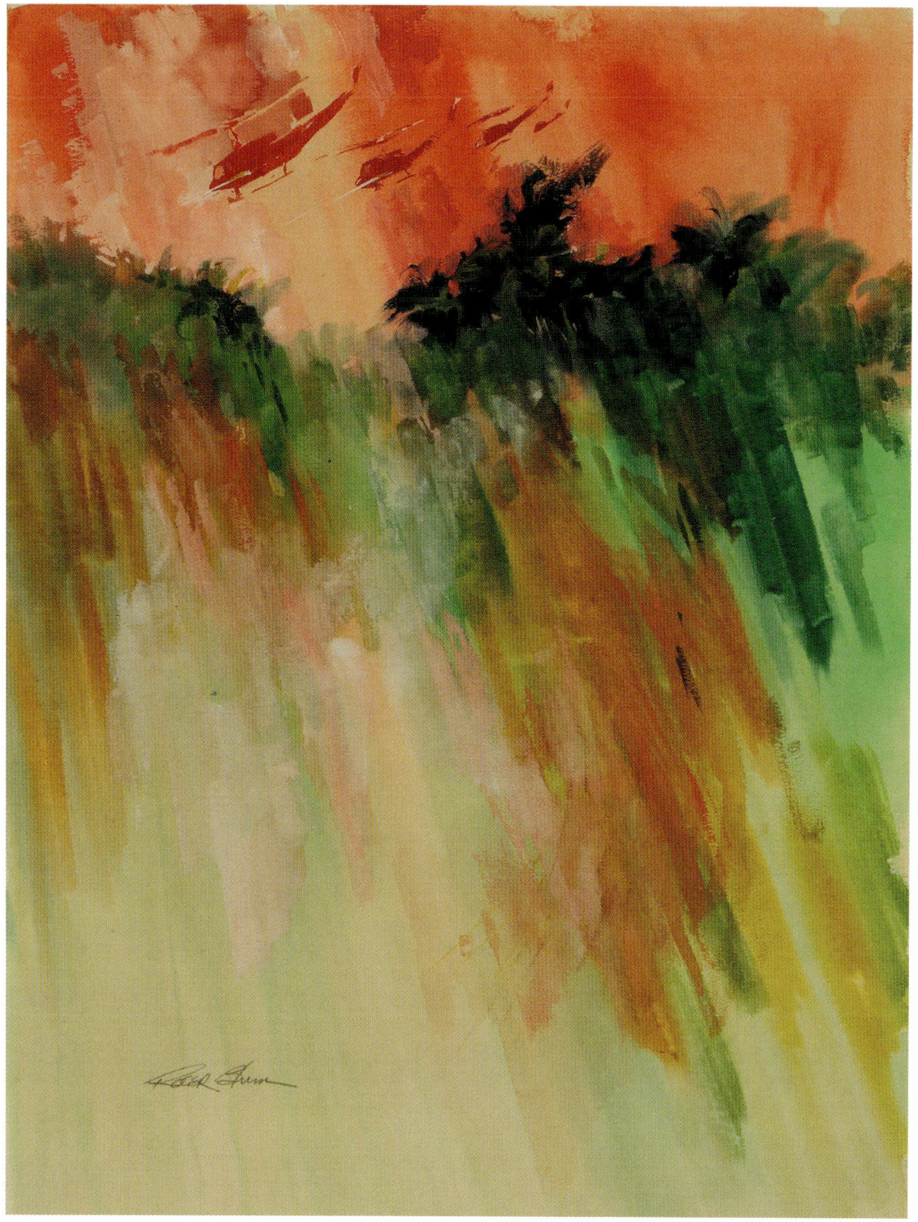

HELICOPTER ASSAULT
Vietnam
Roger Blum, 1966
Watercolor on paper, 24" x 18"

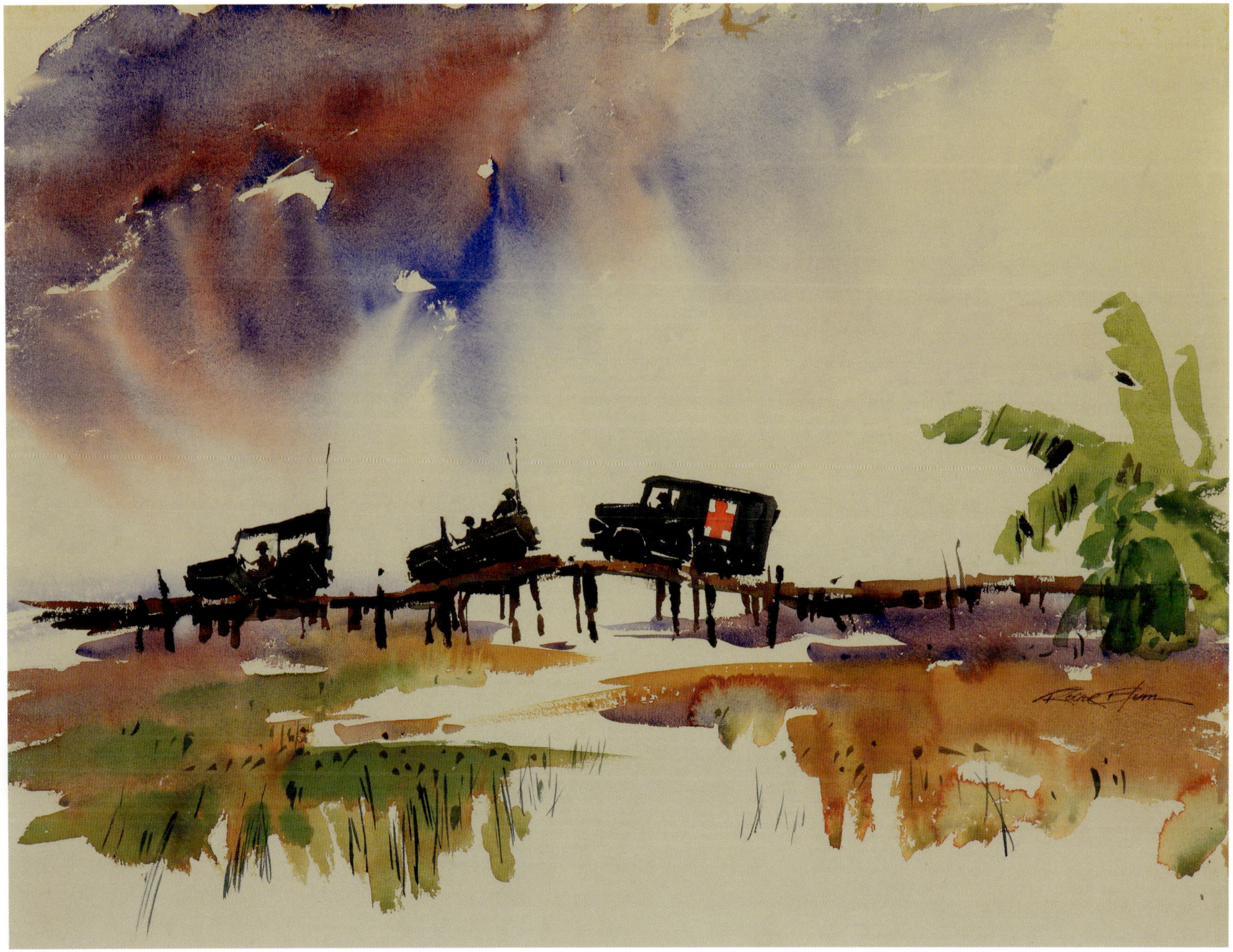

CONVOY
Vietnam
Roger Blum, 1966
Watercolor on paper, 22½" x 29"

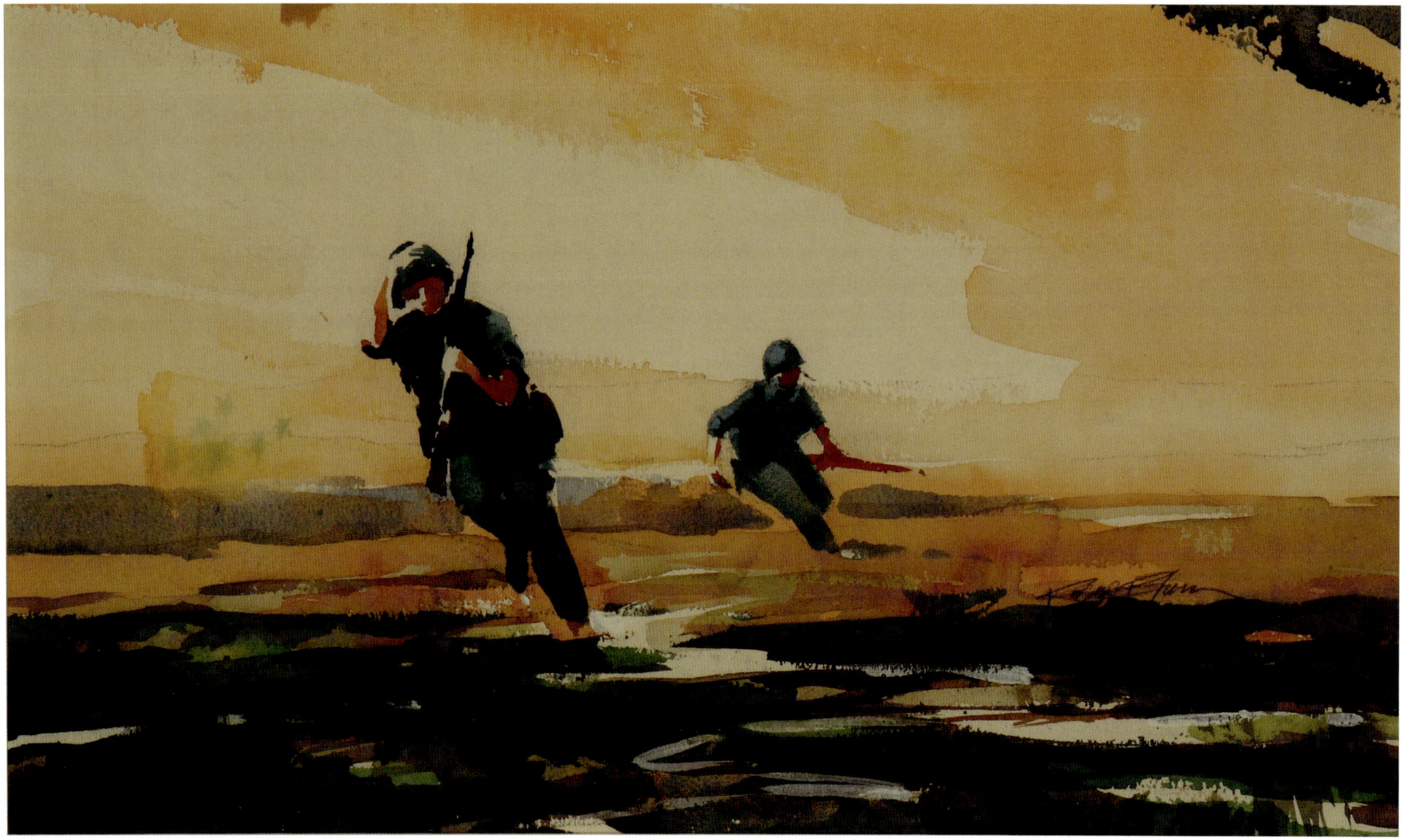

INFANTRY SOLDIER
Vietnam
Roger Blum, 1966
Watercolor on paper, 11" x 19"

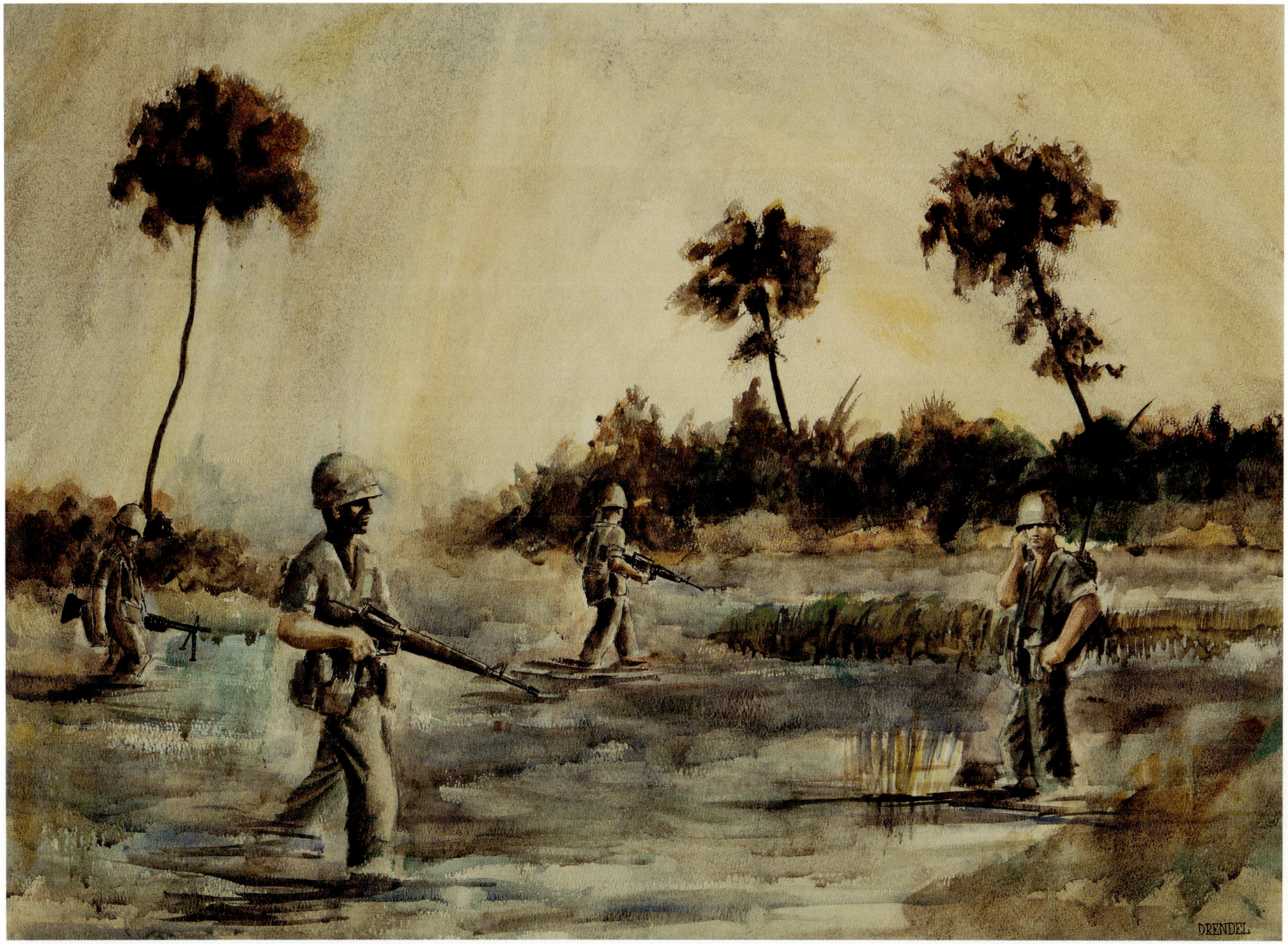

SWAMP FIGHTERS
Vietnam
Theodore E. Drendel, 1966
Watercolor on paper, 18" x 24"

"I saw a beautiful country, culture and people caught up and entangled in a complicated war. I saw in the people a great strength and resilience. . . . The overall Vietnam experience introduced me to the harsh reality of war first hand. I saw the every day images of war through the poverty, pain, fear, and frustration and loss of lives that it creates. I also saw the difficult and dangerous situations that our soldiers found themselves in."
—Robert Knight

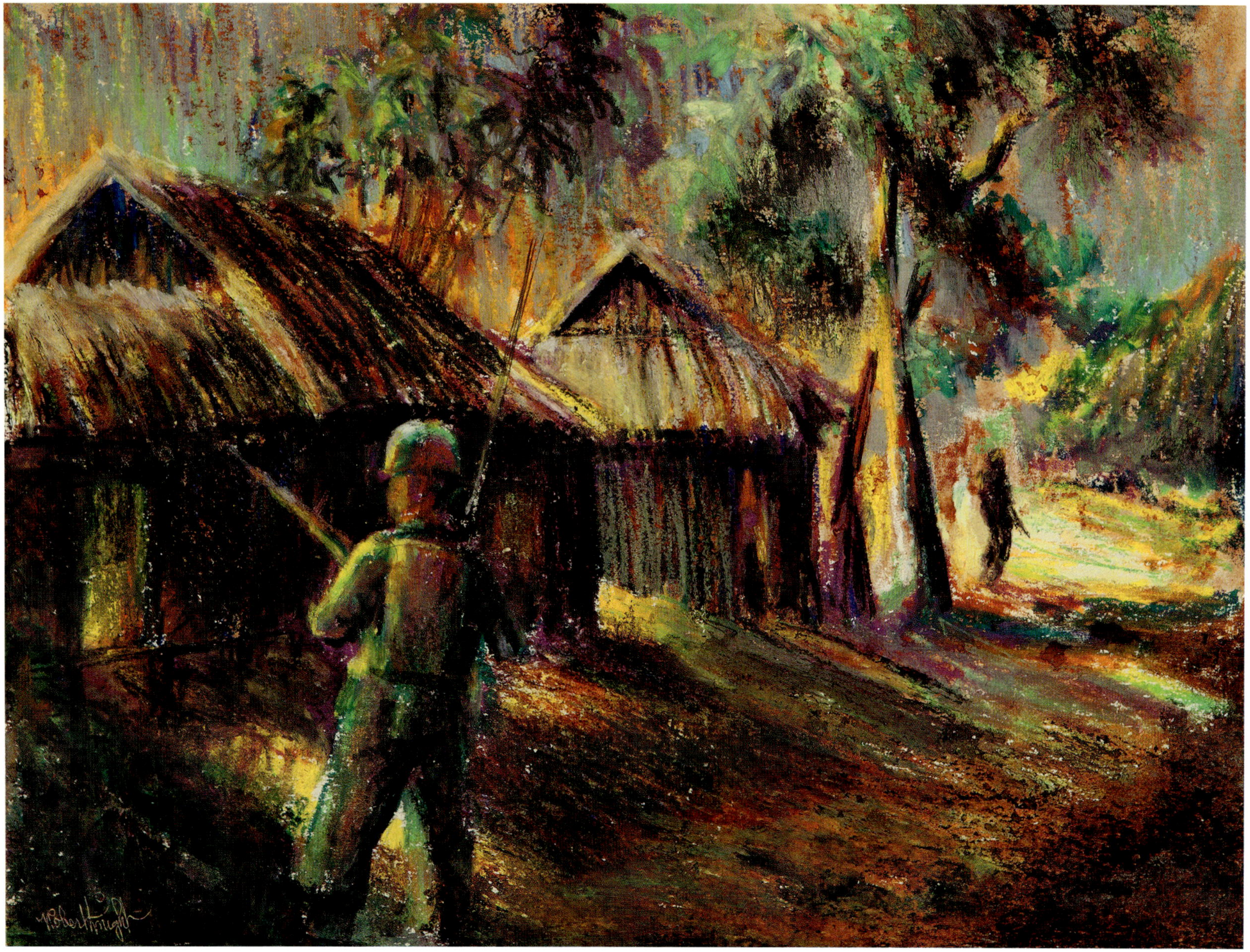

SEARCHING A VILLAGE
Vietnam
Robert Knight, 1966
Crayon on paper, 9" x 11¾"

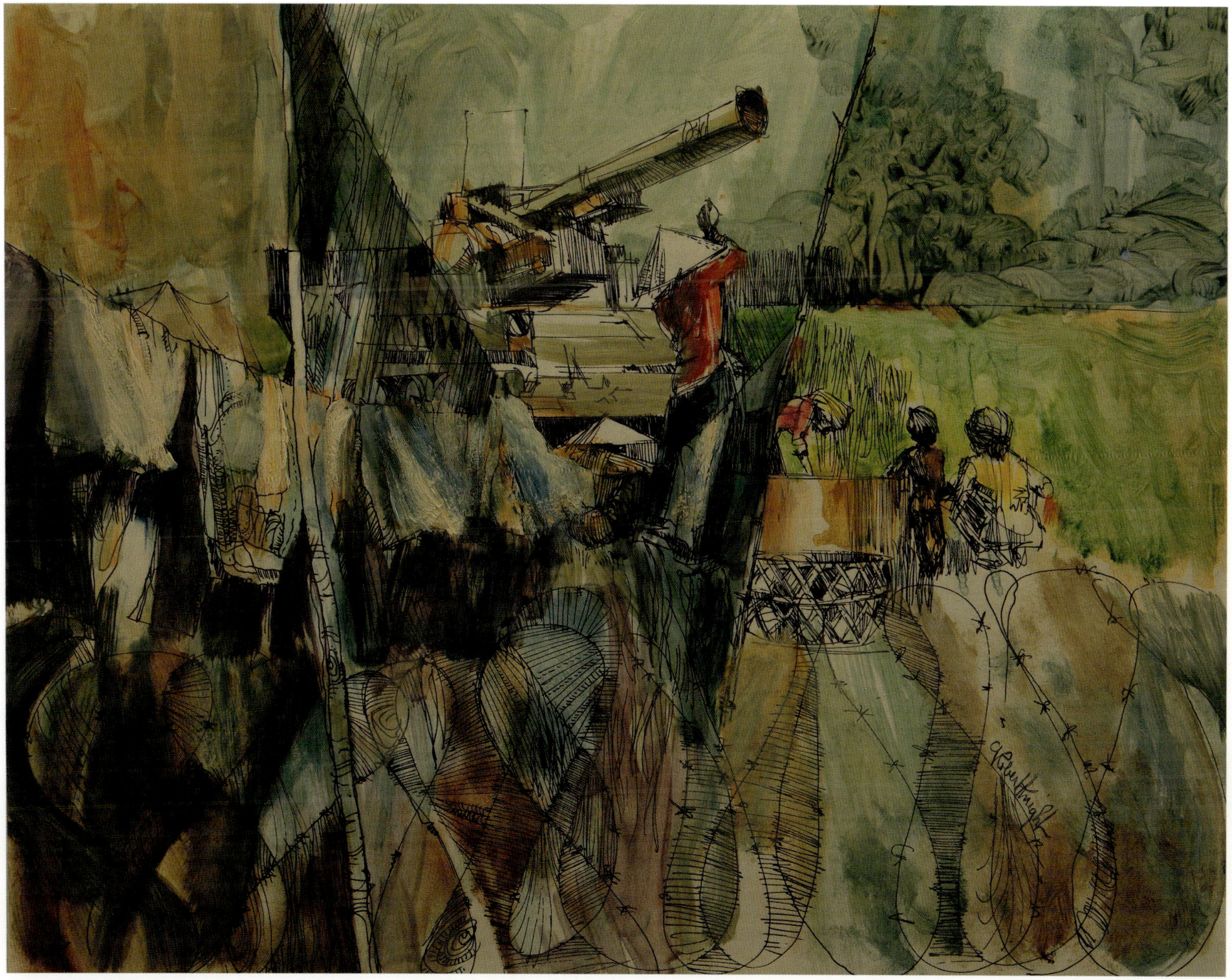

SCENE NEAR CHU CHI
Chu Chi, Vietnam
Robert Knight, 1966
Ink and oil on paper, 11" x 14"

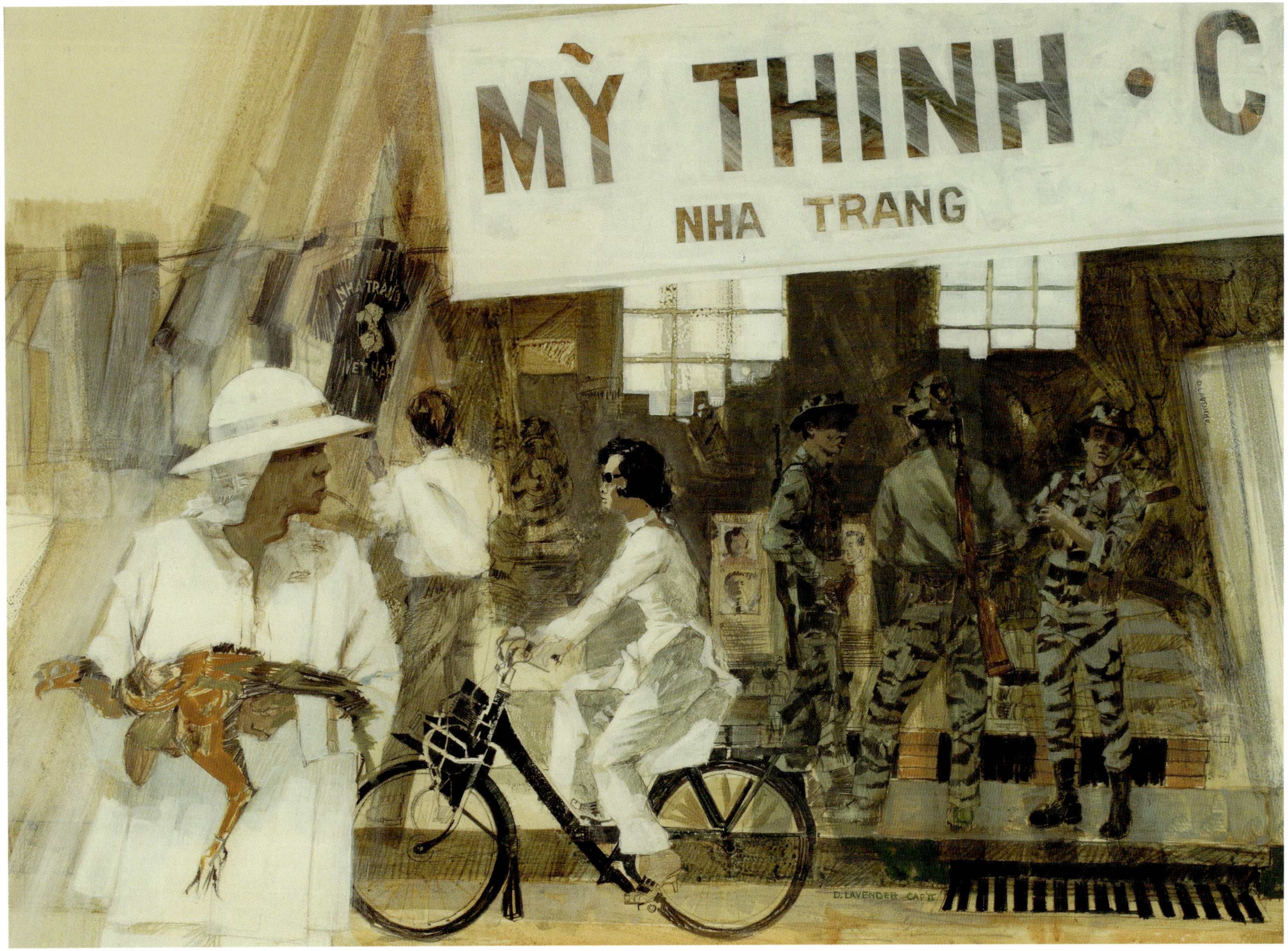

STREET SCENE
Nha Trang, Vietnam
David M. Lavender, 1966
Acrylic on board, 21½" x 25¾"

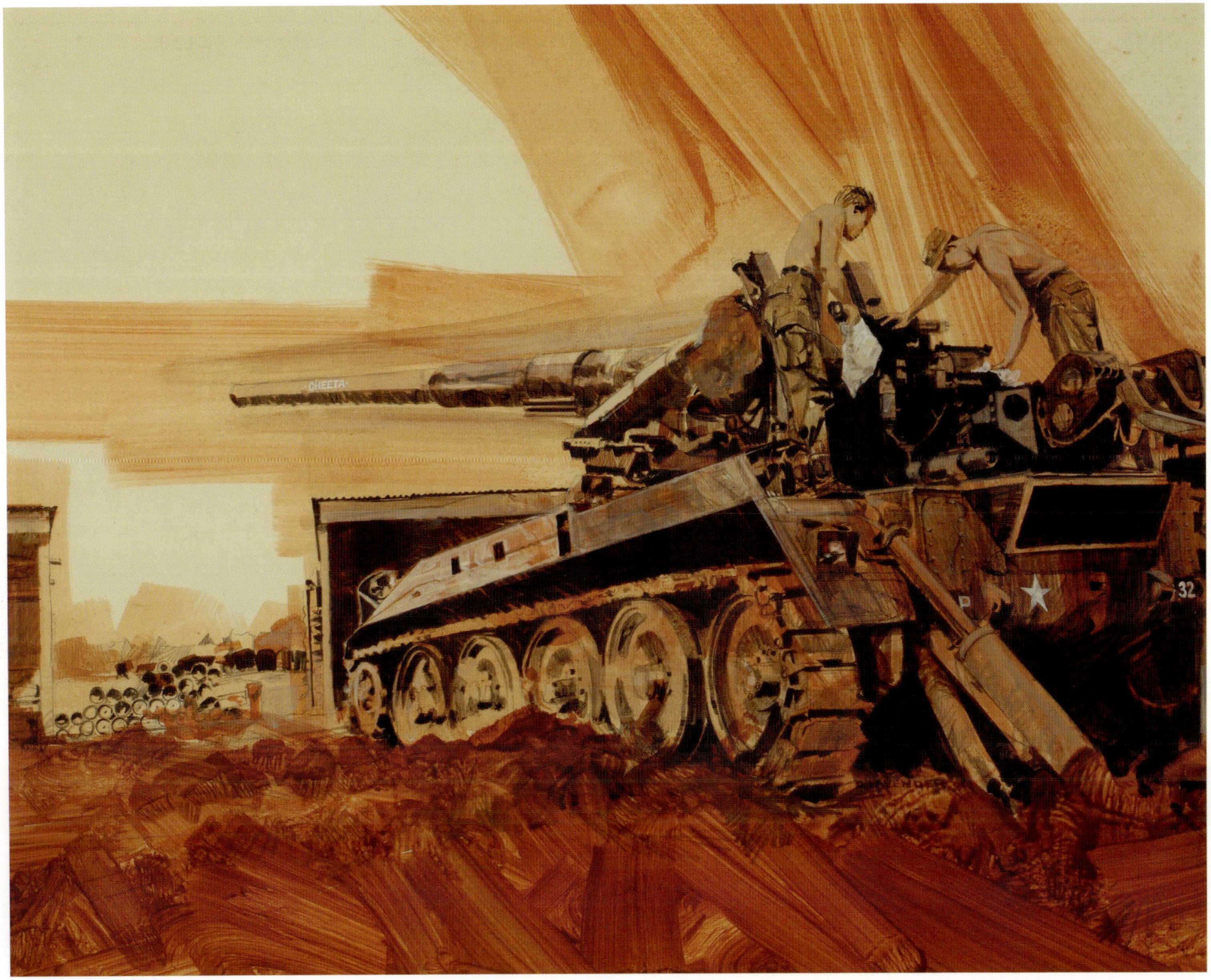

CHEETA GETS A BATH
Cu Chi, Vietnam
David M. Lavender, 1966
Acrylic and pencil on board, 26" x 34"

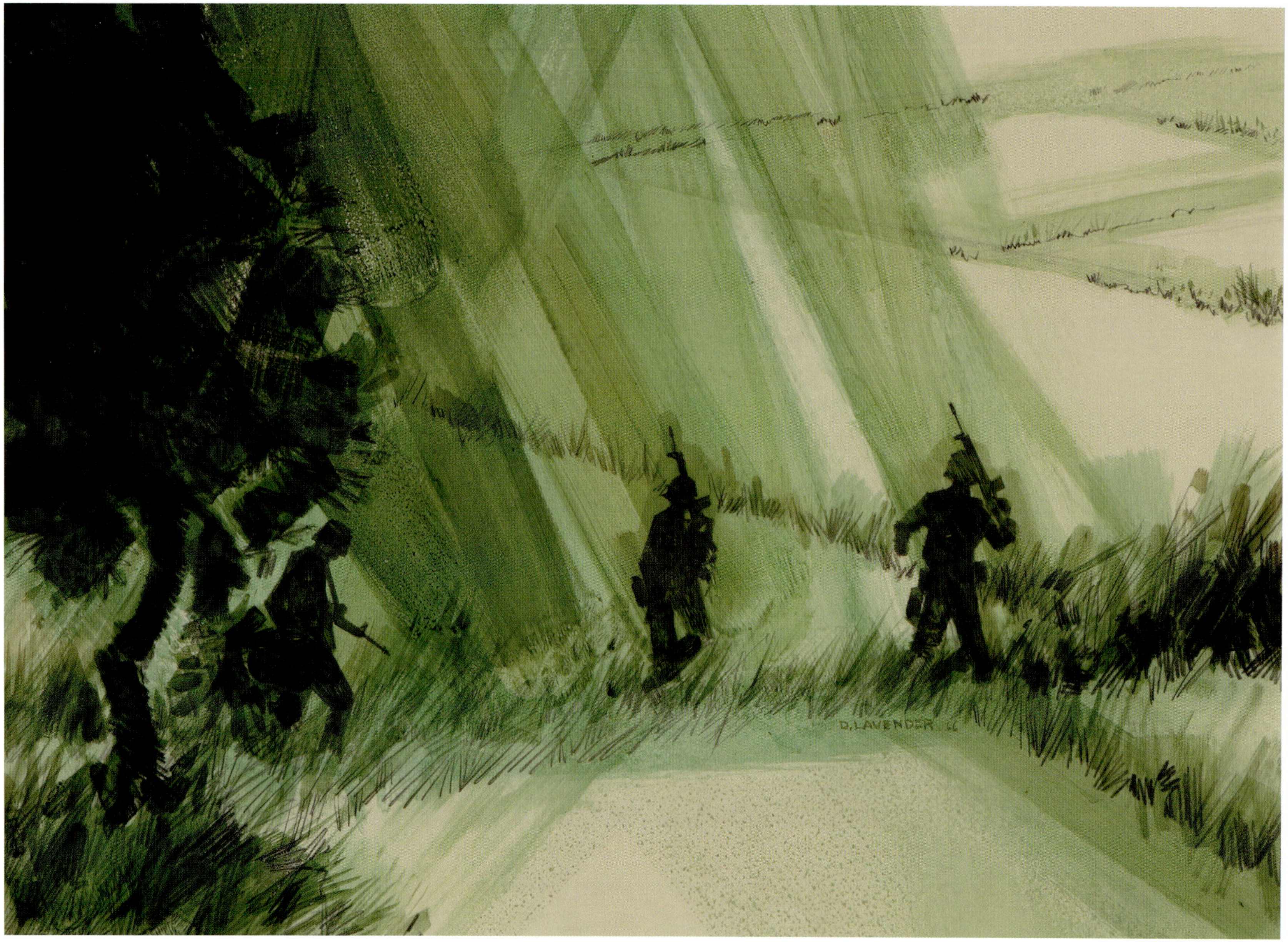

ALWAYS THE RICE PADDIES
Tay Ninh, Vietnam
David M. Lavender, 1966
Acrylic on illustration board, 23" x 28½"

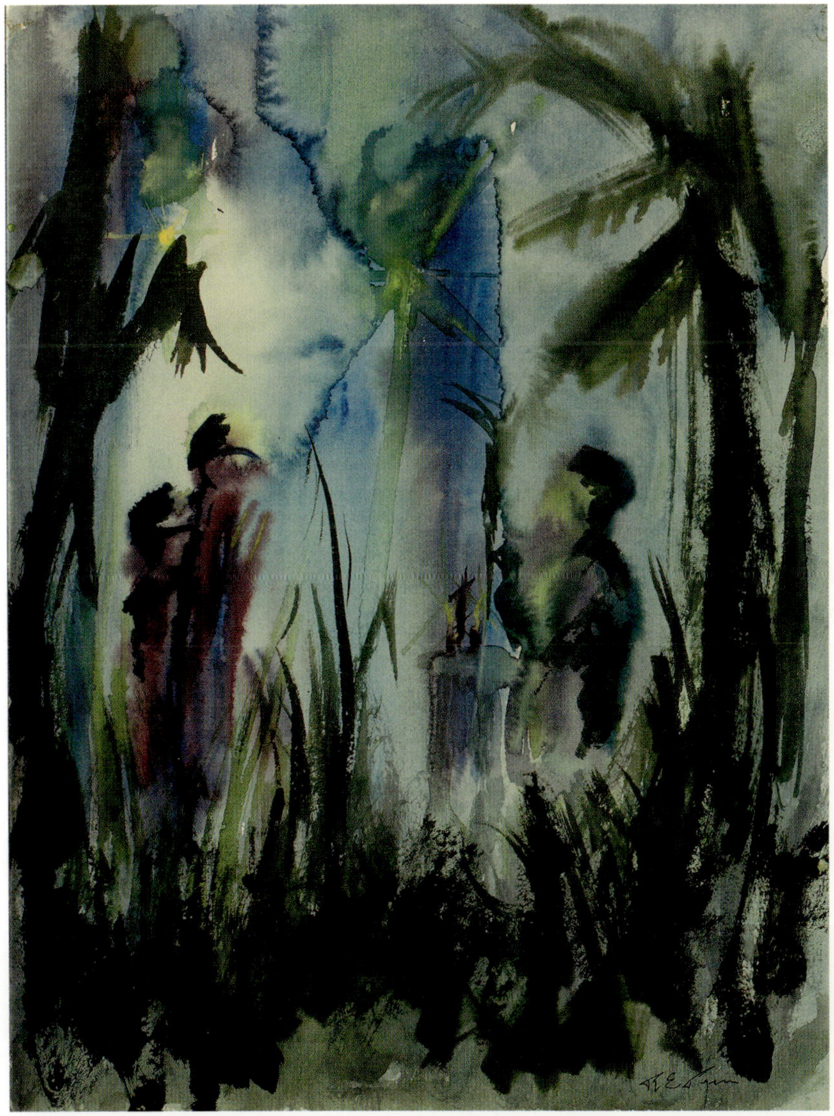

CHURCH SERVICES IN THE JUNGLE
An Khe, Vietnam
Ronald E. Pepin, 1966
Watercolor on paper, 11¾" x 9"

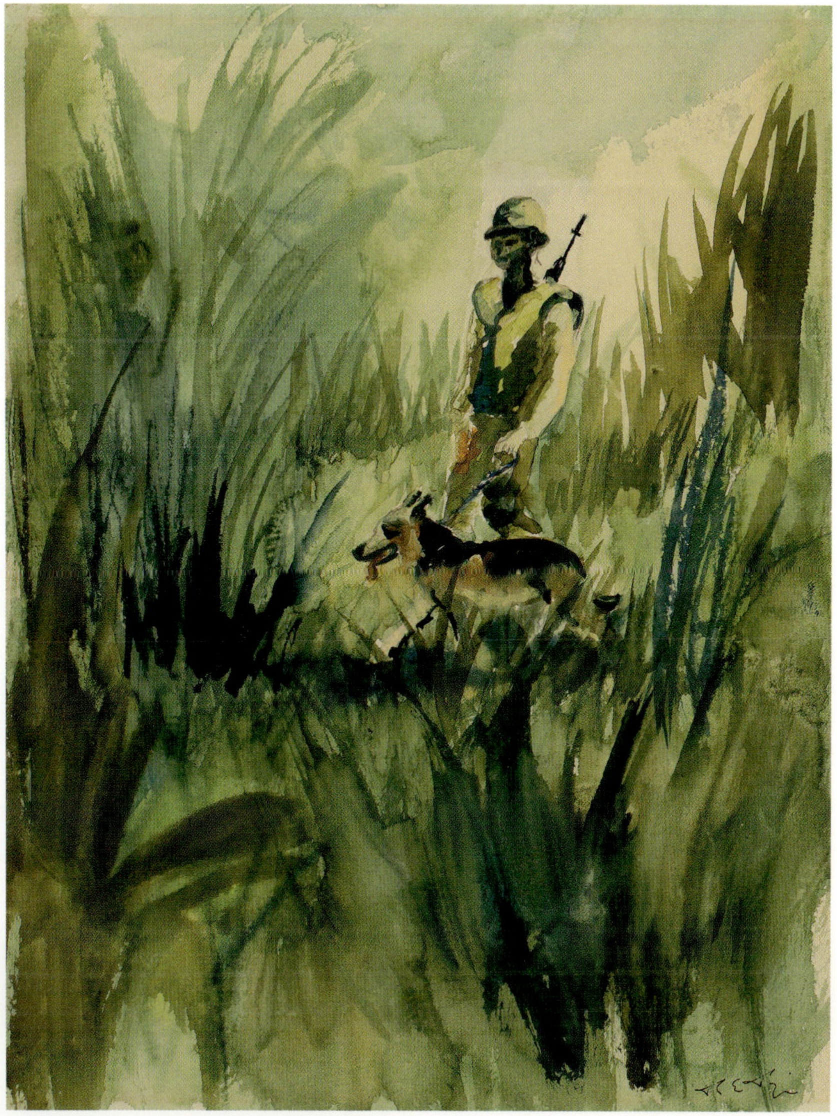

SCOUT DOG
Vietnam
Ronald E. Pepin, 1966
Watercolor on paper, 11¾" x 9"

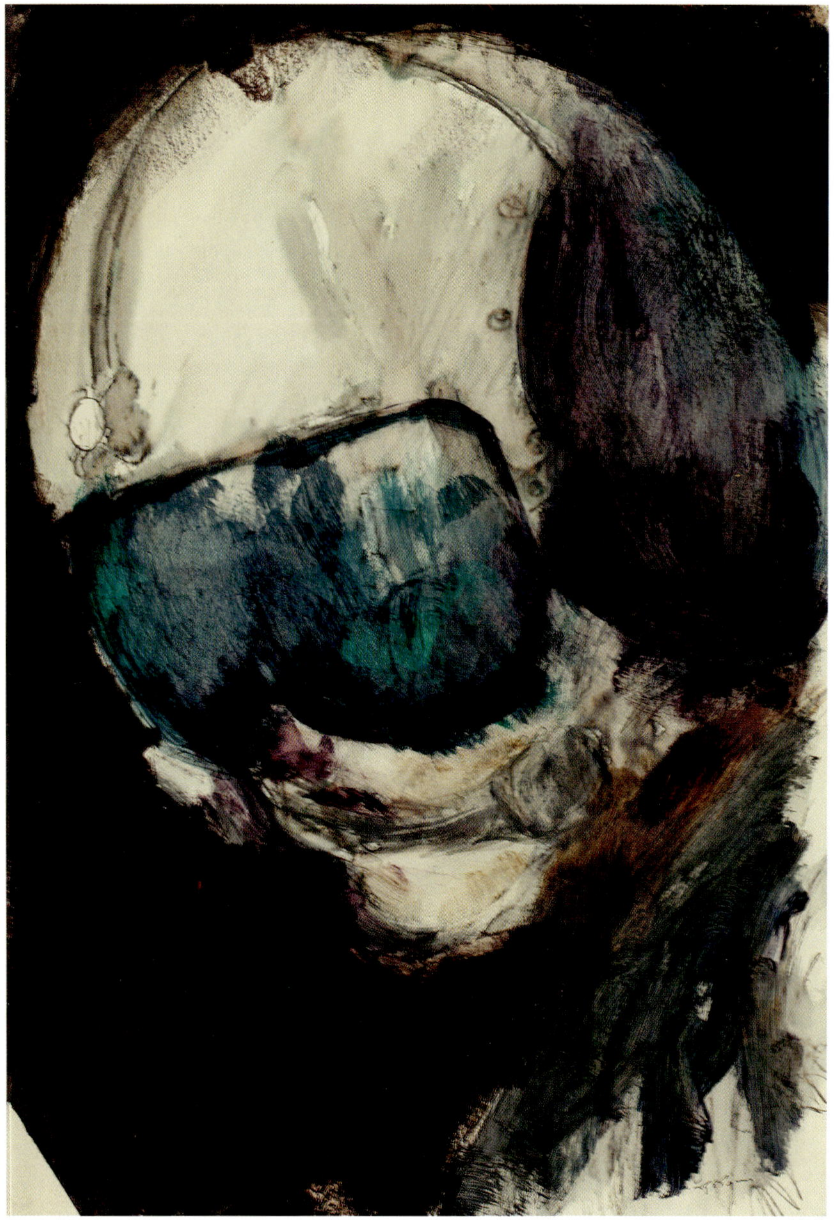

HELICOPTER MACHINE GUNNER
Vietnam
Ronald E. Pepin, 1966
Oil on paper, 17¾" x 12"

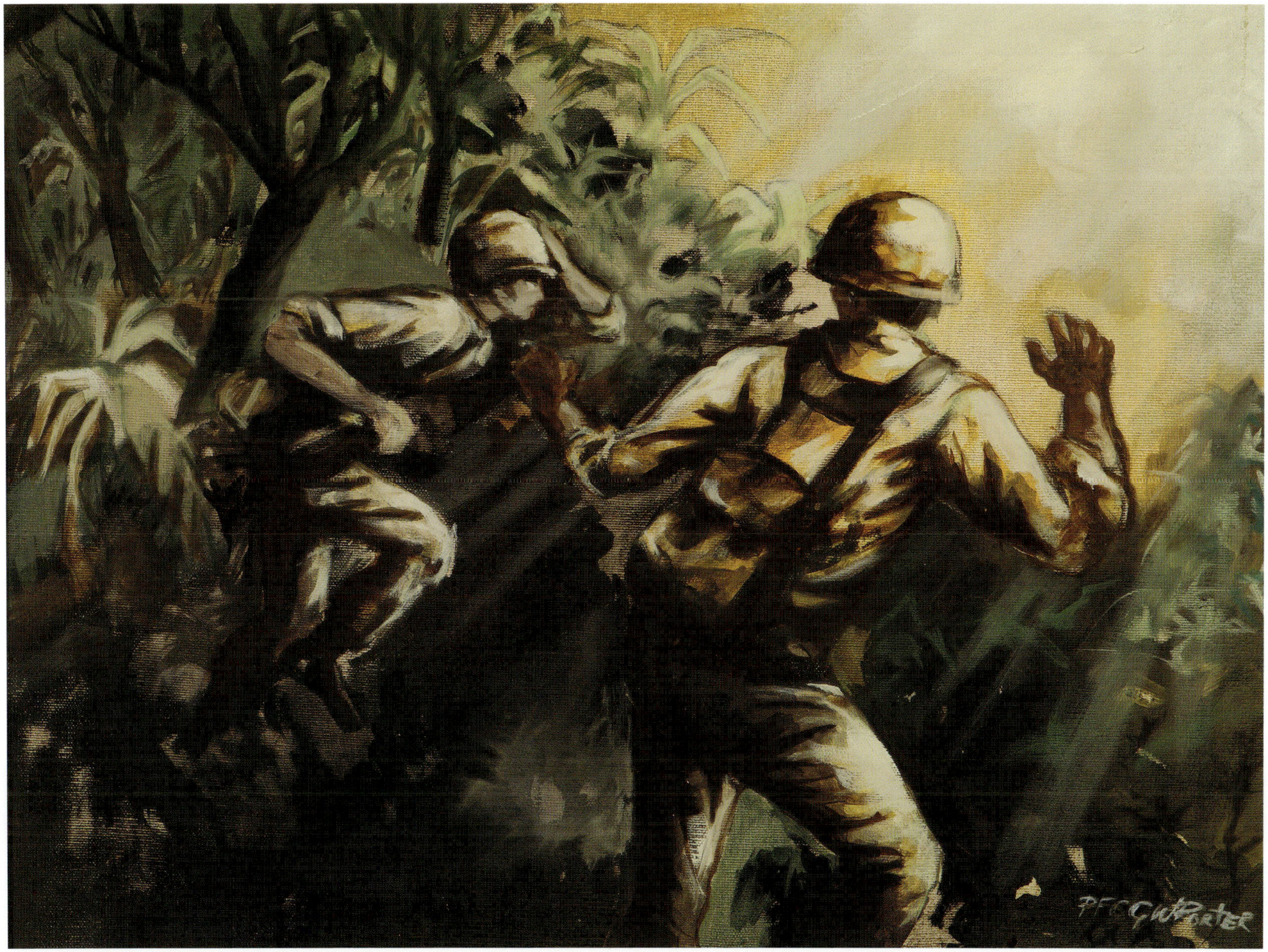

"It's all happening very fast. I would just stick up my head and take a few shots very quickly and get back to business staying alive and keep moving with guys."
—Gary W. Porter

SECOND PLATOON ASSAULT
Vietnam
Gary W. Porter, 1966
Oil on canvas, 18" x 24"

"I think I grew up a little out there; as an artist—and as a man."
—Paul Rickert

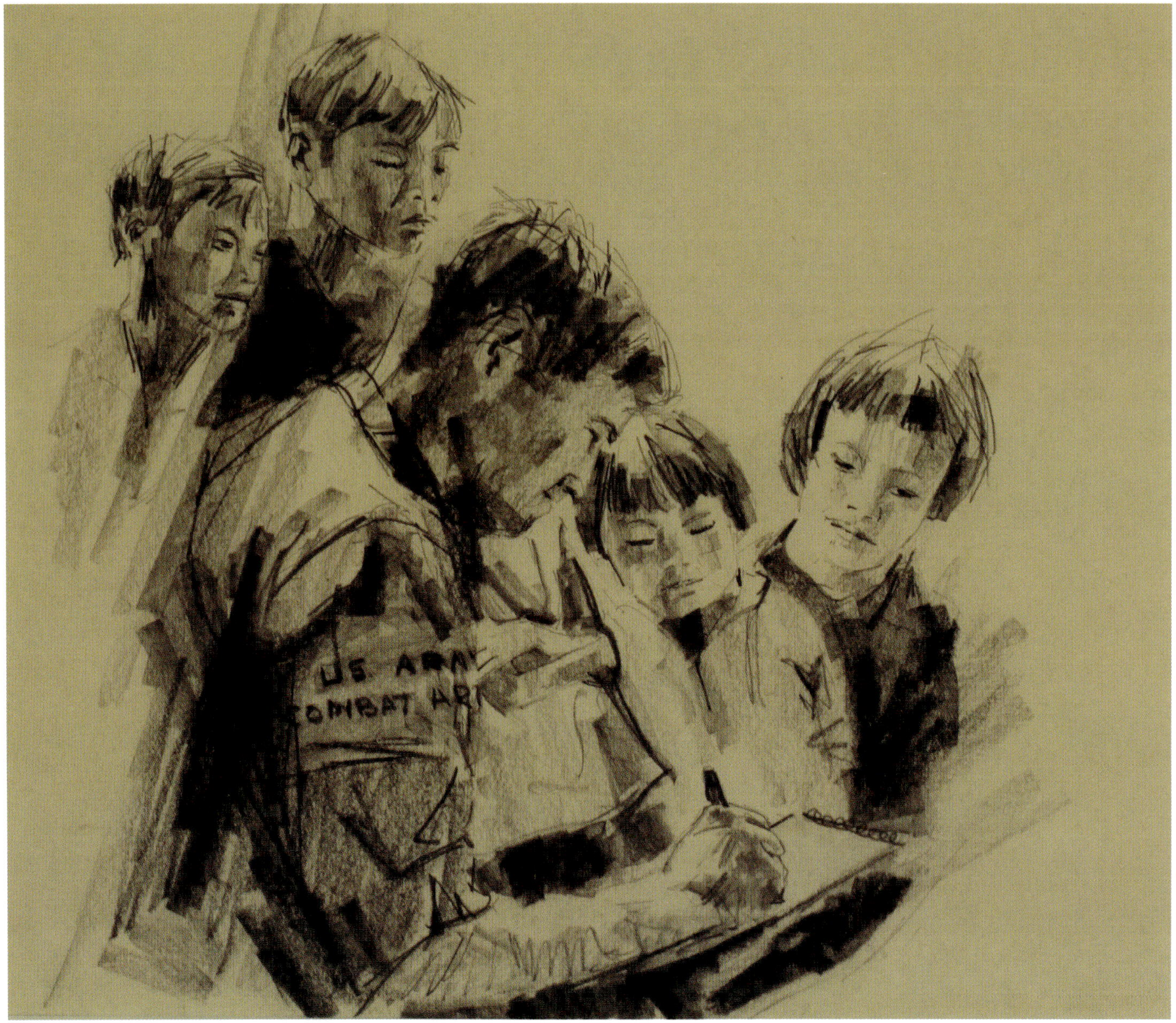

COMBAT ARTIST AT WORK
Duc An, Vietnam
Paul Rickert, 1966
Pencil on paper, 9" x 12"

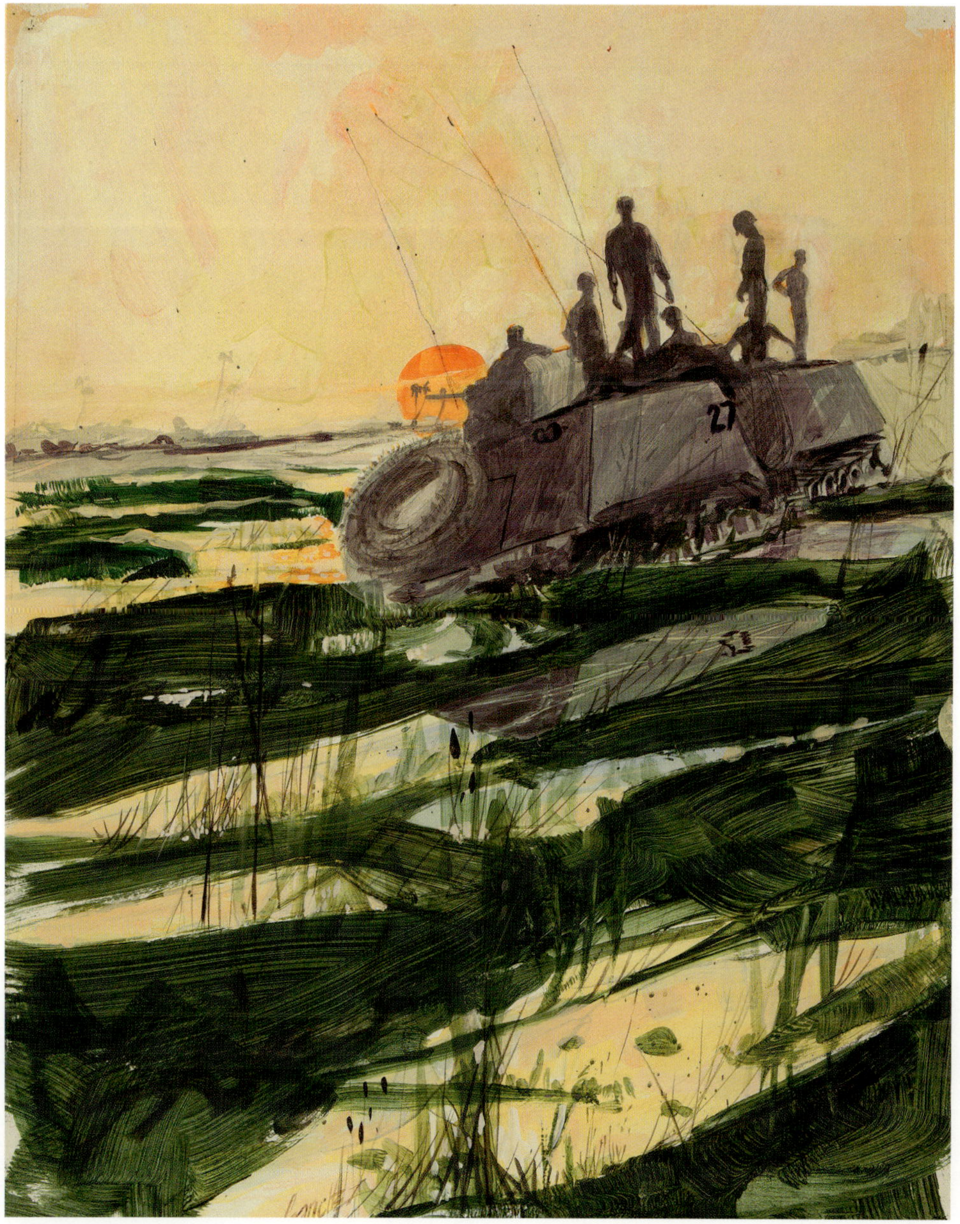

BOG-DOWN
Cu Chi, Vietnam
Felix R. Sanchez, 1966
Acrylic on paper, 29" x 23"

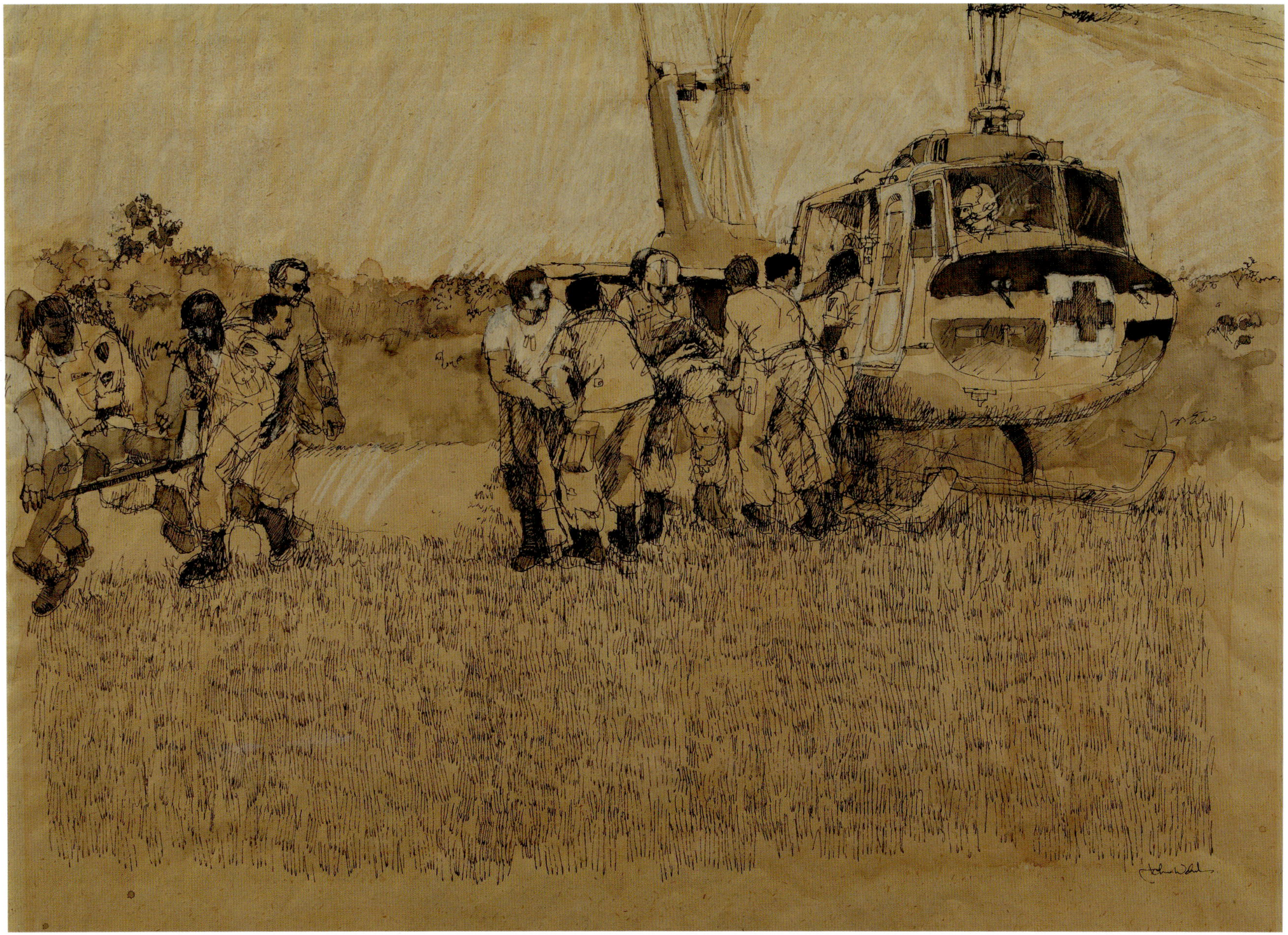

DUSTOFF AT TAN SON NHUT
Tan Son Nhut, Vietnam
John O. Wehrle, 1966
Crayon, ink, watercolor, and wash on paper, 17¾" x 24"

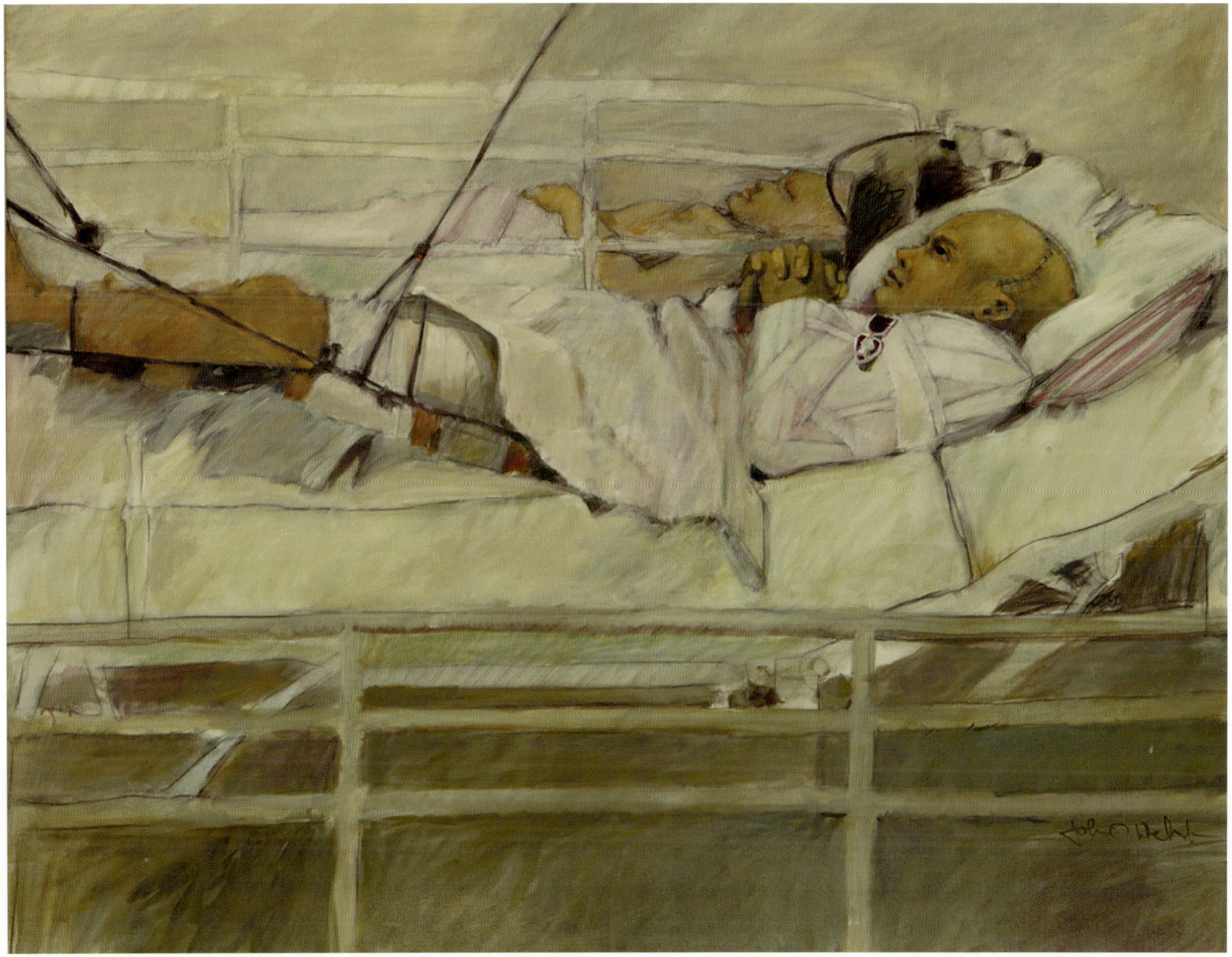

PURPLE HEART
Saigon, Vietnam
John O. Wehrle, 1966
Oil on canvas, 36" x 46"

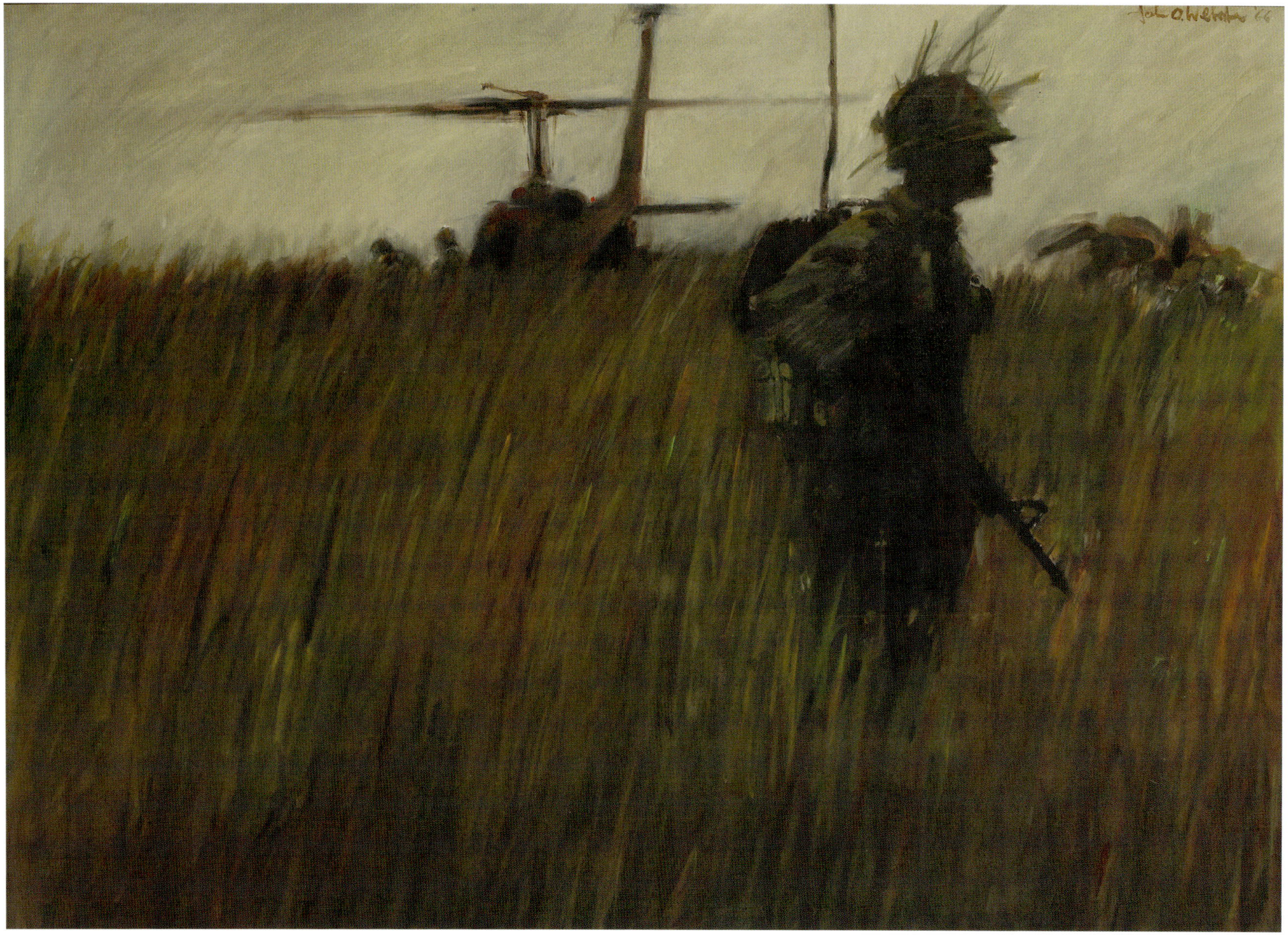

LANDING ZONE
Vietnam
John O. Wehrle, 1966
Oil on canvas, 32" x 46"

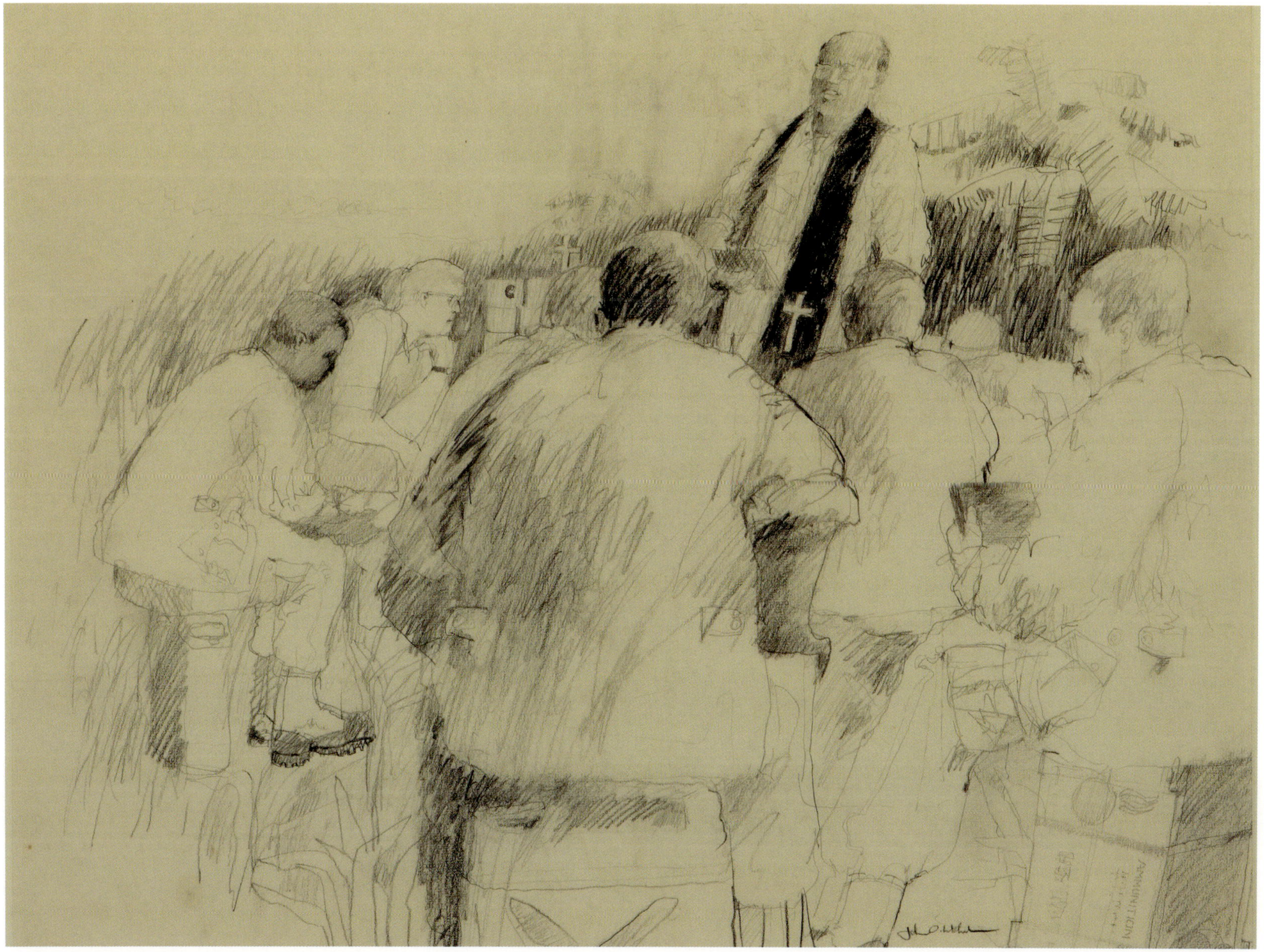

RELIGIOUS SERVICES AT CU CHI
Cu Chi, Vietnam
John O. Wehrle, 1966
Pencil on paper, 19" x 24½"

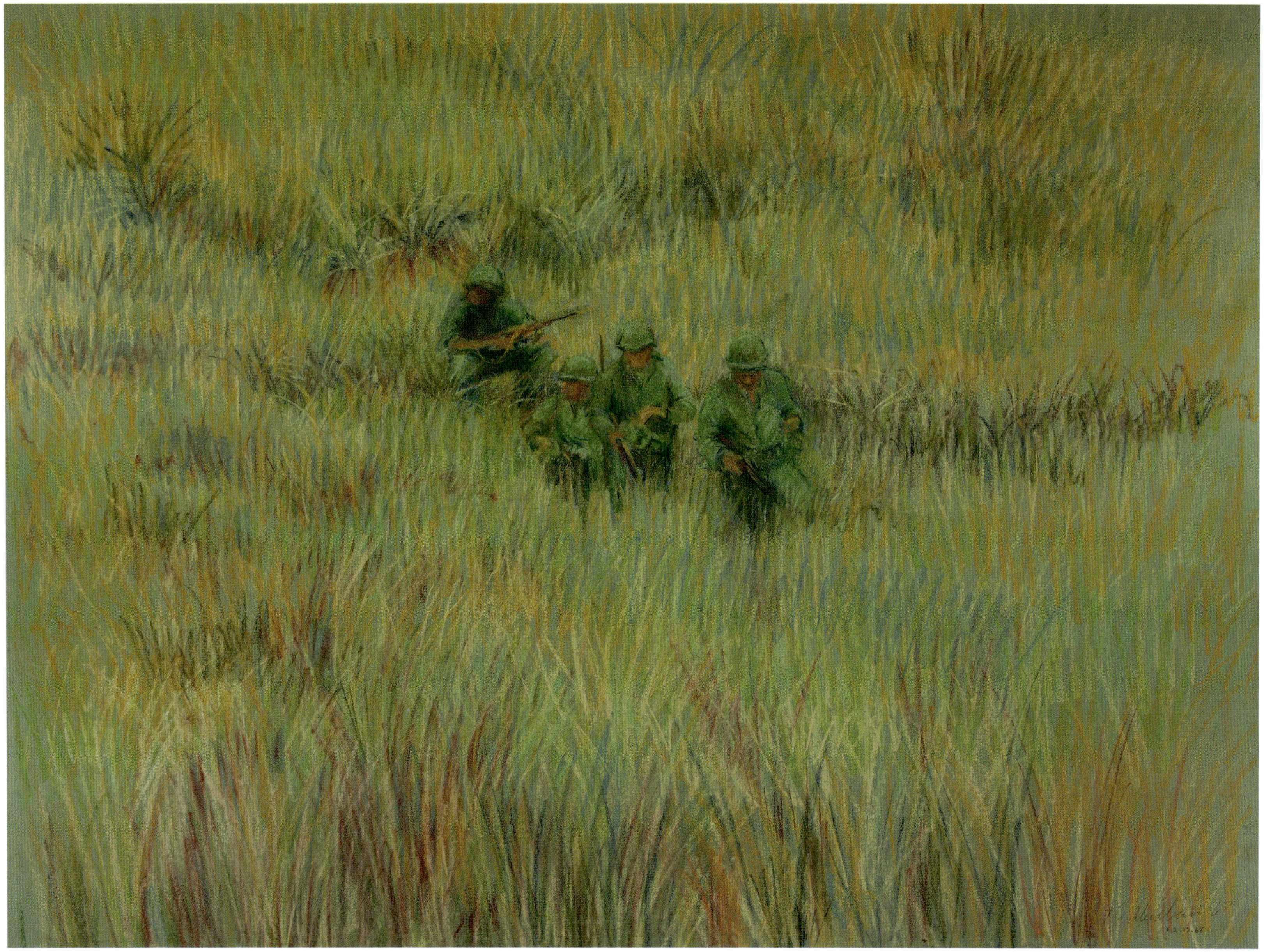

RECONNAISSANCE PATROL
Central Highlands, Vietnam
Theodore J. Abraham, 1967
Pastel on paper, 19" x 25"

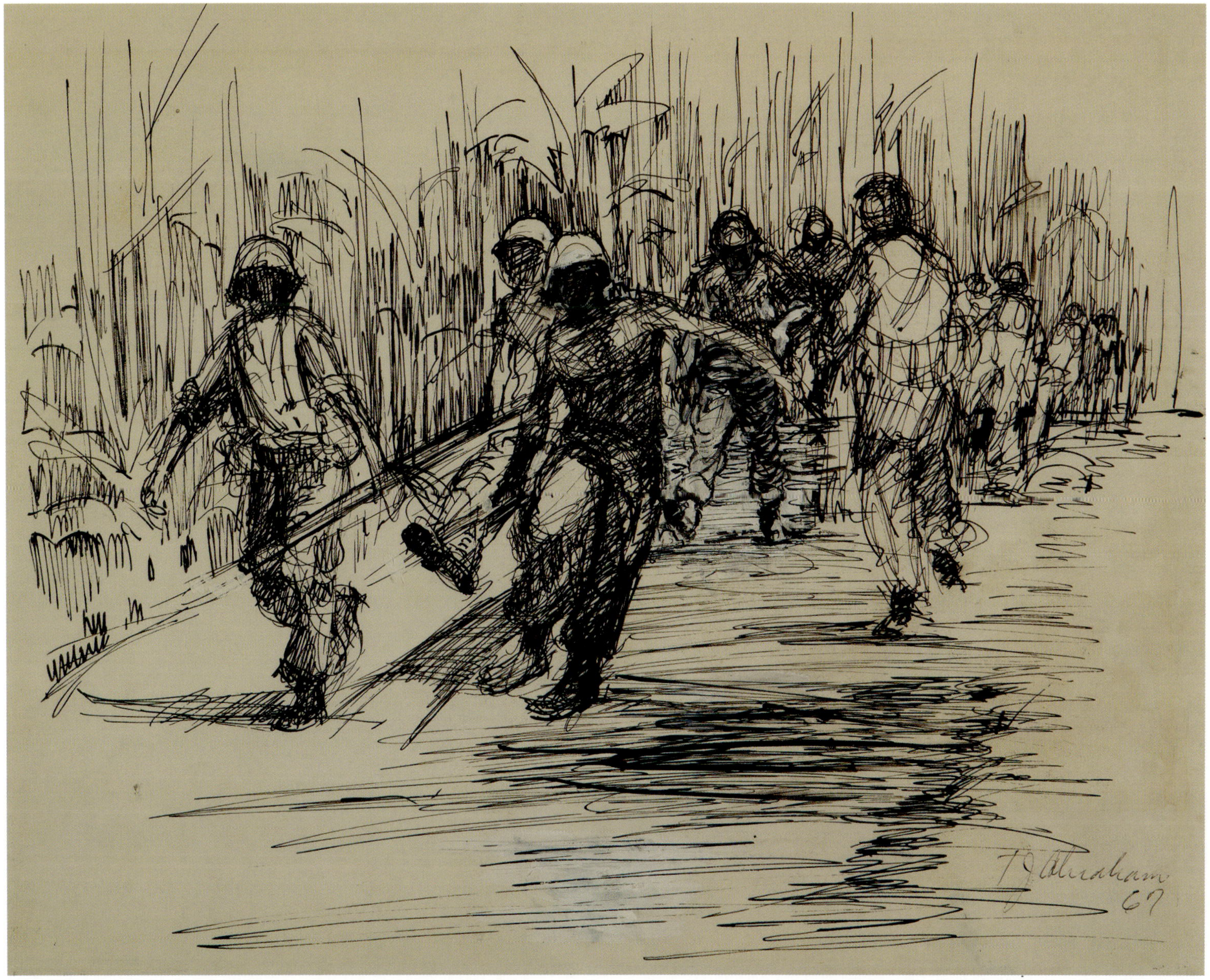

QUICK TRIP (EVACUATING A MINE CASUALTY)
Vietnam
Theodore J. Abraham, 1967
Ink on paper, 11" x 13½"

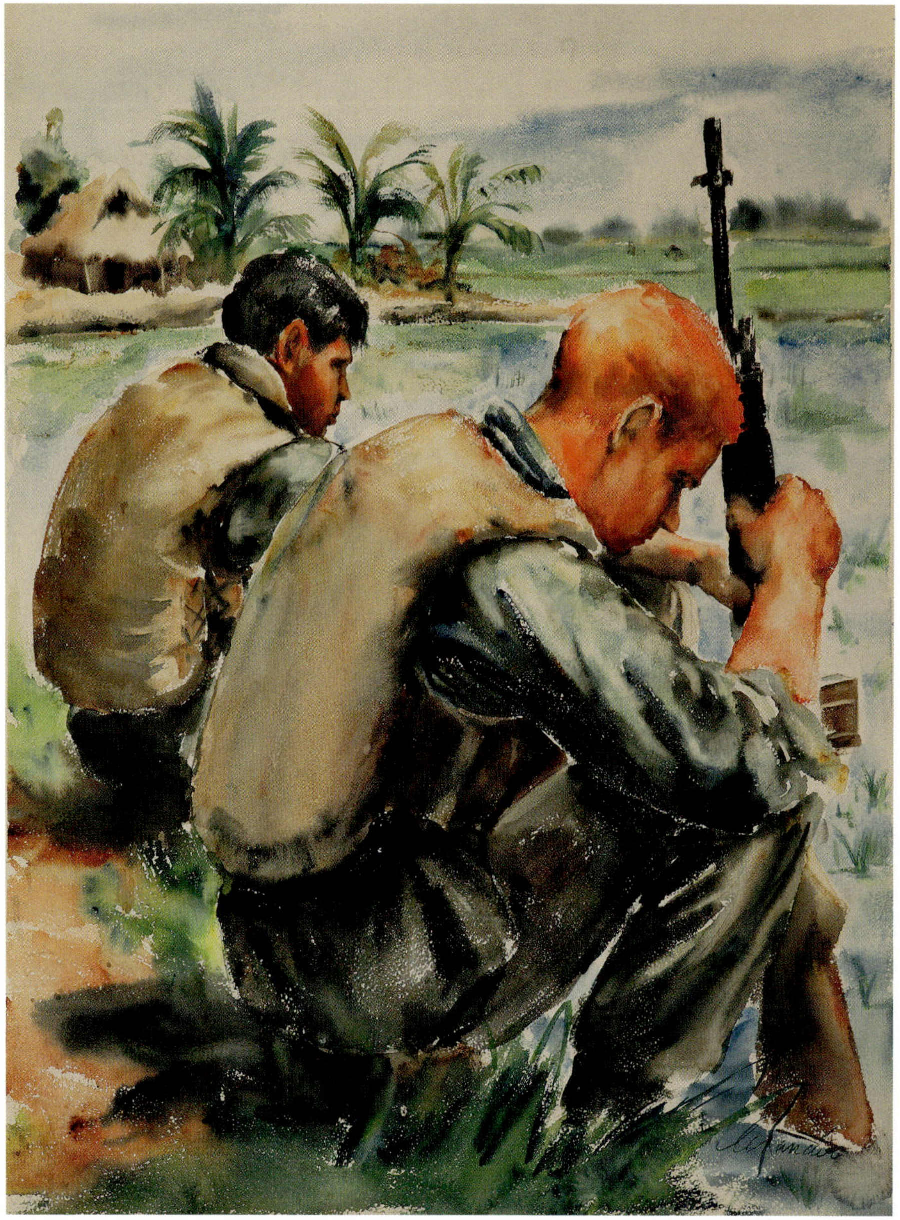

SOLDIERS RESTING (TEN MINUTE BREAK)
Thach Tru, Vietnam
Samuel E. Alexander, 1968
Watercolor on paper, 29½" x 22"

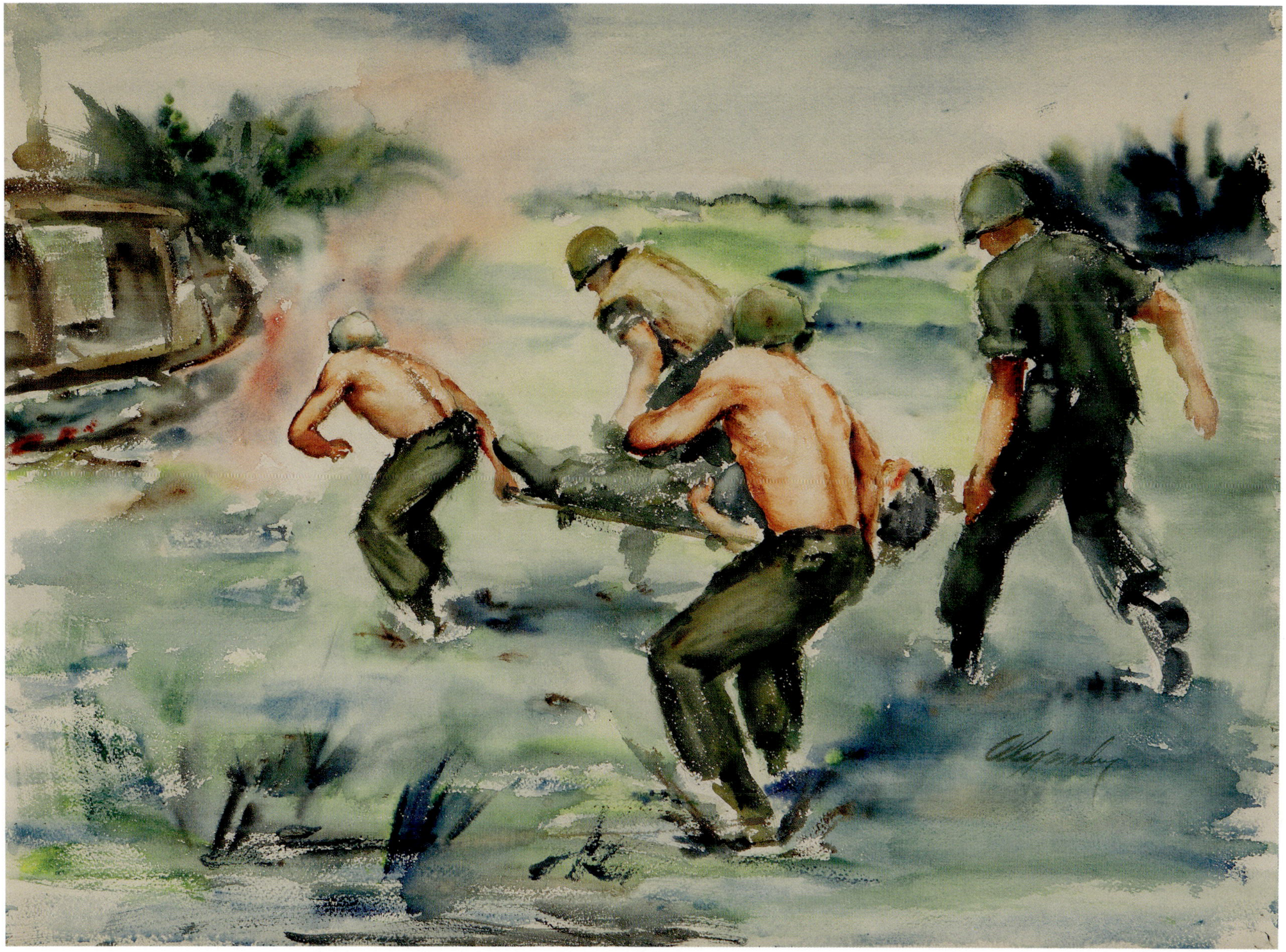

DUST OFF
Dong Tam, Vietnam
Samuel E. Alexander, 1967
Watercolor on paper, 22" x 29½"

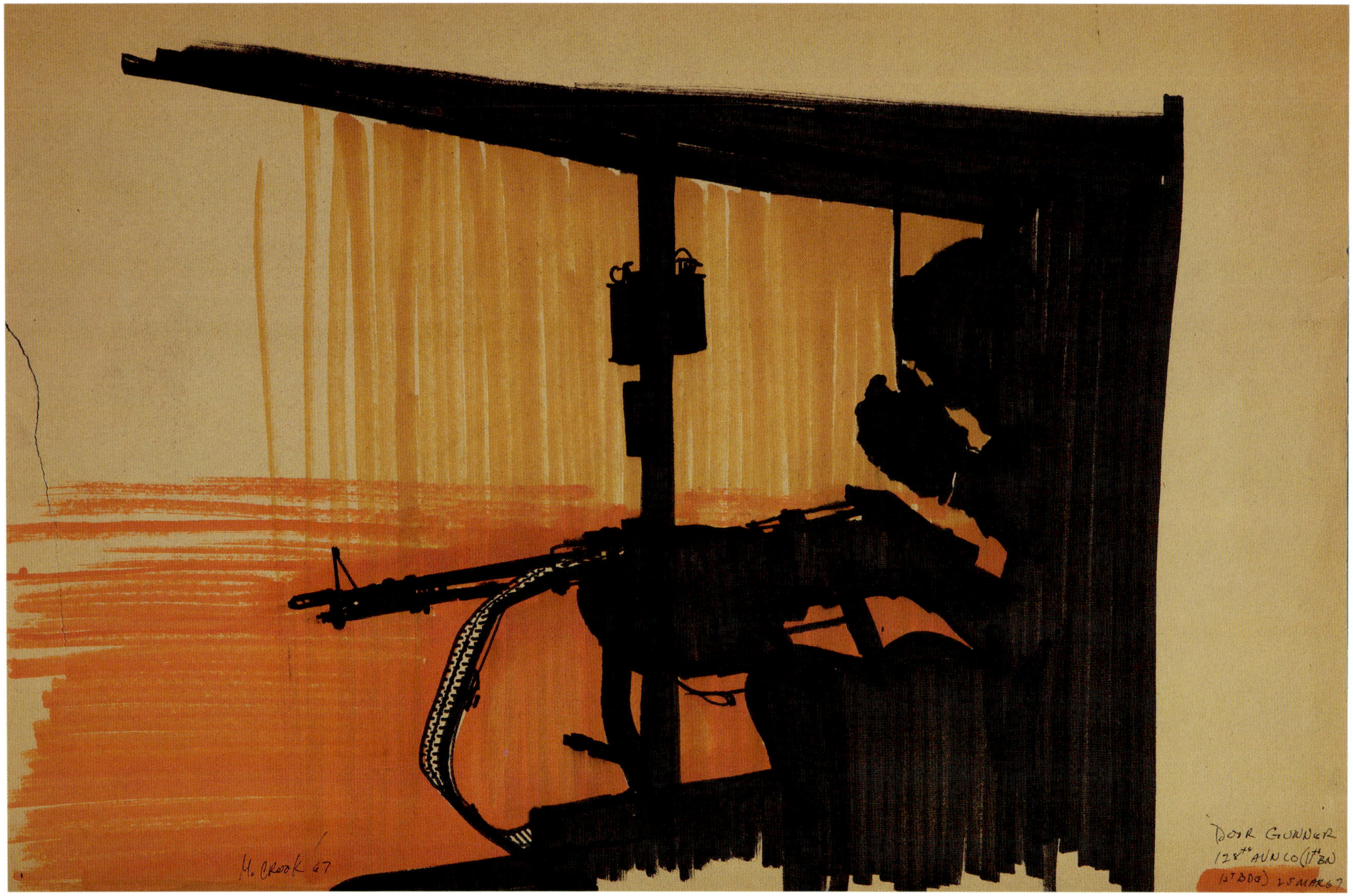

DOOR GUNNER
Vietnam
Michael R. Crook, 1967
Ink on paper, 17¾" x 13"

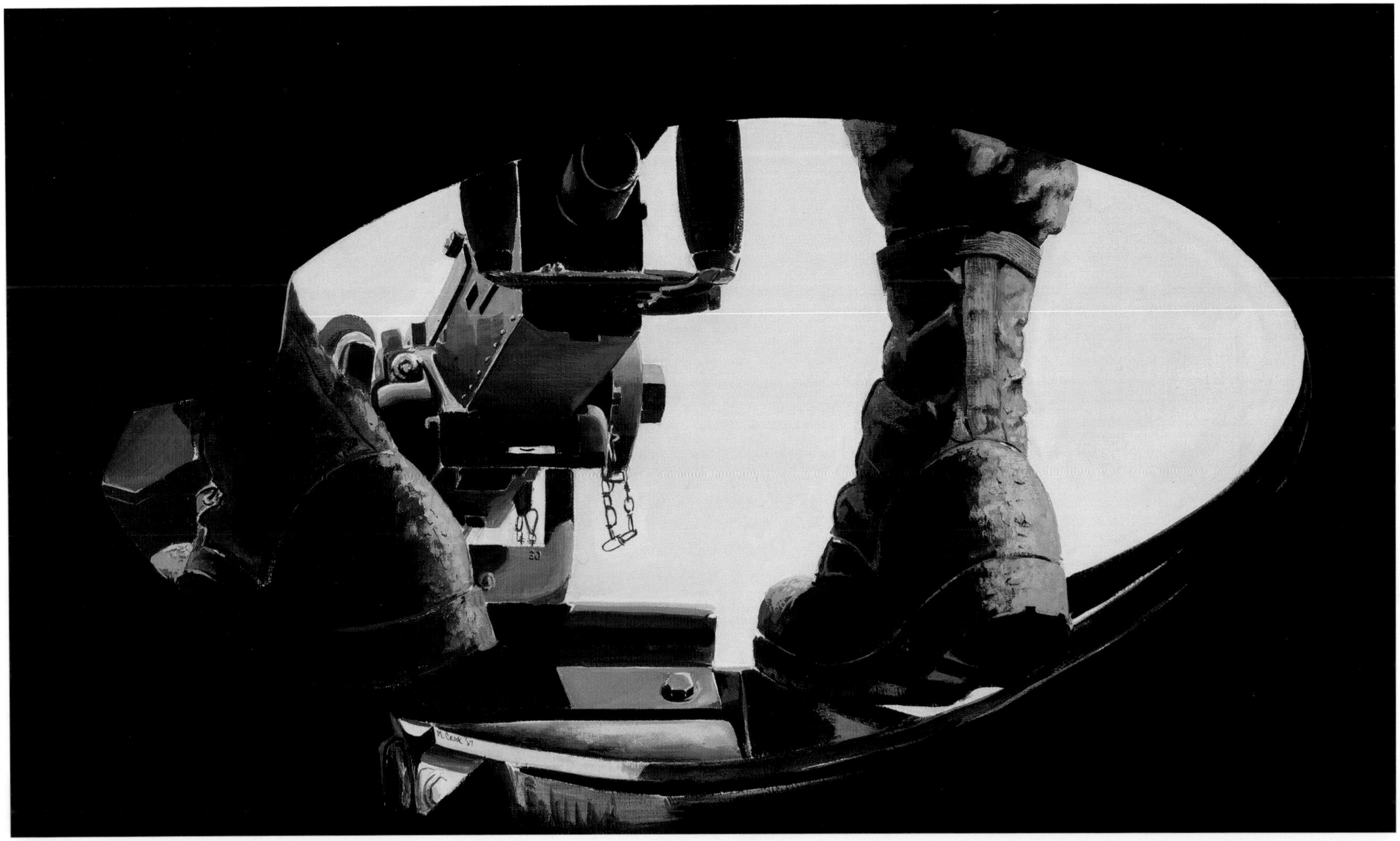

JONAH '67
Tan Hiep, Vietnam
Michael R. Crook, 1967
Acrylic on canvas, 23¾" x 34"

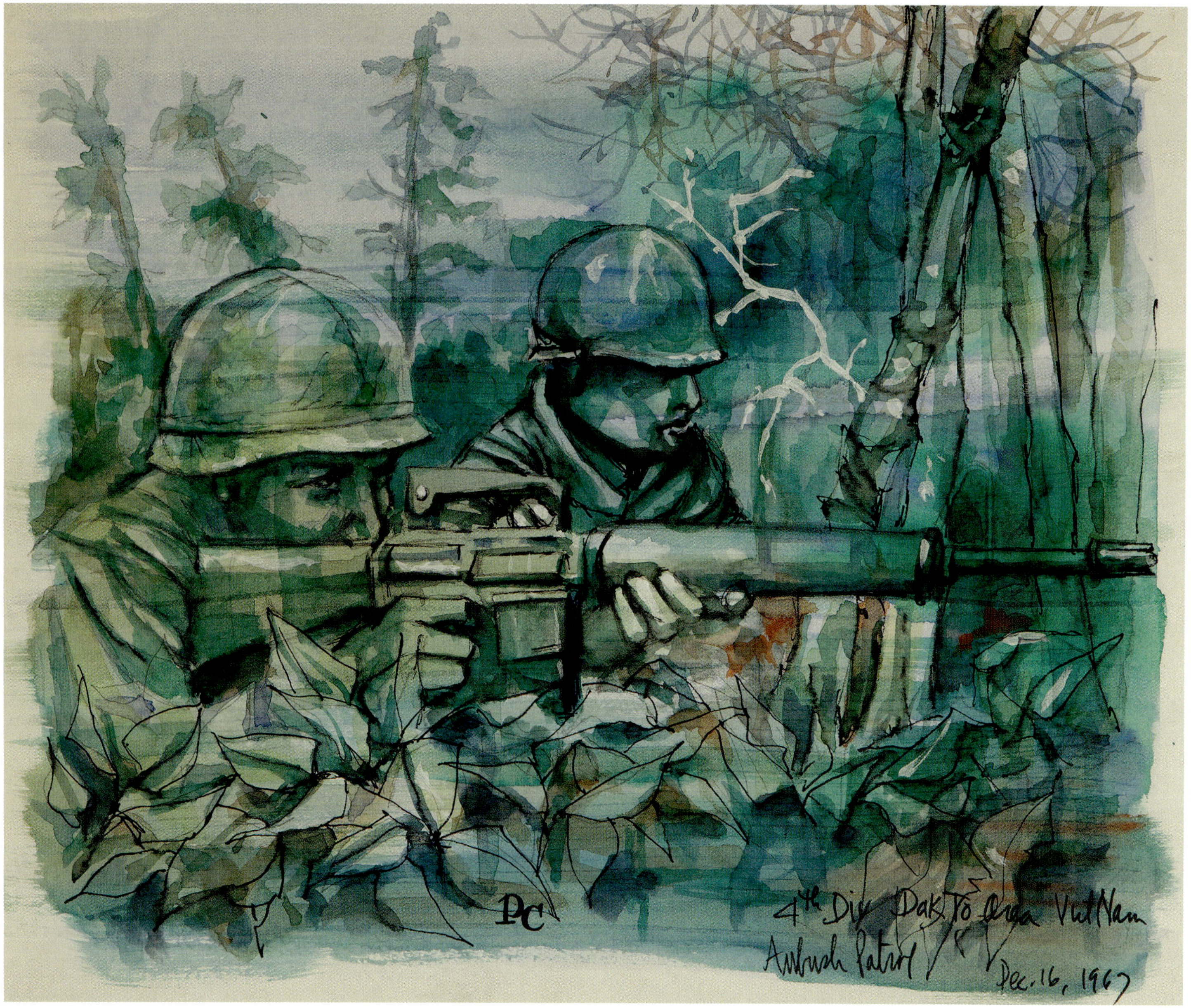

AMBUSH PATROL
Vietnam
Peter F. Copeland, 1967
Watercolor on paper, 11" x 14"

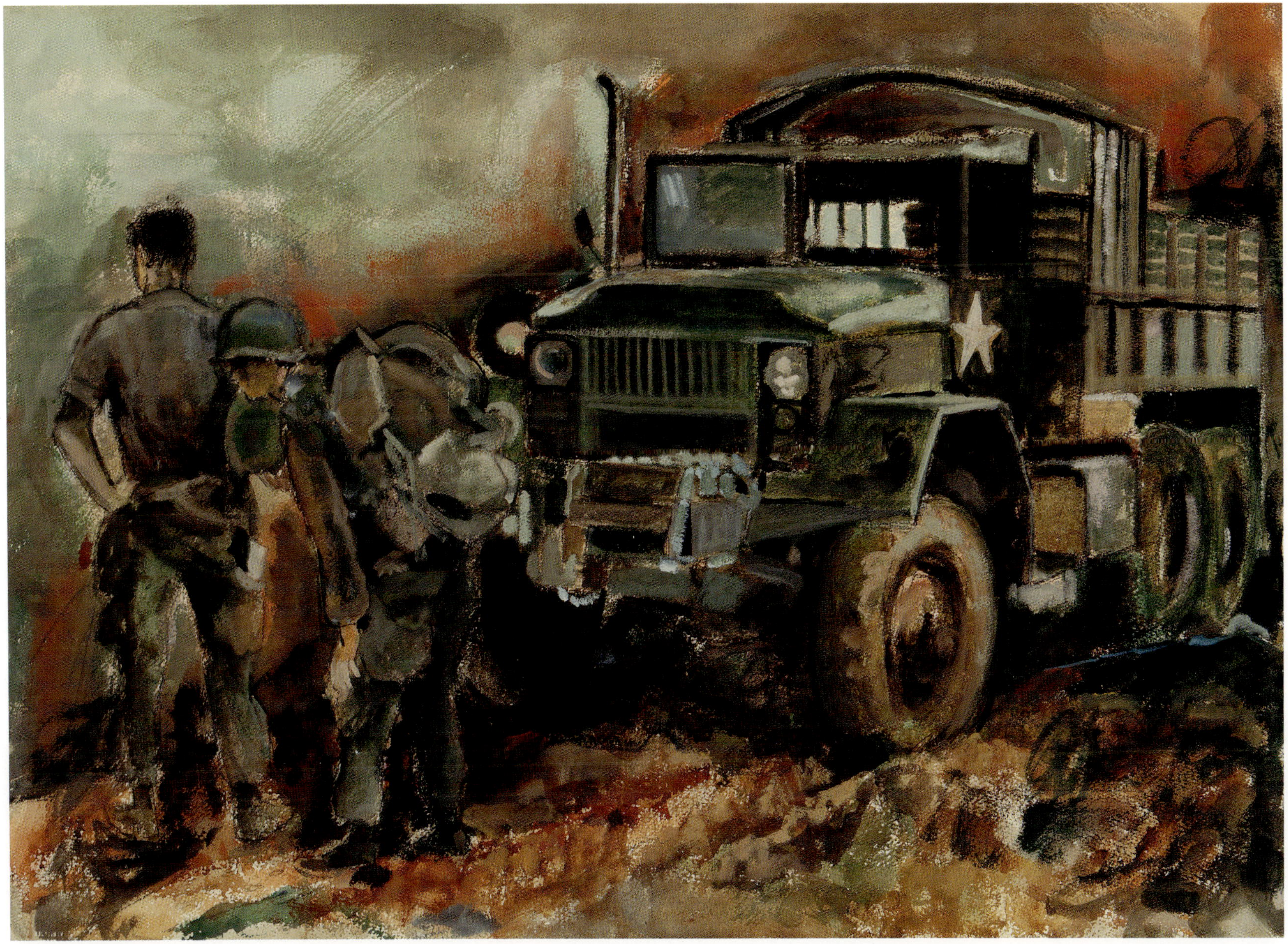

"We were supposed to show the Army in action in our paintings—at least that's what I understood. . . . I enjoyed being on the team, however, because I was allowed to do my own thing. Being a Combat Artist and being a civilian artist is like apples and oranges."
—Phillip W. Jones

BREAKDOWN
Vietnam
Phillip W. Jones, 1967
Acrylic on paper, 22" x 30"

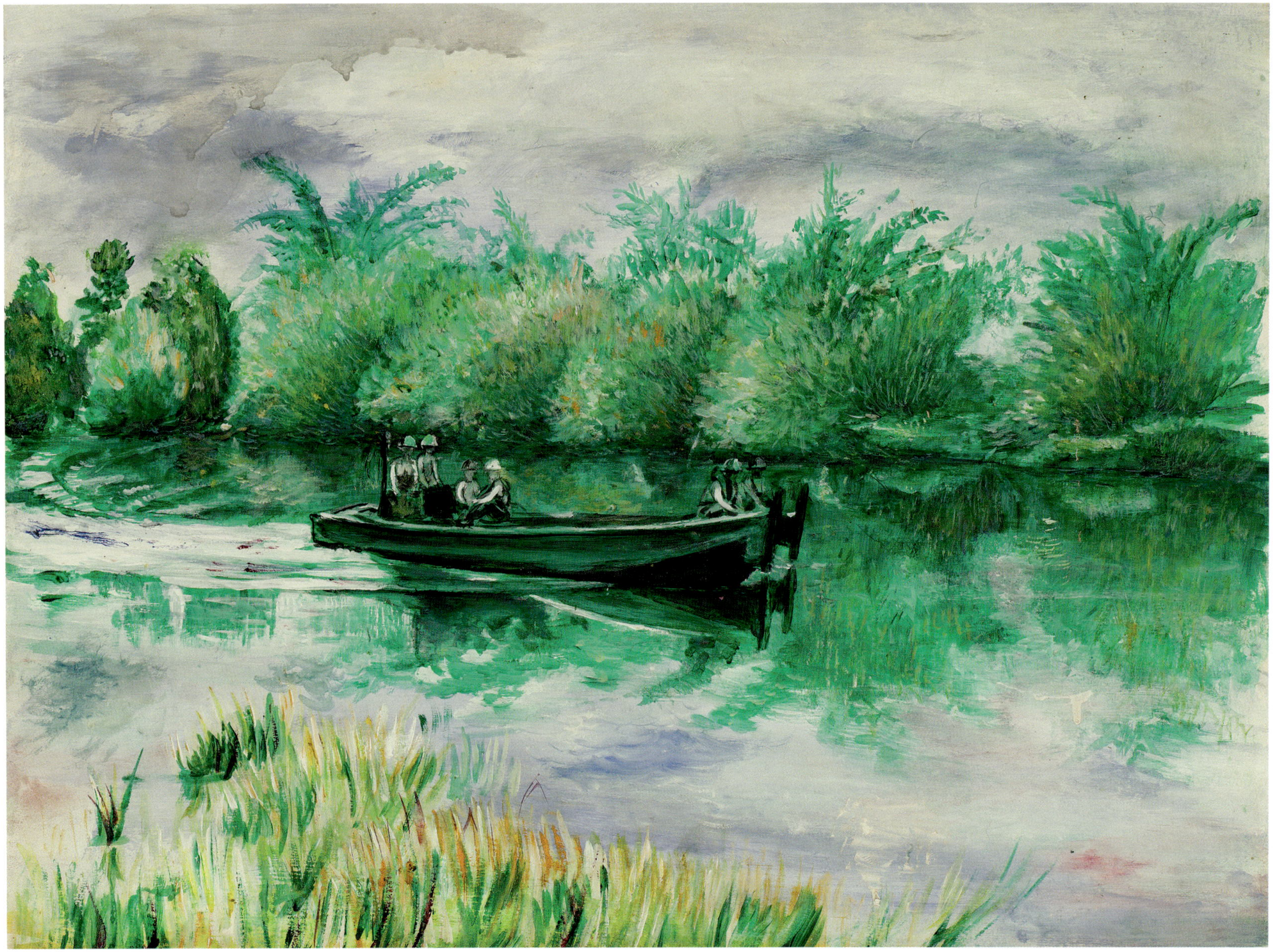

RIVER PATROL
Long Binh, Vietnam
Daniel T. Lopez, 1967
Tempera on paper, 15" x 20"

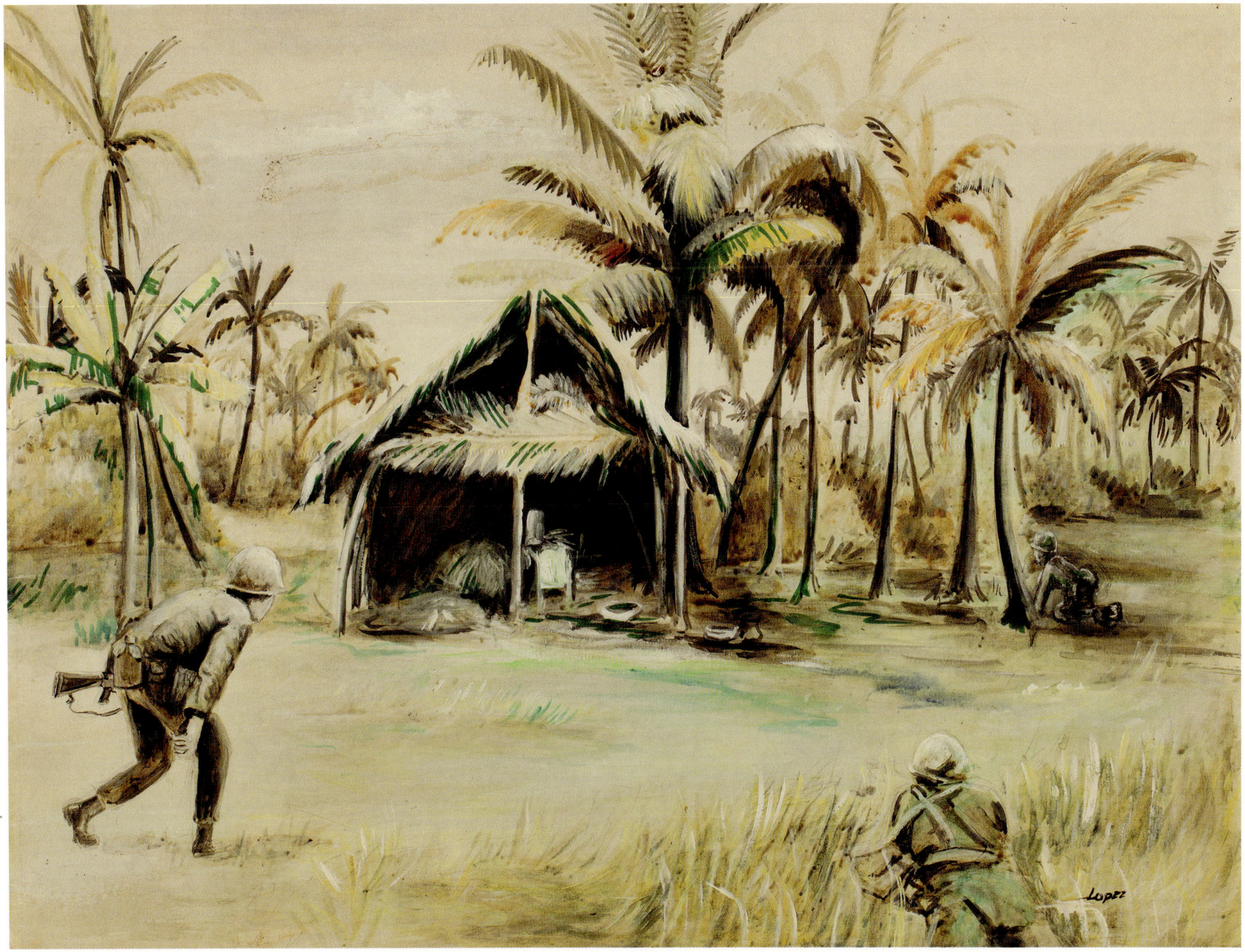

HUNTED HUT
Vietnam
Daniel T. Lopez, 1967
Watercolor on board, 19" x 24¾"

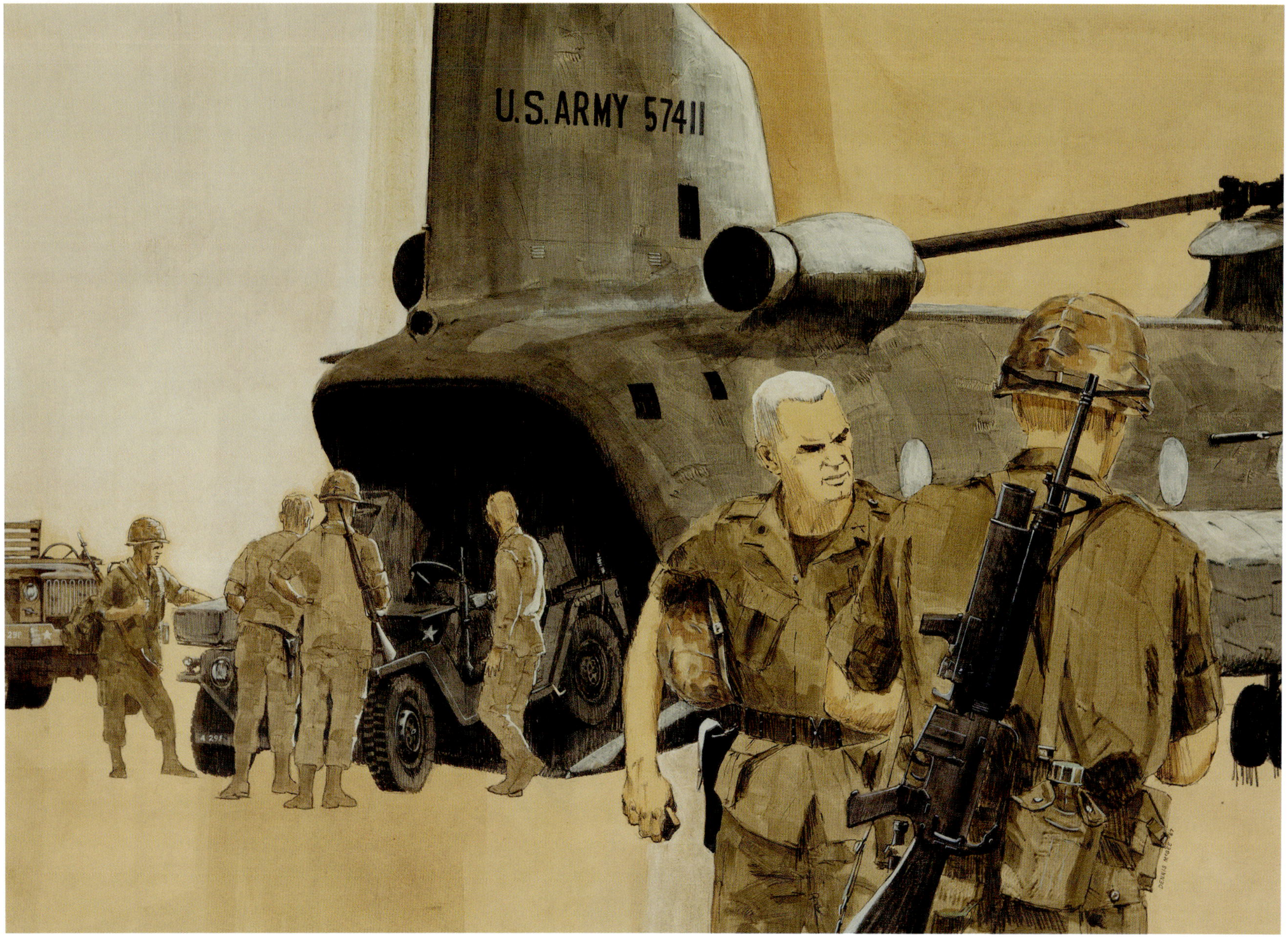

JOLLY GREEN GIANT
Pleiku, Vietnam
Dennis O. McGee, 1967
Acrylic on board, 28" x 40"

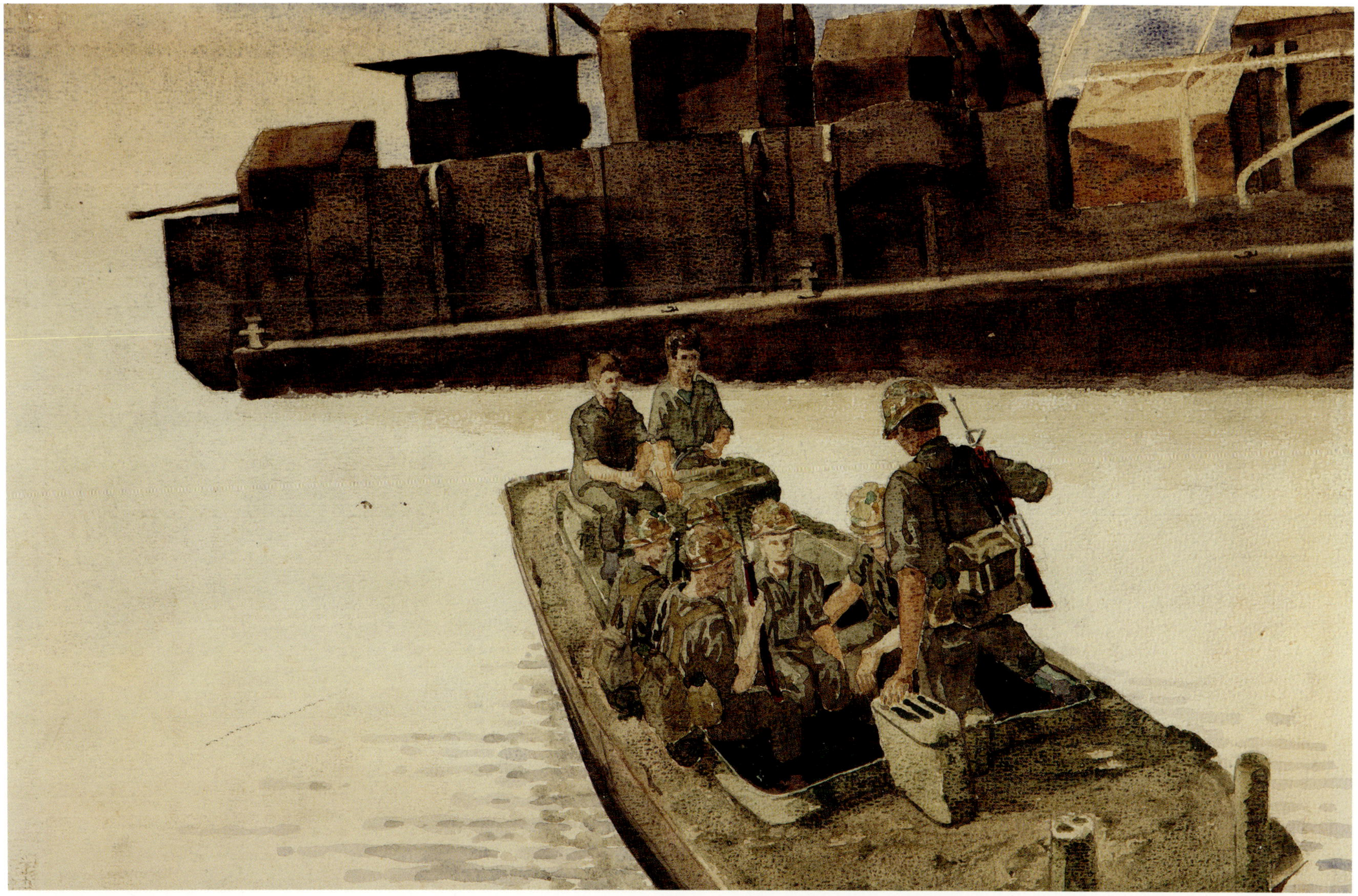

LEAVING FOR PATROL
Rung Sat, Vietnam
Robert T. Myers, 1967
Watercolor on paper, 11¾" x 17¾"

"I remember the hard lives and unsung sacrifices made by those young men of 19 and 20 years of age."
—Burdell Moody

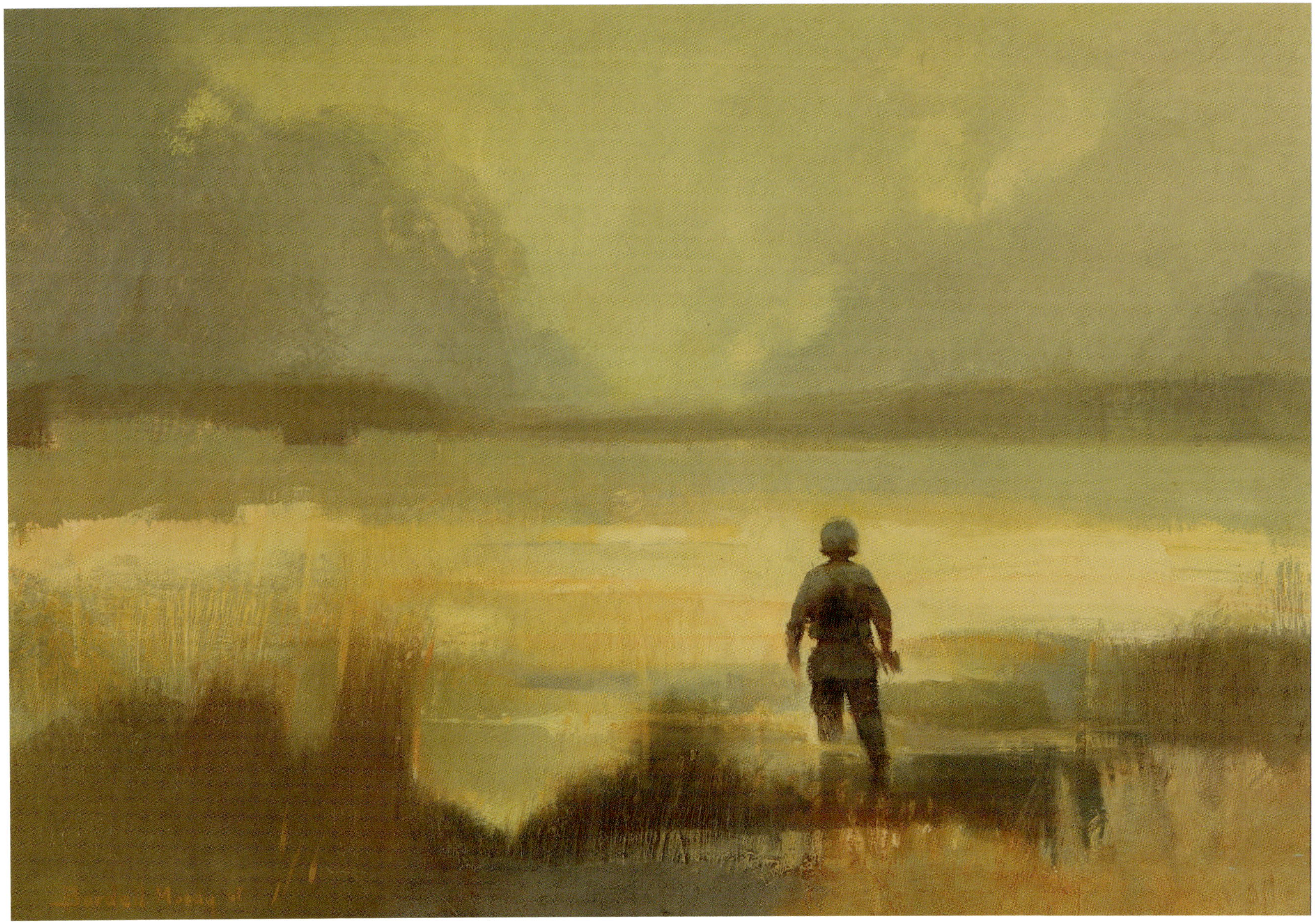

POINT MAN
Phuoc Vinh, Vietnam
Burdell Moody, 1967
Oil on canvas, 20" x 28"

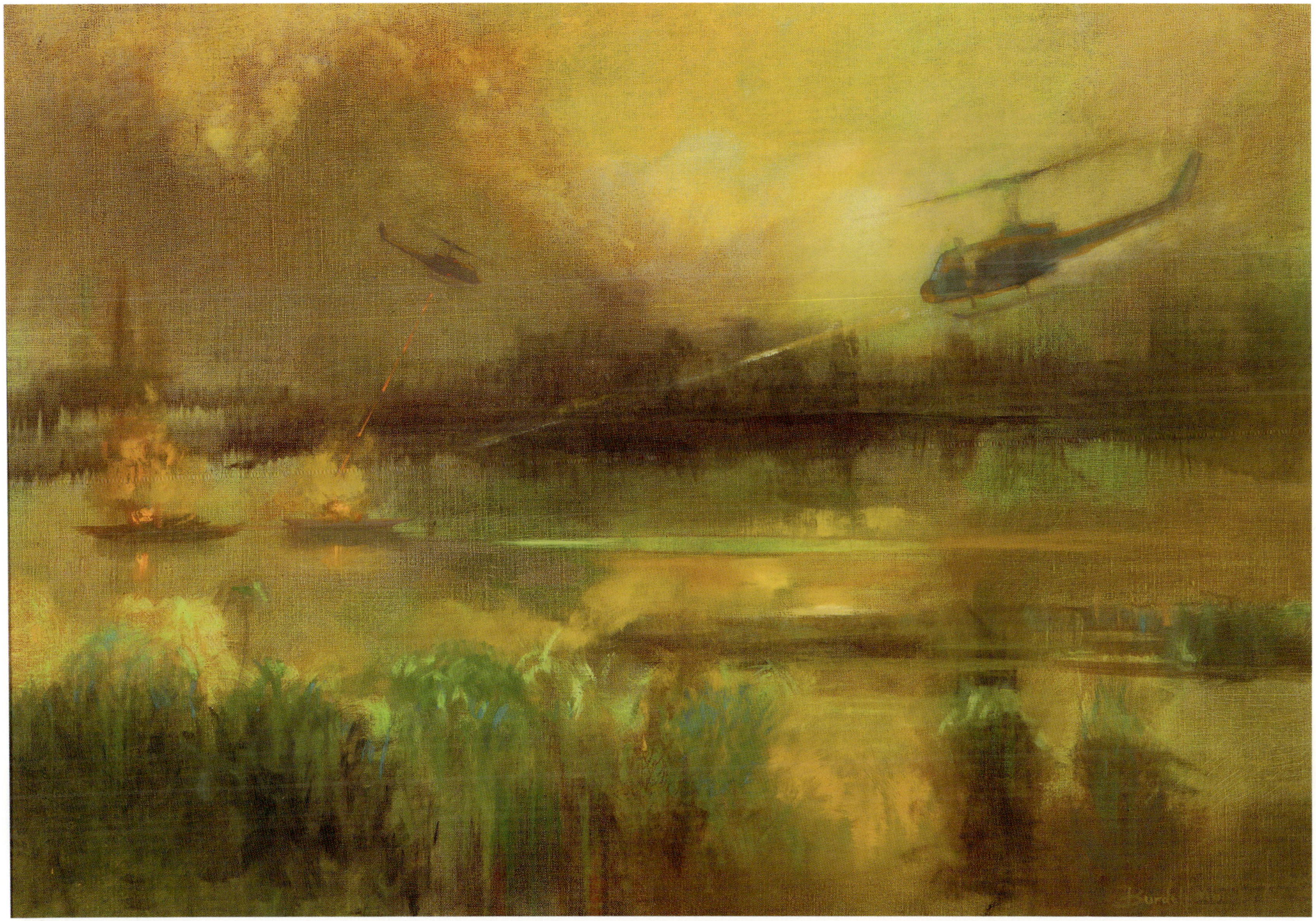

FIREFLY MISSION
Rung Sat, Vietnam
Burdell Moody, 1967
Oil on canvas, 24" x 34"

"It was early in my experience as a soldier artist that I came to understand the hardship ordinary soldiers in Vietnam experienced on a day-to-day basis."
—James Pollock

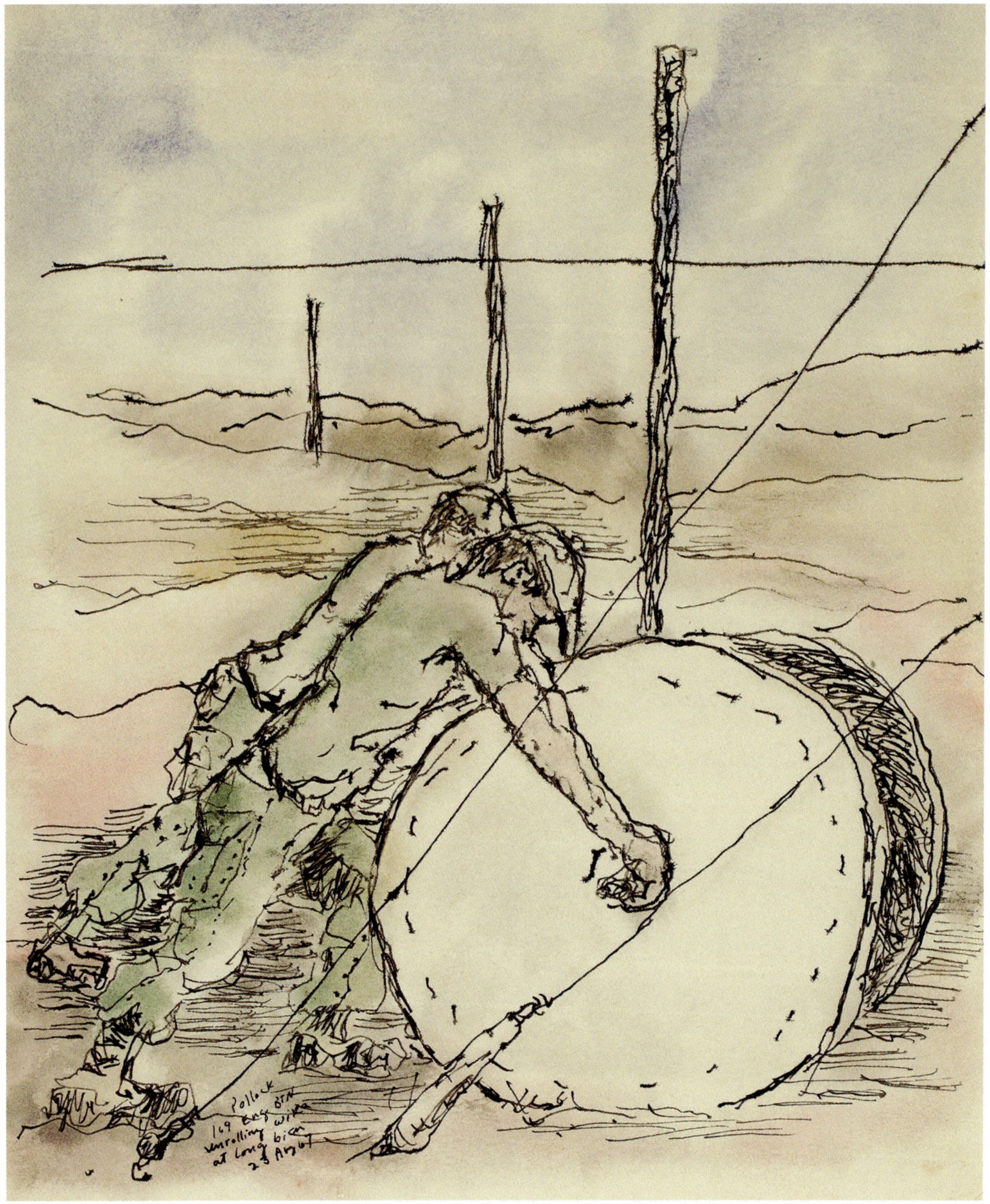

UNROLLING WIRE
Long Binh, Vietnam
James Pollock, 1967
Watercolor on paper, 12" x 14¼"

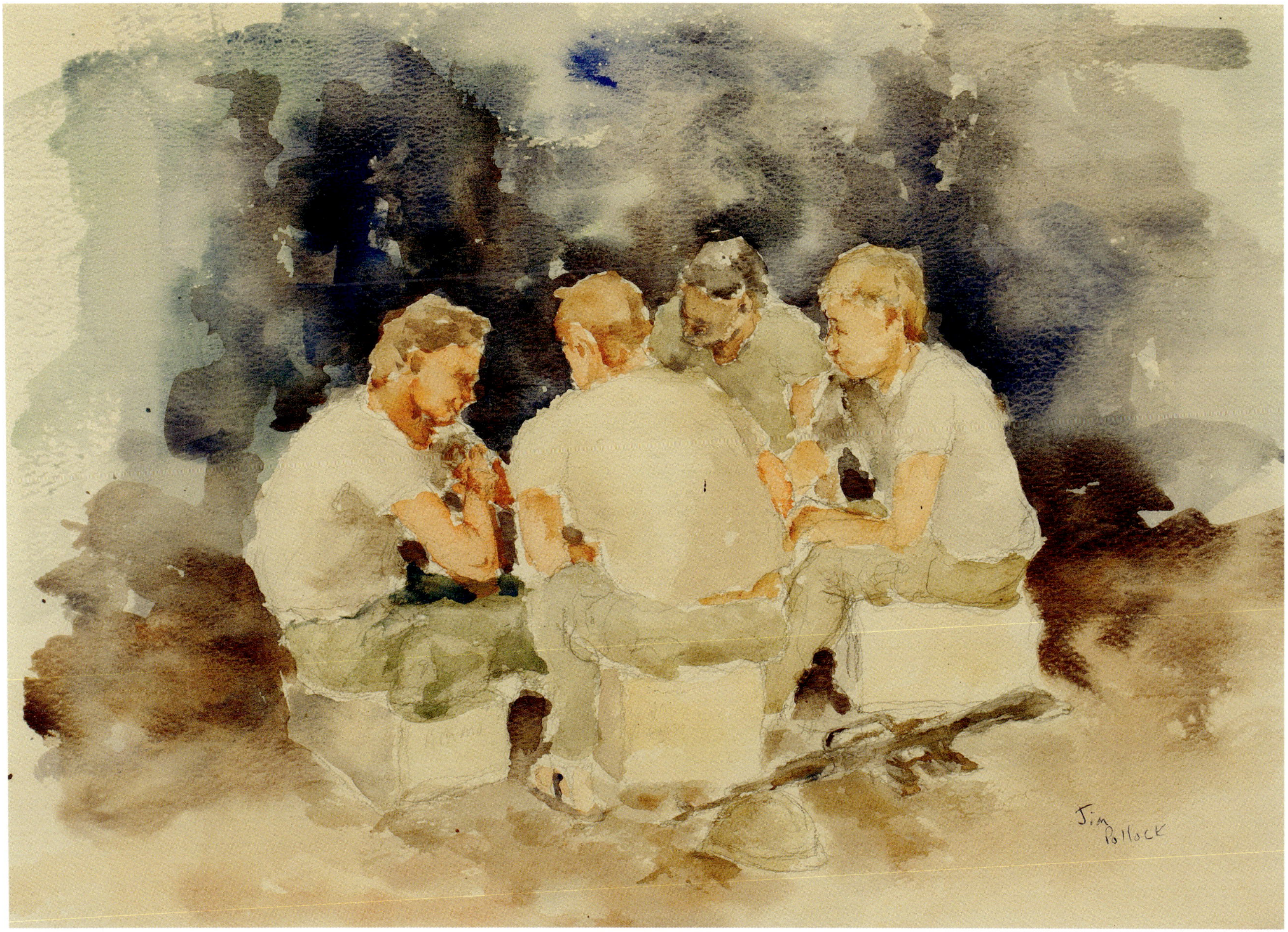

CARD GAME
Vietnam
James Pollock, 1967
Watercolor on paper, 13" x 16"

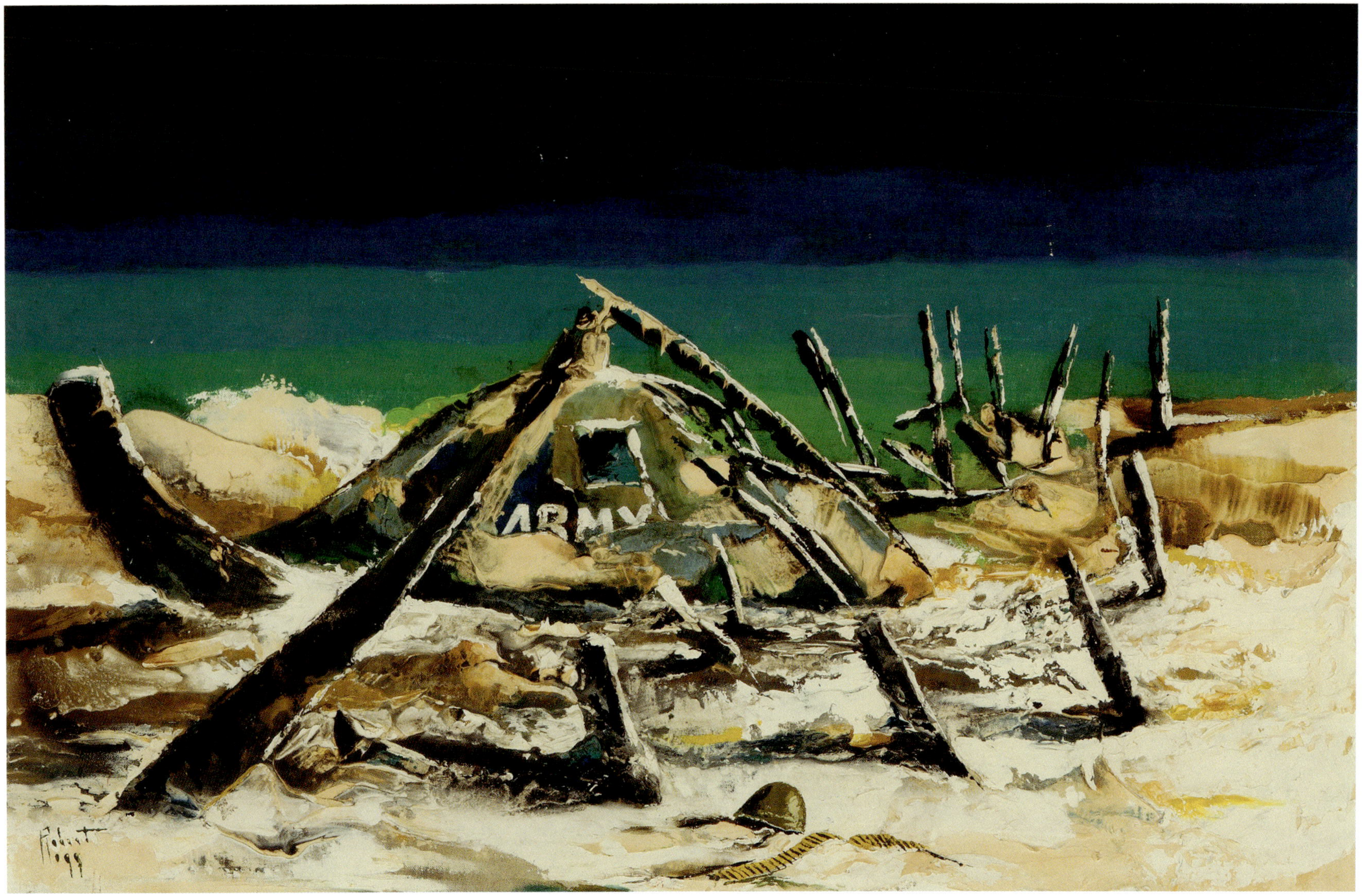

RUINS OF COMBAT
Vietnam
Robert B. Rigg, 1967
Oil on canvas, 24" x 36"

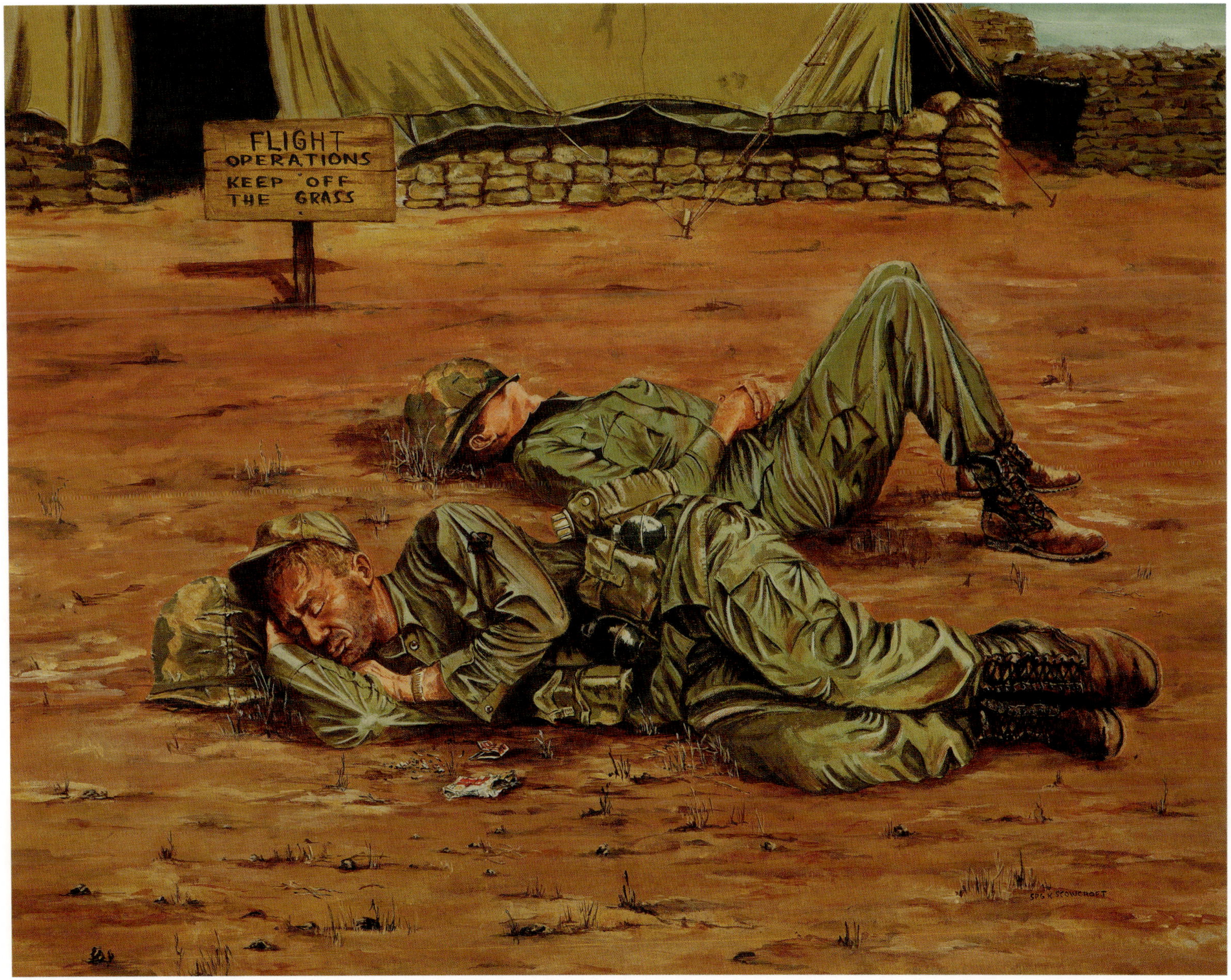

THE WAITING GAME
Dragon Mountain Heliport, Pleiku, Vietnam
Kenneth J. Scowcroft, 1967
Acrylic on canvas, 24" x 30"

"One thing I could not portray with paint was the feeling of the unreality of it all. The Vietnam War I experienced was so different and far removed from all the other experiences in my life that it felt like being on Mars."

—Ronald A. Wilson

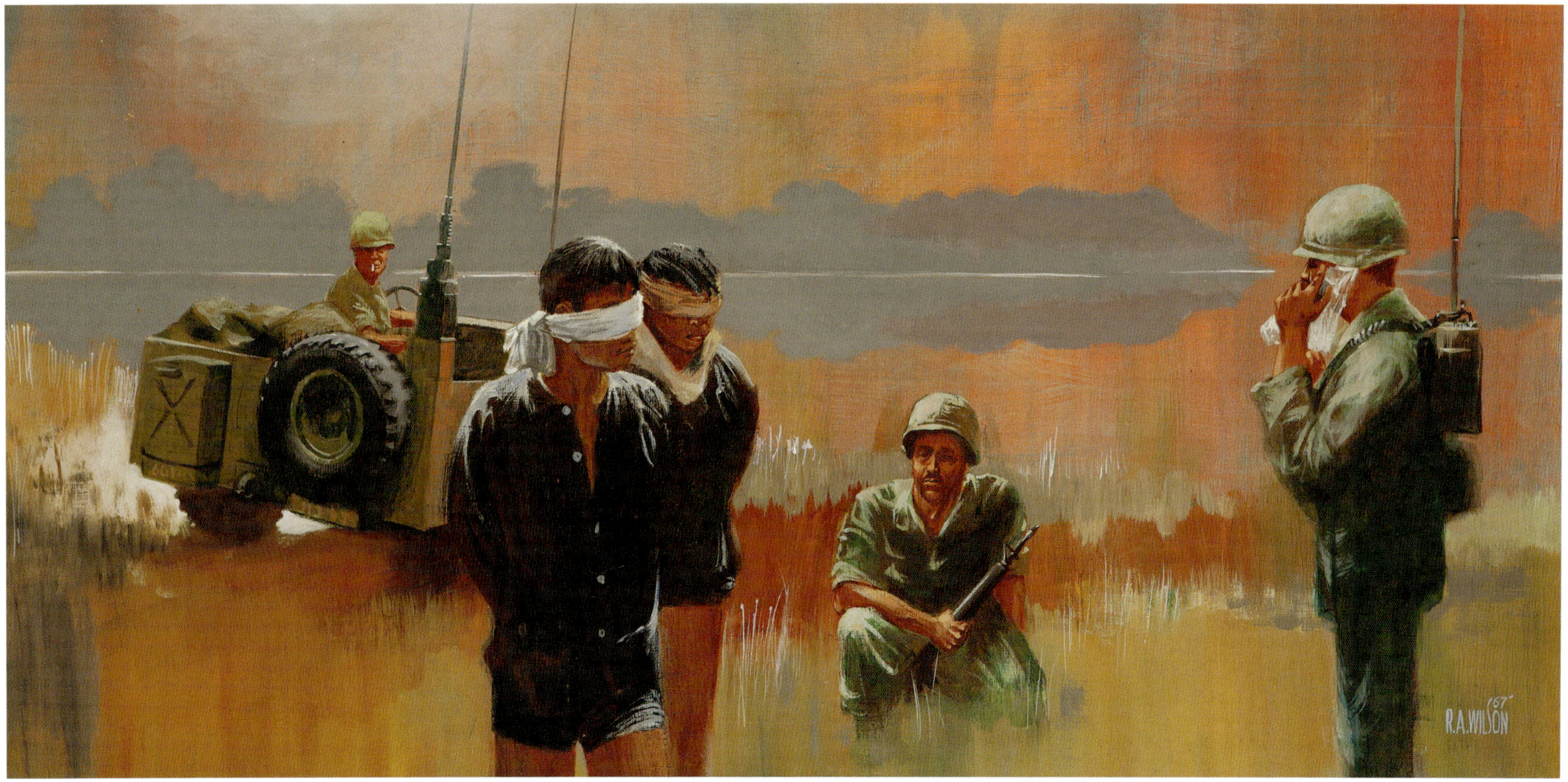

APPREHENDED
Vietnam
Ronald A. Wilson, 1967
Acrylic on masonite, 20" x 40"

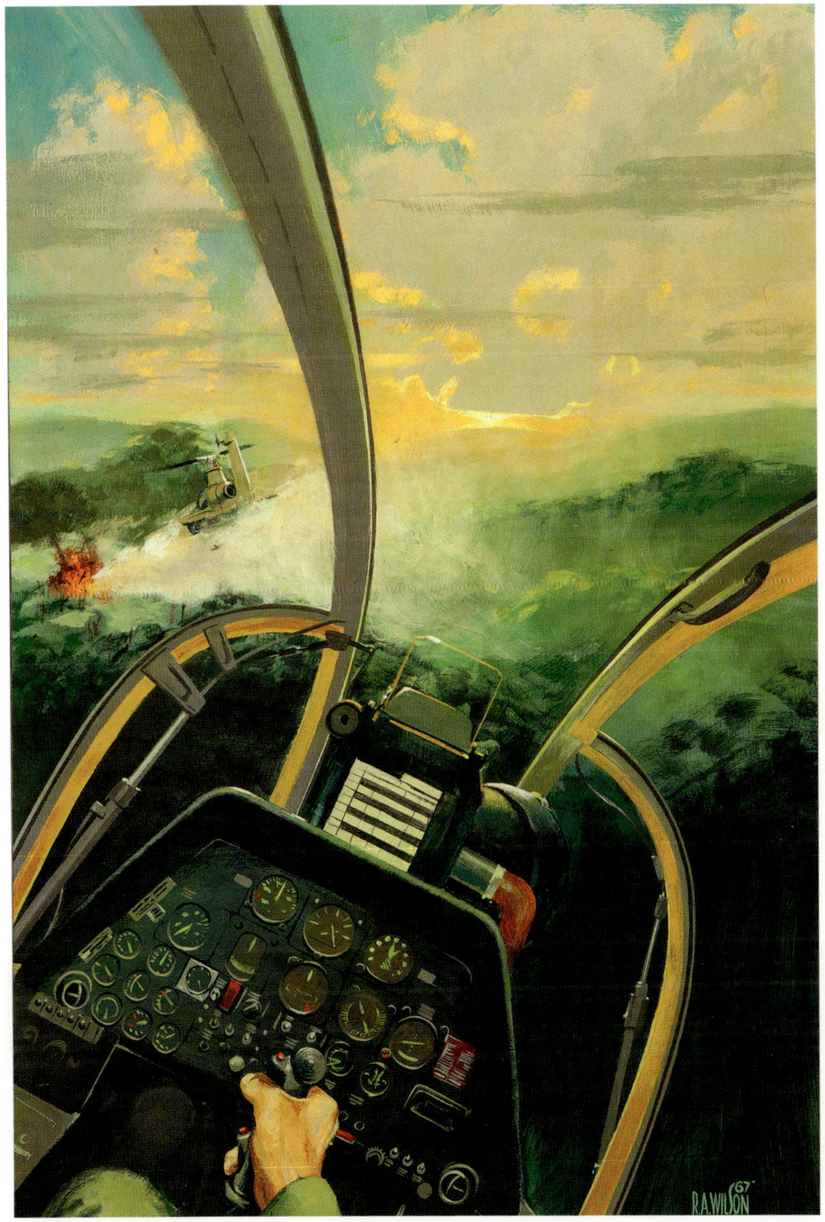

FIRING RUN
Vietnam
Ronald A. Wilson, 1967
Acrylic on masonite, 36" x 24"

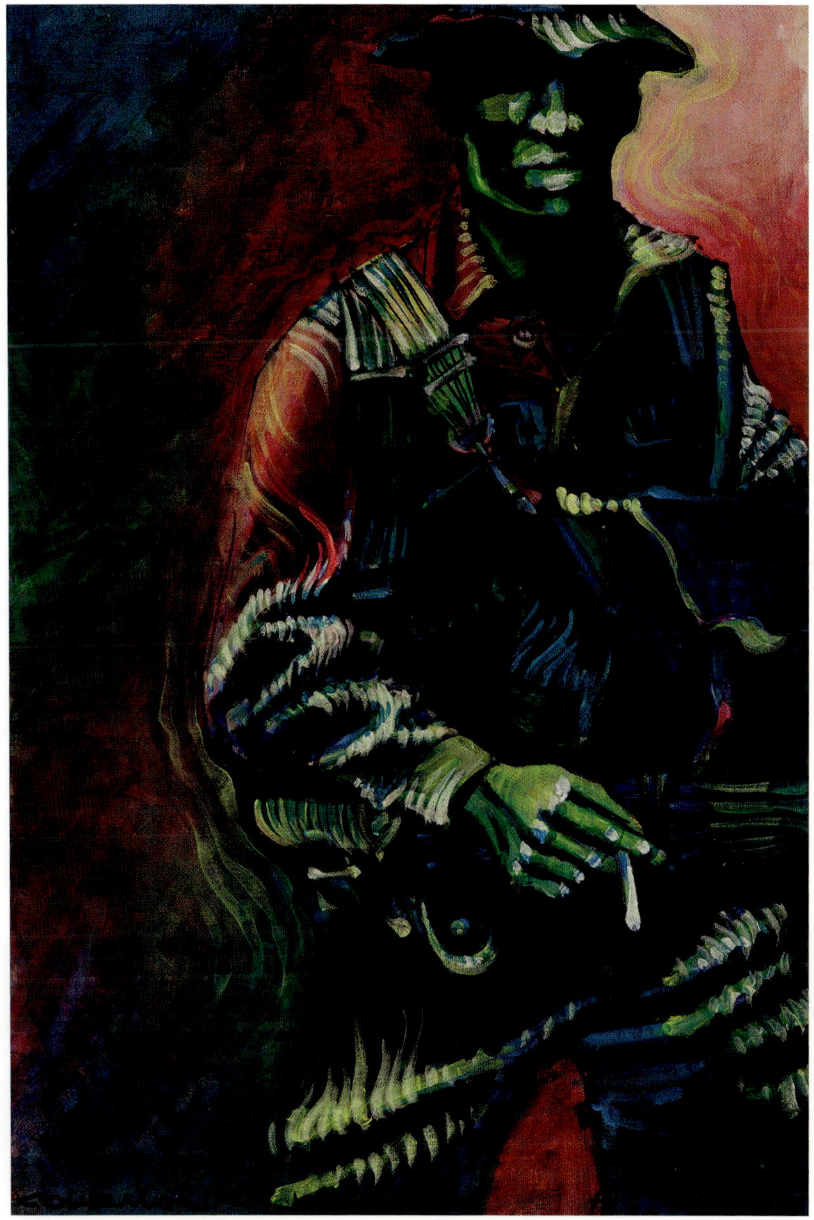

SMOKE BREAK
Vietnam
Robert T. Coleman, 1968
Acrylic on canvas, 30" x 20¼"

"Troops on the move try to capture the constant uncertainty of life at its rawest, never sure of what chaos and the necessary response might be next. And always, the omnipresent chopper."
—Brian H. Clark

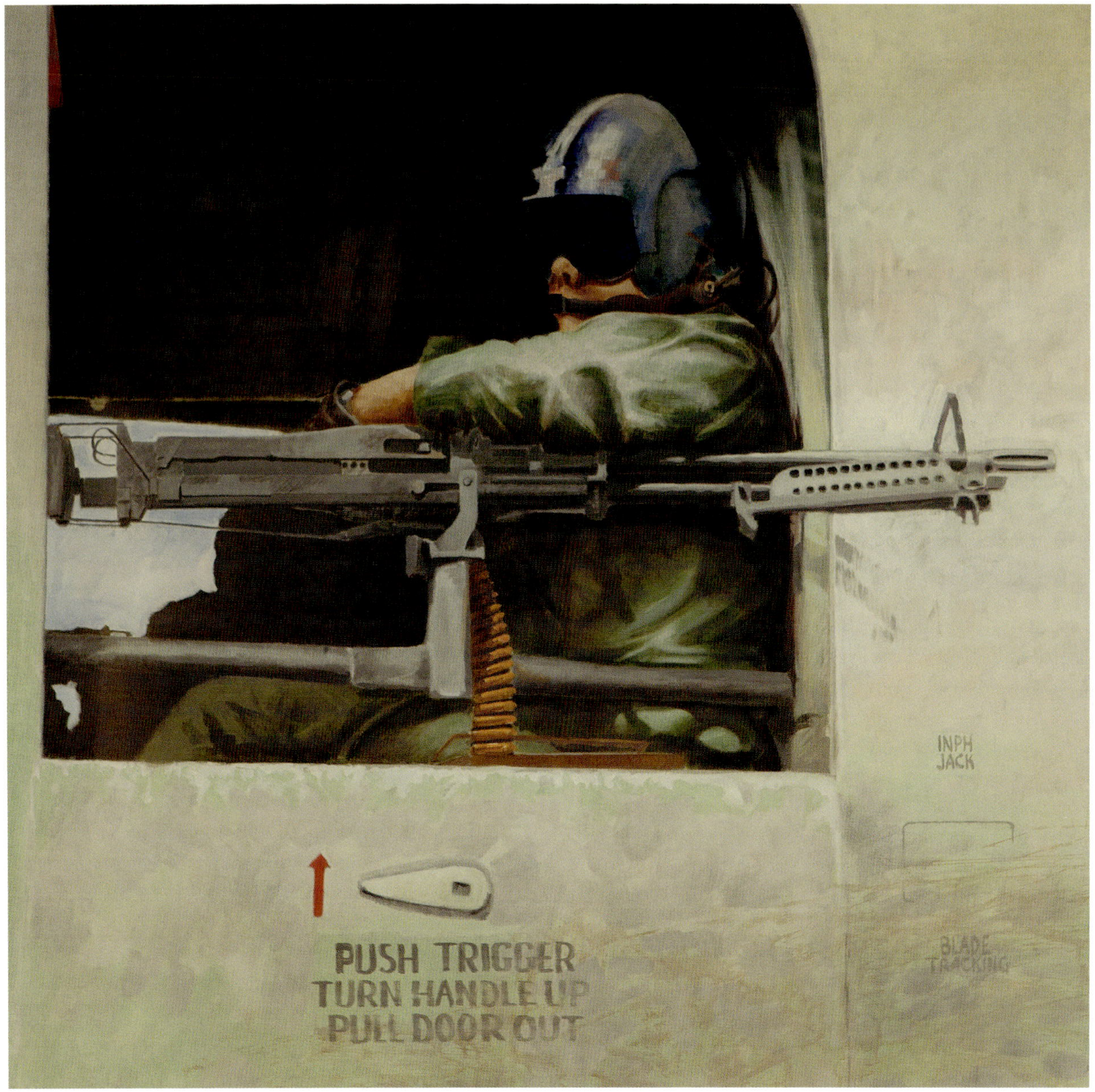

CHINOOK GUNNER
Vietnam
Brian H. Clark, 1968
Acrylic on canvas, 40" x 40"

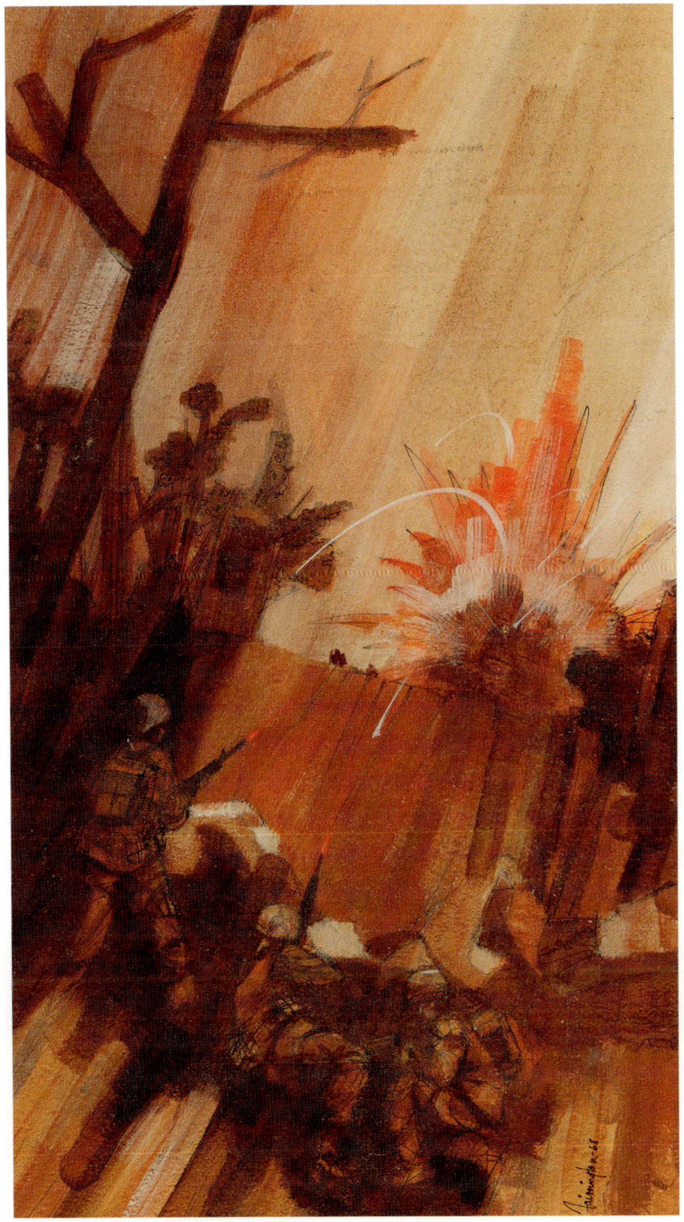

FIRFIGHT AT DAKTO

Dak To, Vietnam

David N. Fairrington, 1968

Acrylic on board, 20¾" x 11¼"

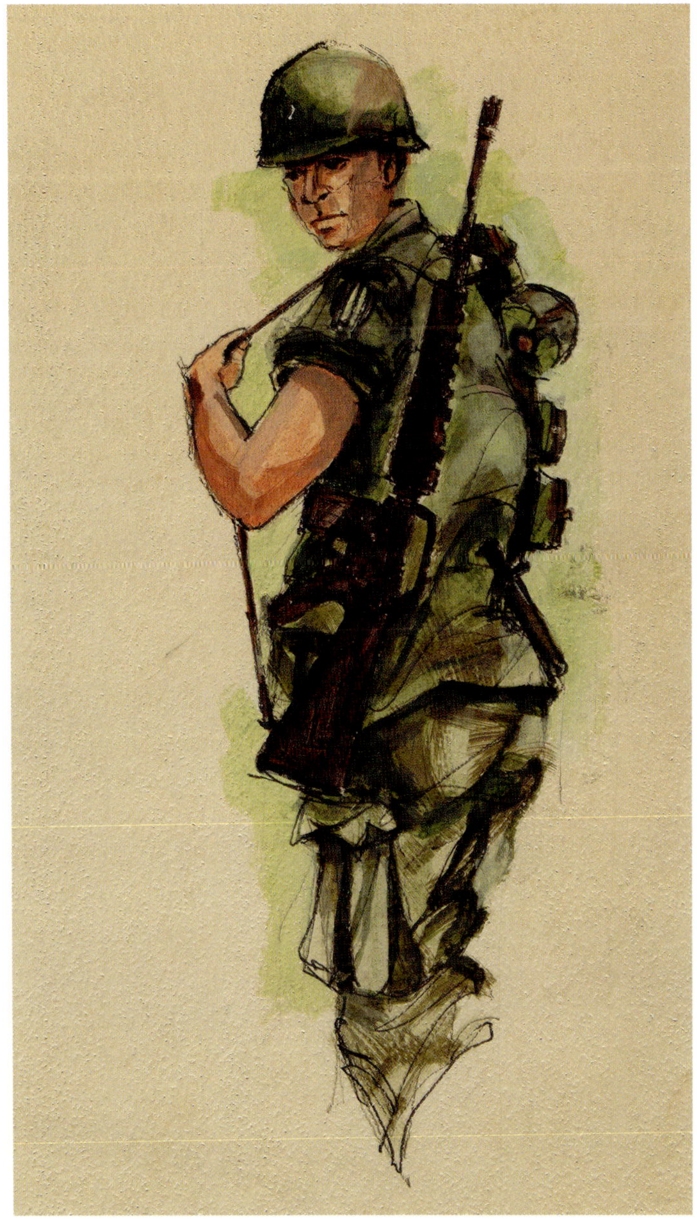

SOLDIER SKETCH

Vietnam

David N. Fairrington, 1968

Acrylic on board, 18¼" x 11½"

"I would only paint still lives and landscapes when I got back and then I stopped painting altogether for years. I didn't really start painting (again) until 1995. That's when I realized, people are all that matter. It took me a long time to know that in my heart."

—David N. Fairrington

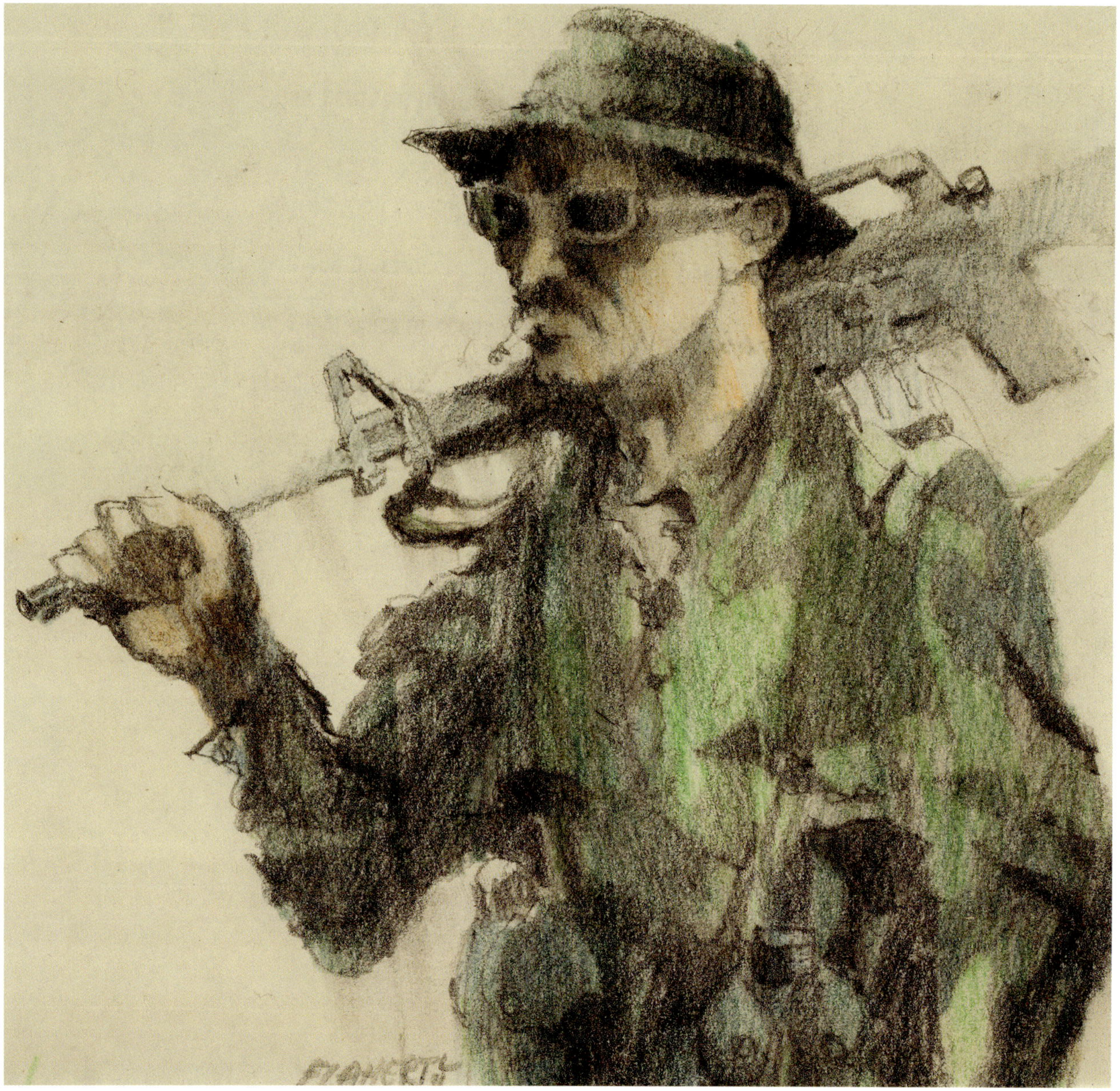

VETERAN
Long Binh Province, Vietnam
William E. Flaherty Jr., 1968
Pencil on paper, 6" x 5¾"

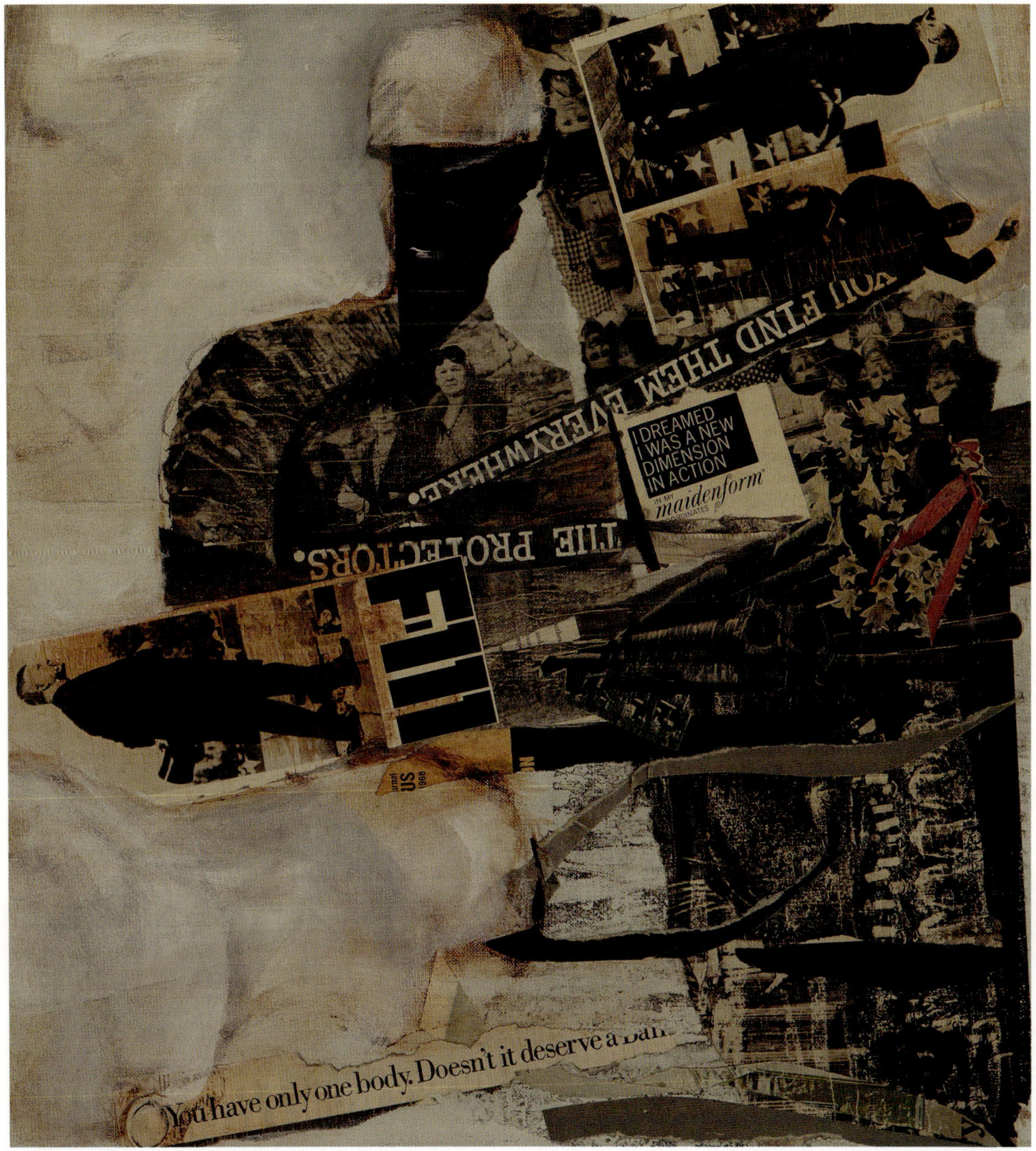

THE PROTECTORS
Vietnam
William C. Harrington, 1968
Acrylic on canvas, 21" x 19"

"The ubiquitous helicopters would have been an obvious subject. They darkened the skies there. To this day, I recoil from the sight and sound of a helicopter."
—John D. Kurtz

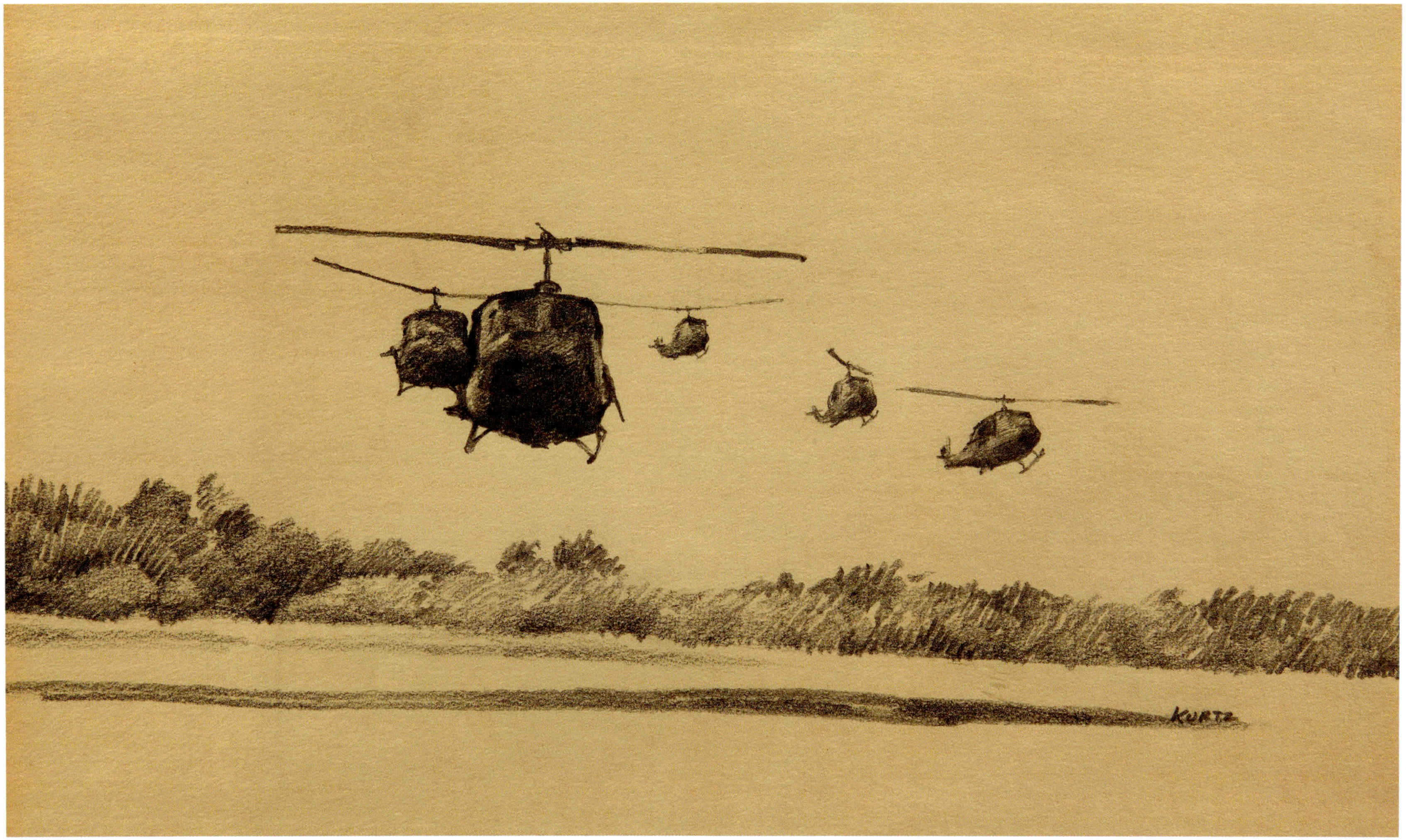

CHOPPERS
Vietnam
John D. Kurtz, 1968
Pencil on paper, 9" x 11½"

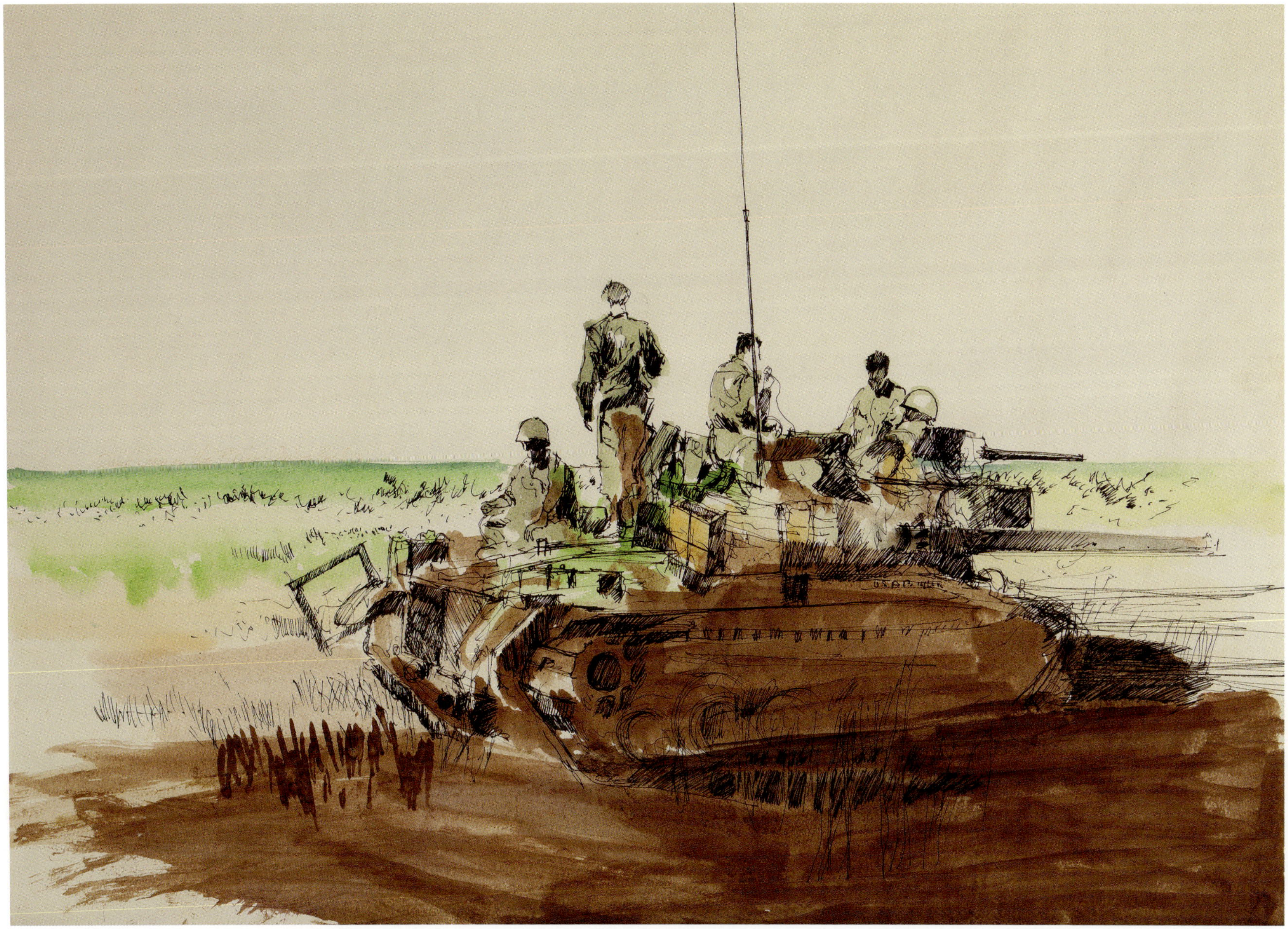

LUNCH BREAK
Vietnam
Stephen J. Matthias, 1968
Watercolor on board, 22" x 30"

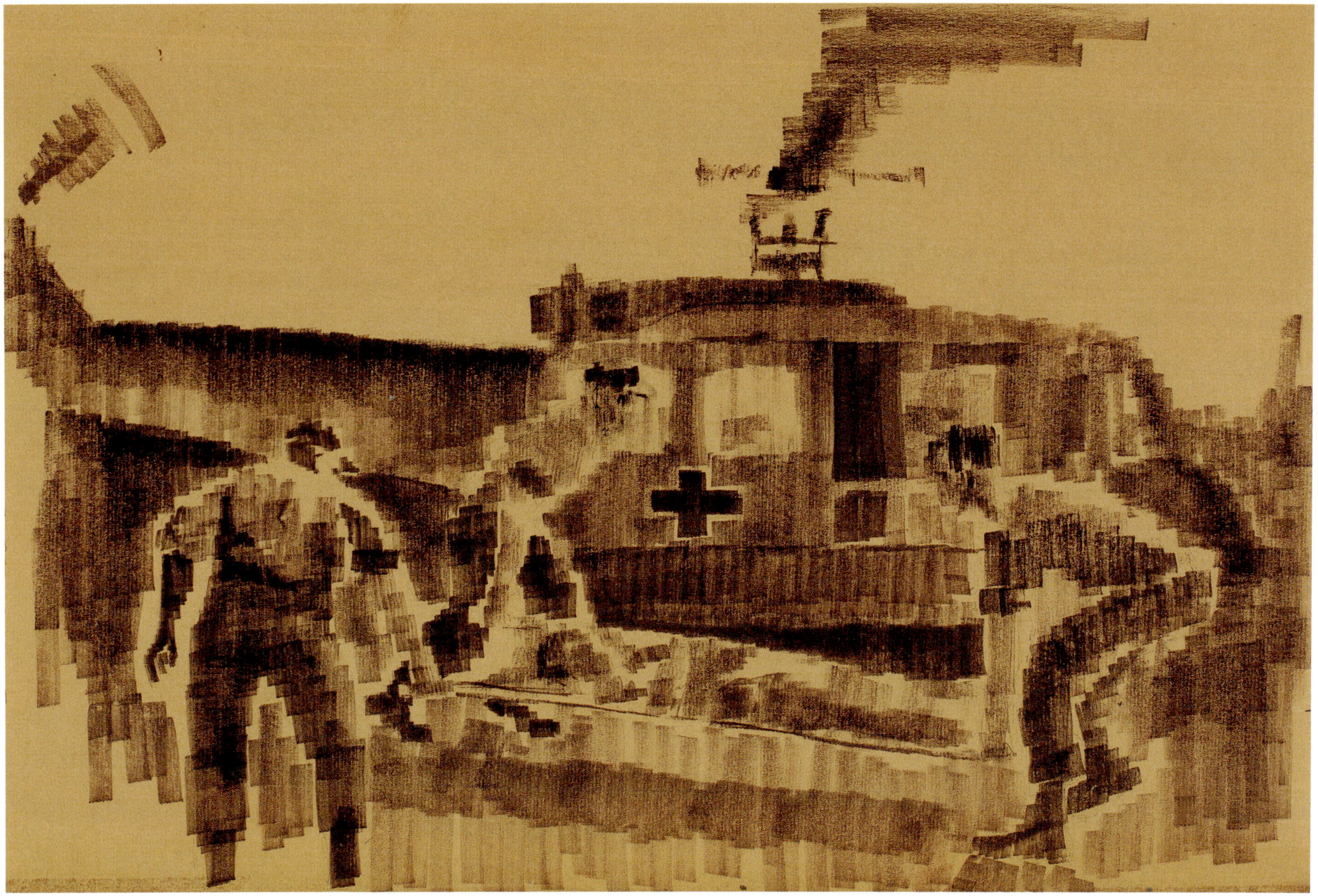

MED EVAC
Vietnam
Michael P. Pala, 1968
Pencil on paper, 17¾" x 11¾"

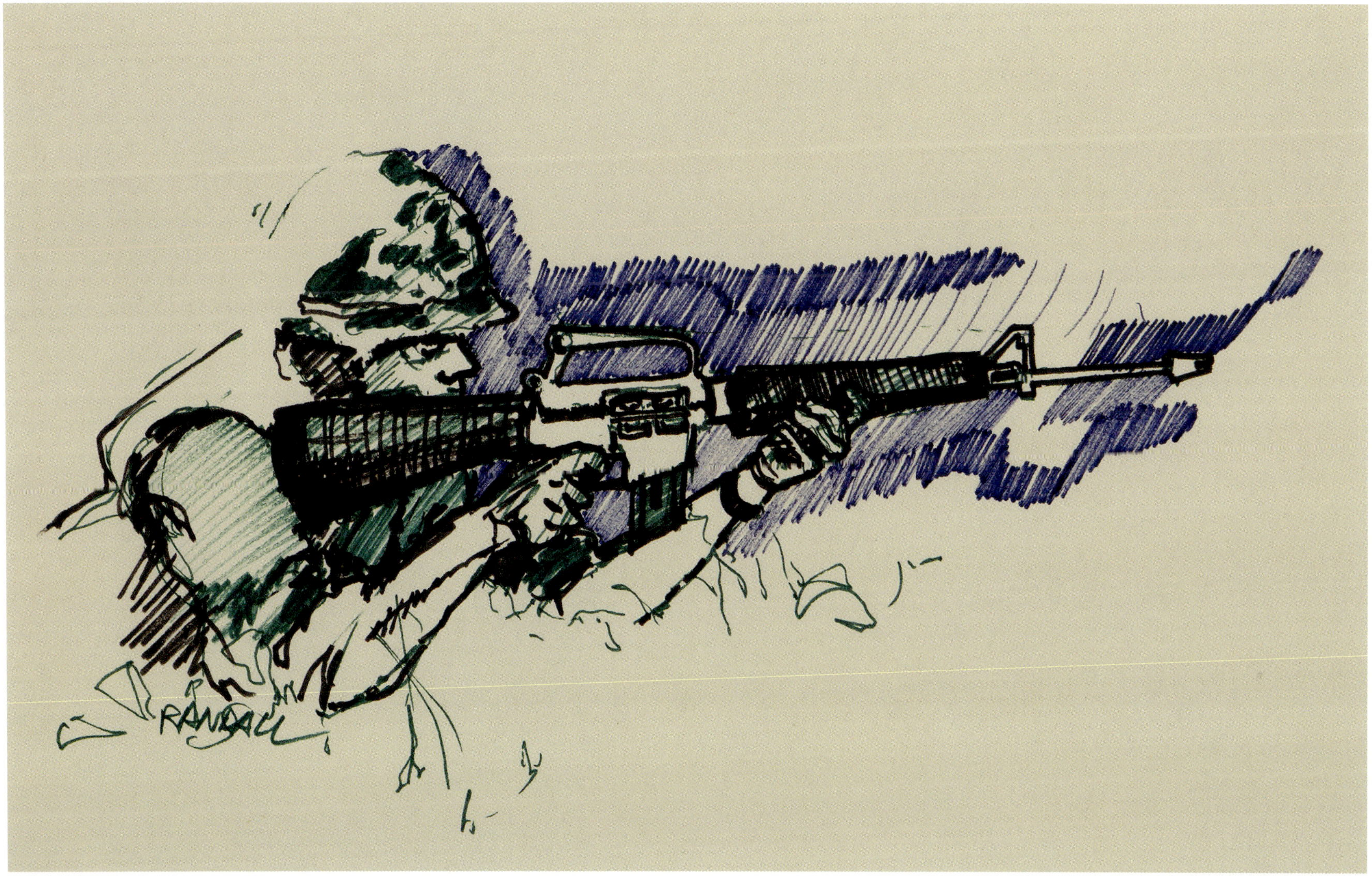

"Although I didn't think about it then, I now think that the circumstances of the units I served with did affect my art in the choice of subjects, media, and techniques. I think I approached support units and services with as much emphasis on equipment as on people."

—Stephen H. Randall

SIGHT PICTURE
Vietnam
Stephen H. Randall, 1968
Felt-tip marker and ink on paper, 9" x 11½"

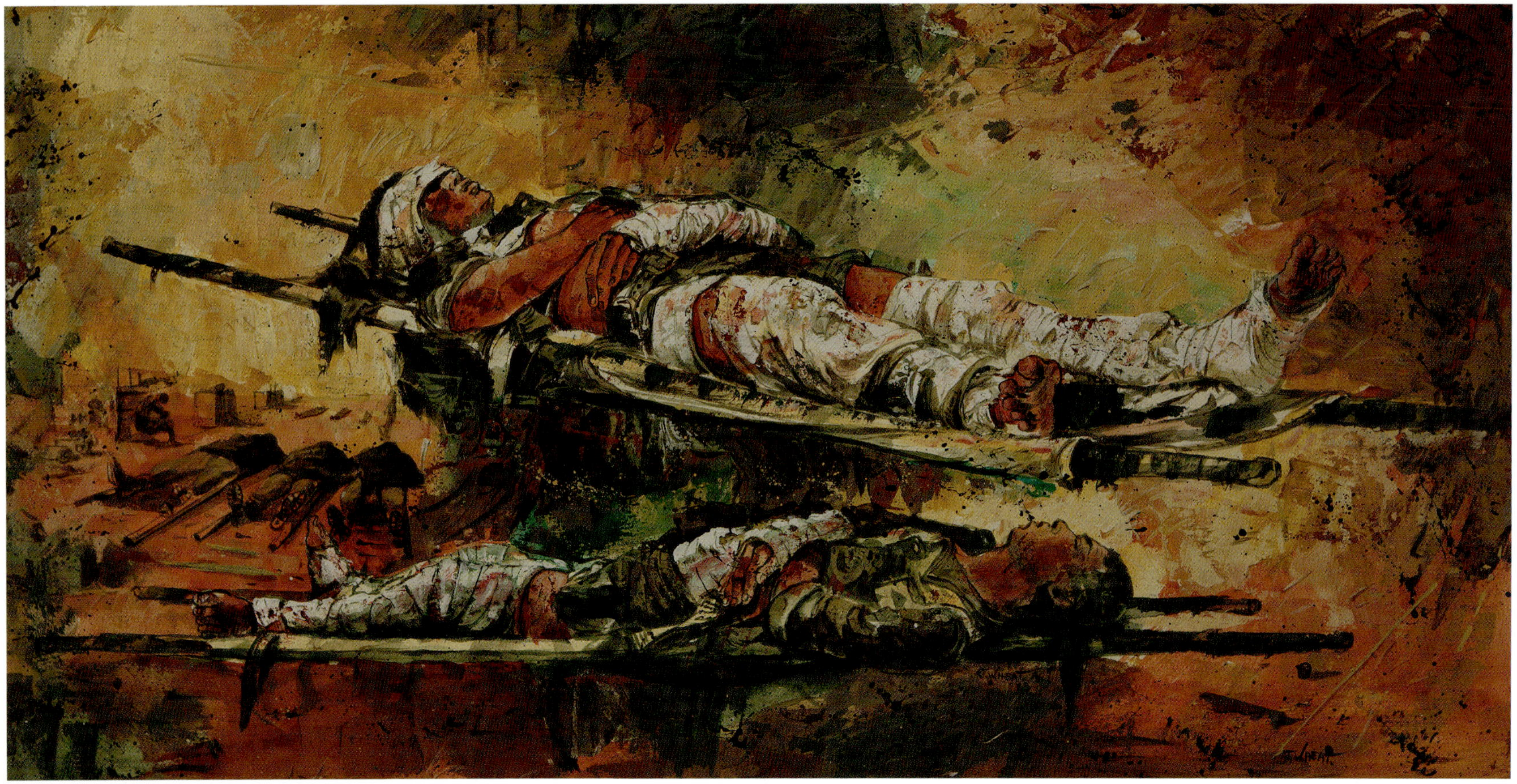

WOUNDED MEN
Qui Nhon, Vietnam
John Potter Wheat, 1968
Acrylic on masonite, 13½" x 25½"

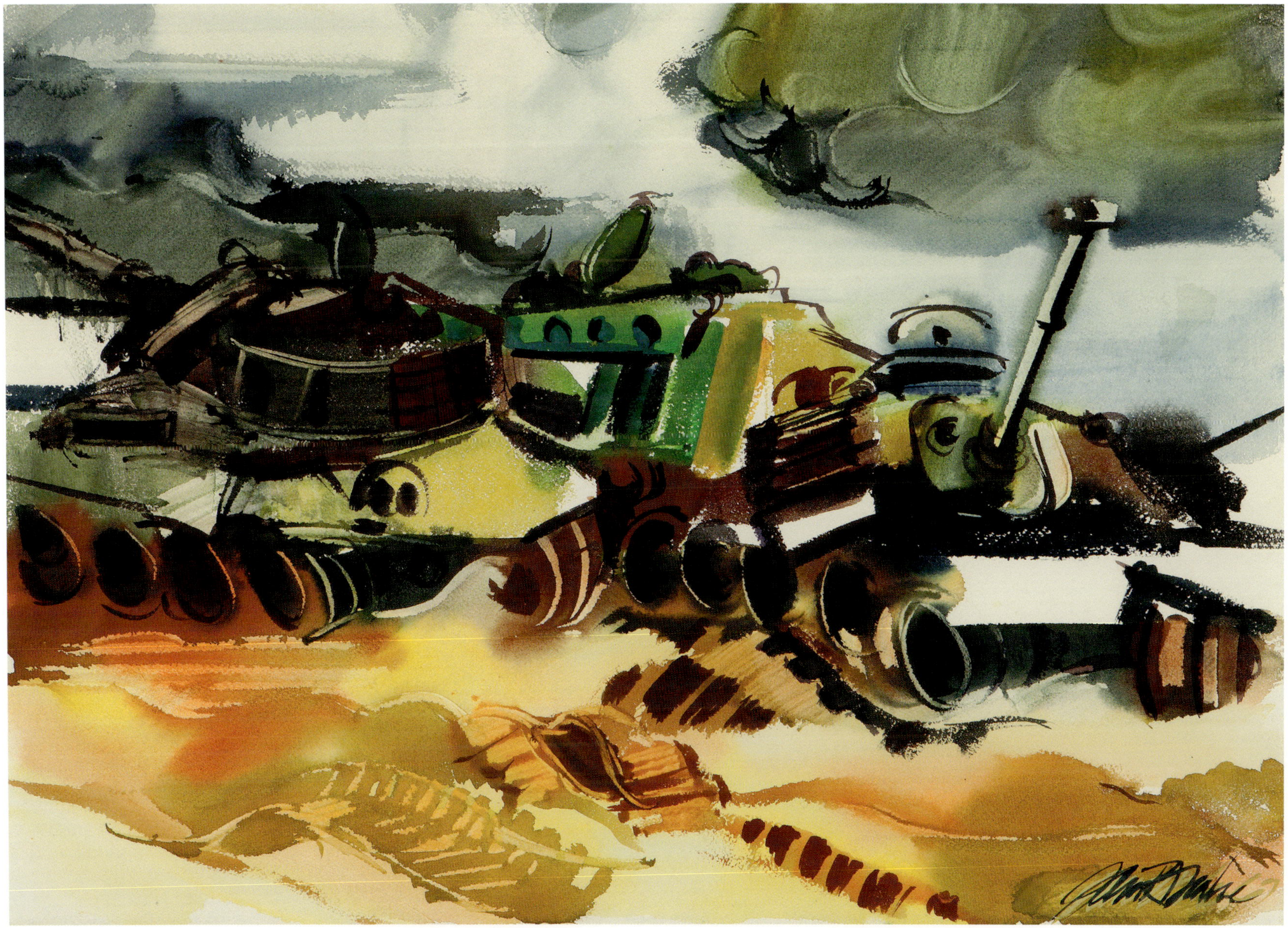

TANK GRAVEYARD
Vietnam
James R. Drake, 1969
Watercolor on paper, 20" x 28½"

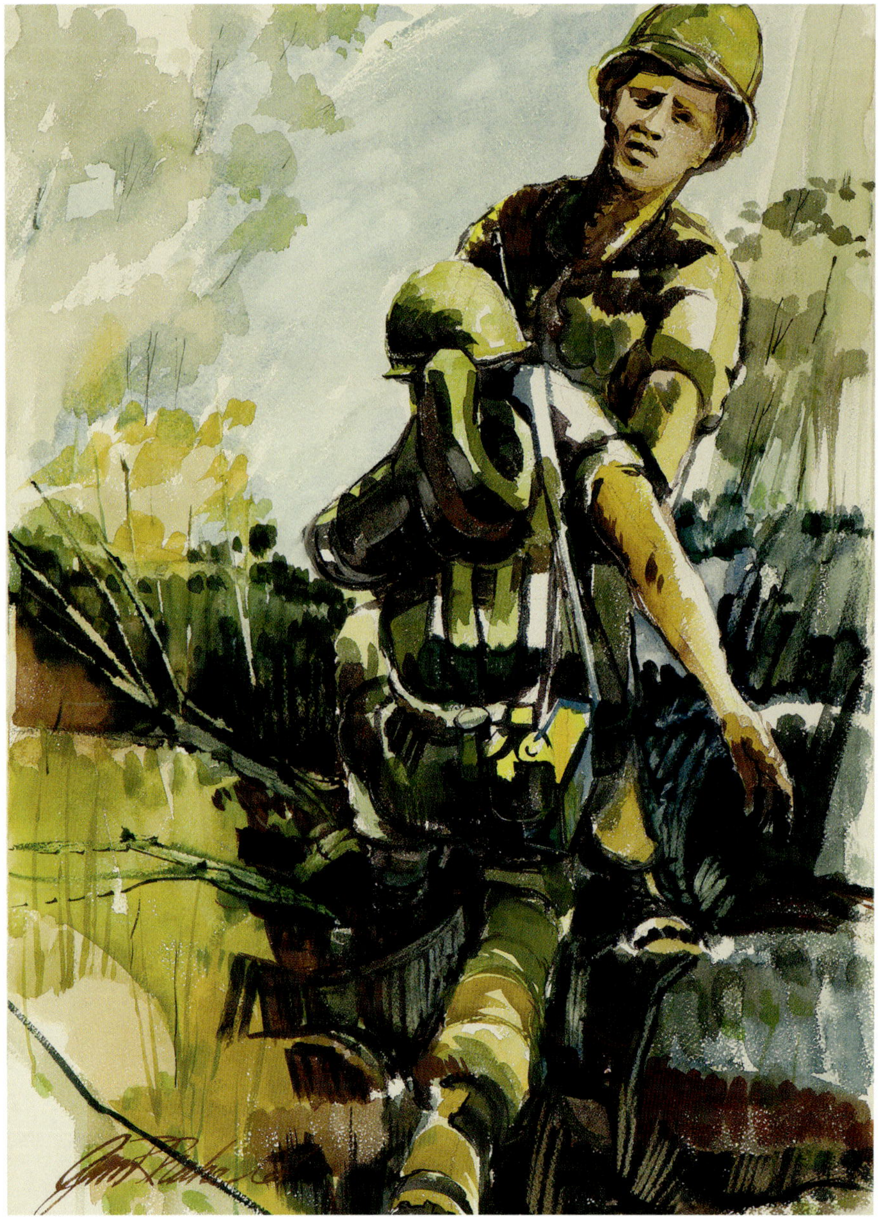

PATROL
Vietnam
James R. Drake, 1969
Watercolor on paper, 20" x 28½"

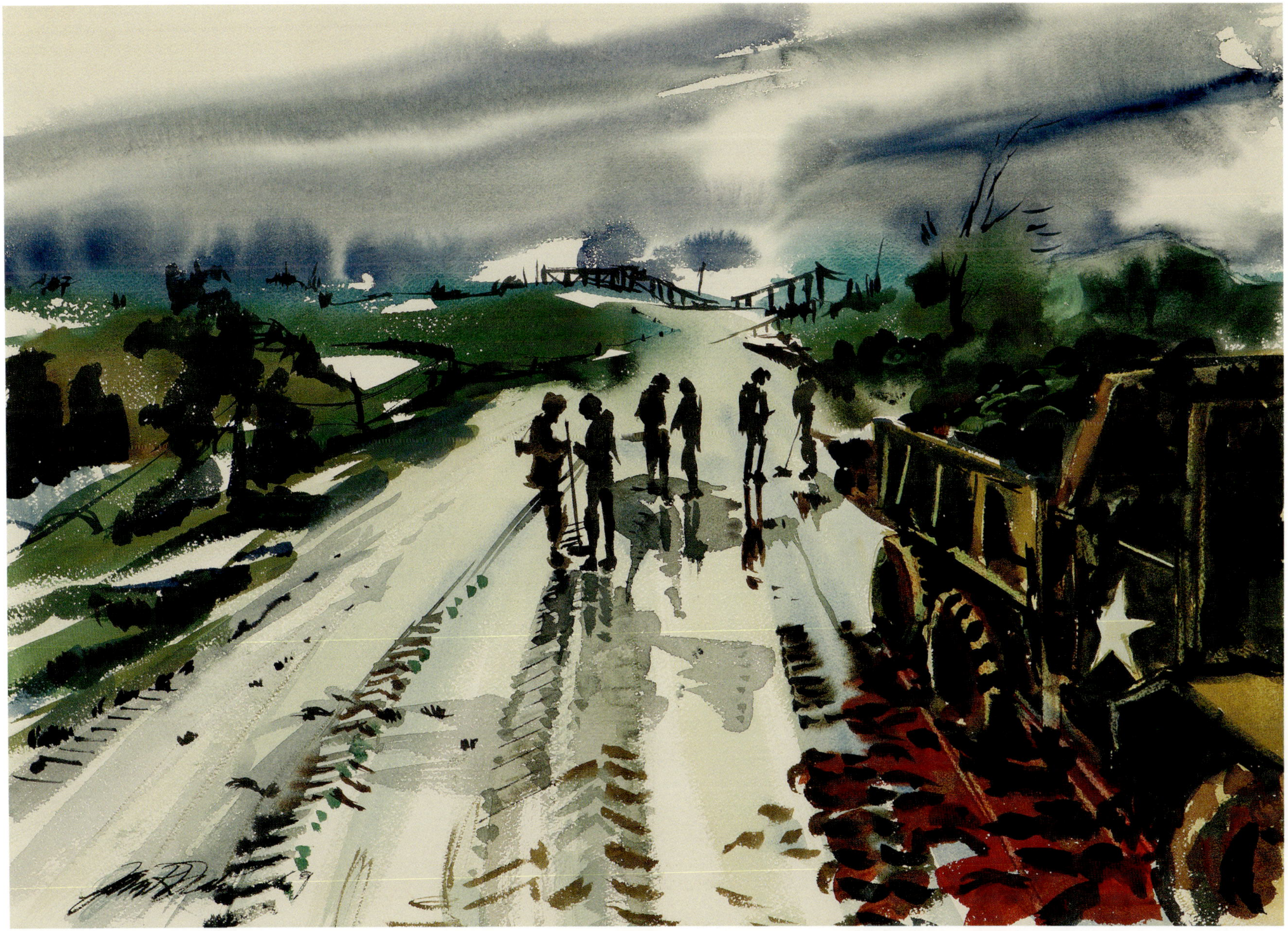

MORNING RITUAL
Vietnam
James R. Drake, 1969
Watercolor on paper, 20" x 28½"

"The portraits and character studies I feel are my strongest works. Never in my life have I been in the companionship of finer people than when I was with those infantry units."
—James S. Hardy

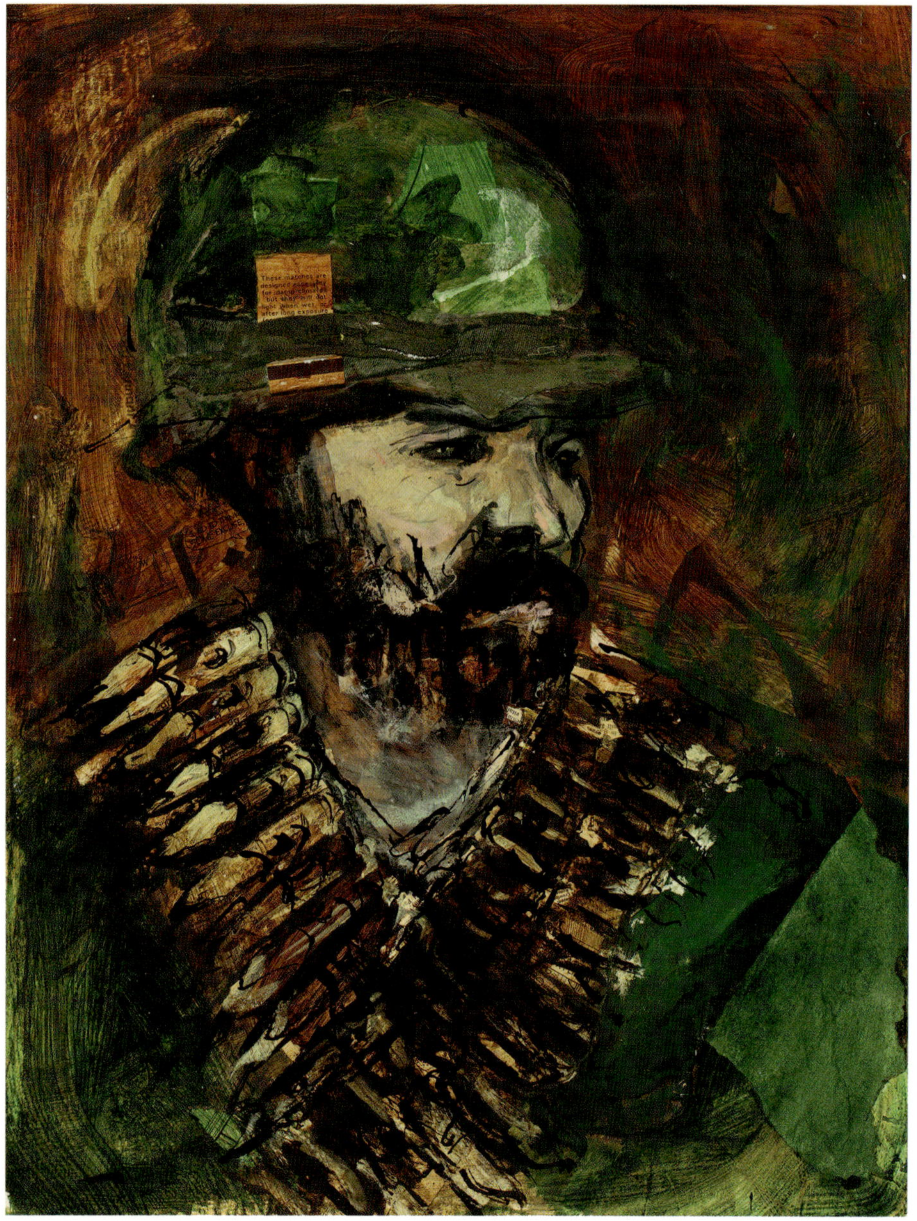

PORTRAIT
Vietnam
James S. Hardy, 1969
Acrylic on board, 18" x 24"

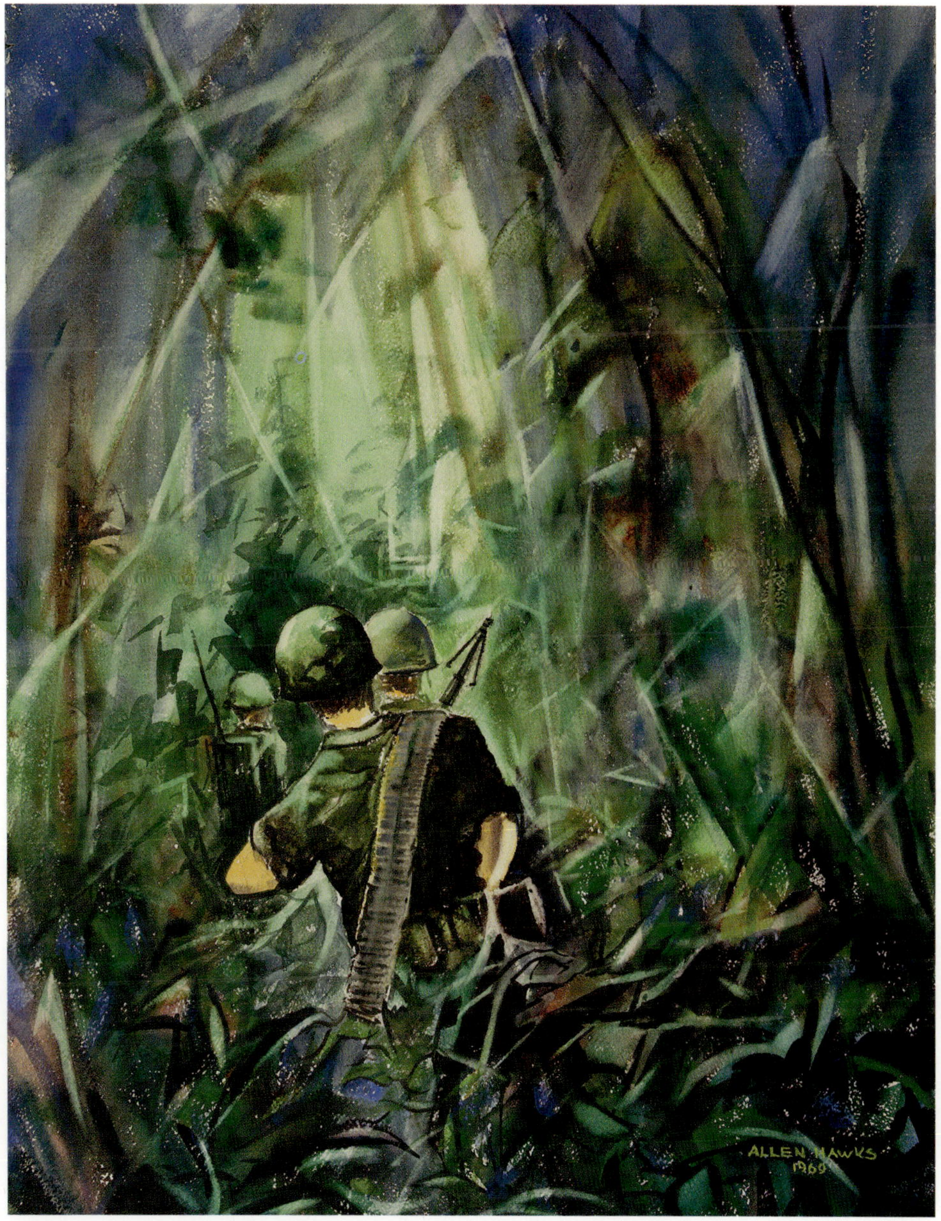

LONG RANGE PATROL
Vietnam
Horatio A. Hawks, 1969
Watercolor on paper, 22" x 30"

"I don't think it (being a Combat Artist) affected my style or what I paint. It did, however, strengthen my drawing ability because of my constant sketching on location. I guess I thought more about adding figures to my compositions."
—Carlton Plummer

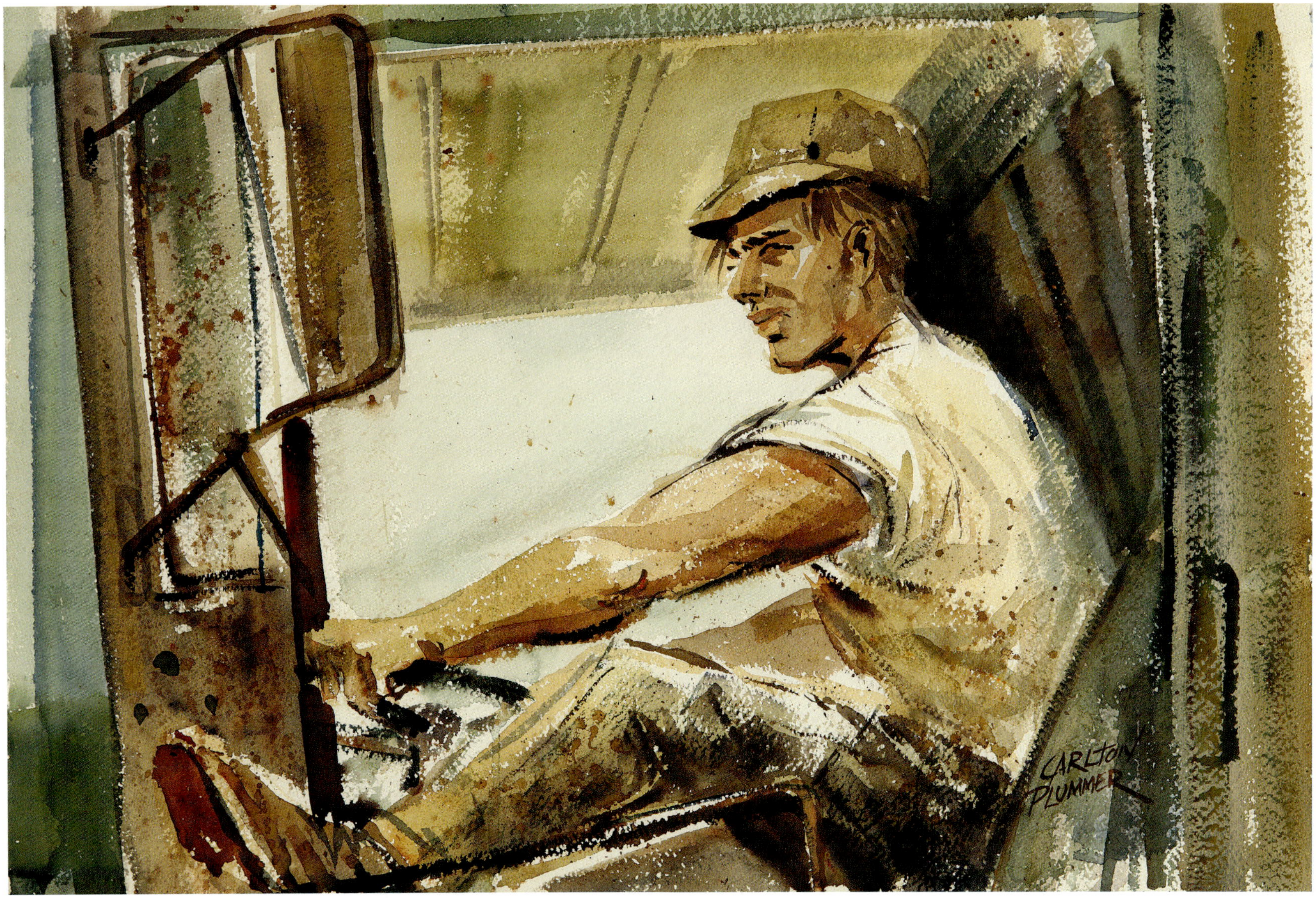

JUNGLE DRIVER
Thailand, Vietnam War
Carlton Plummer, 1969
Watercolor on paper, 14" x 21"

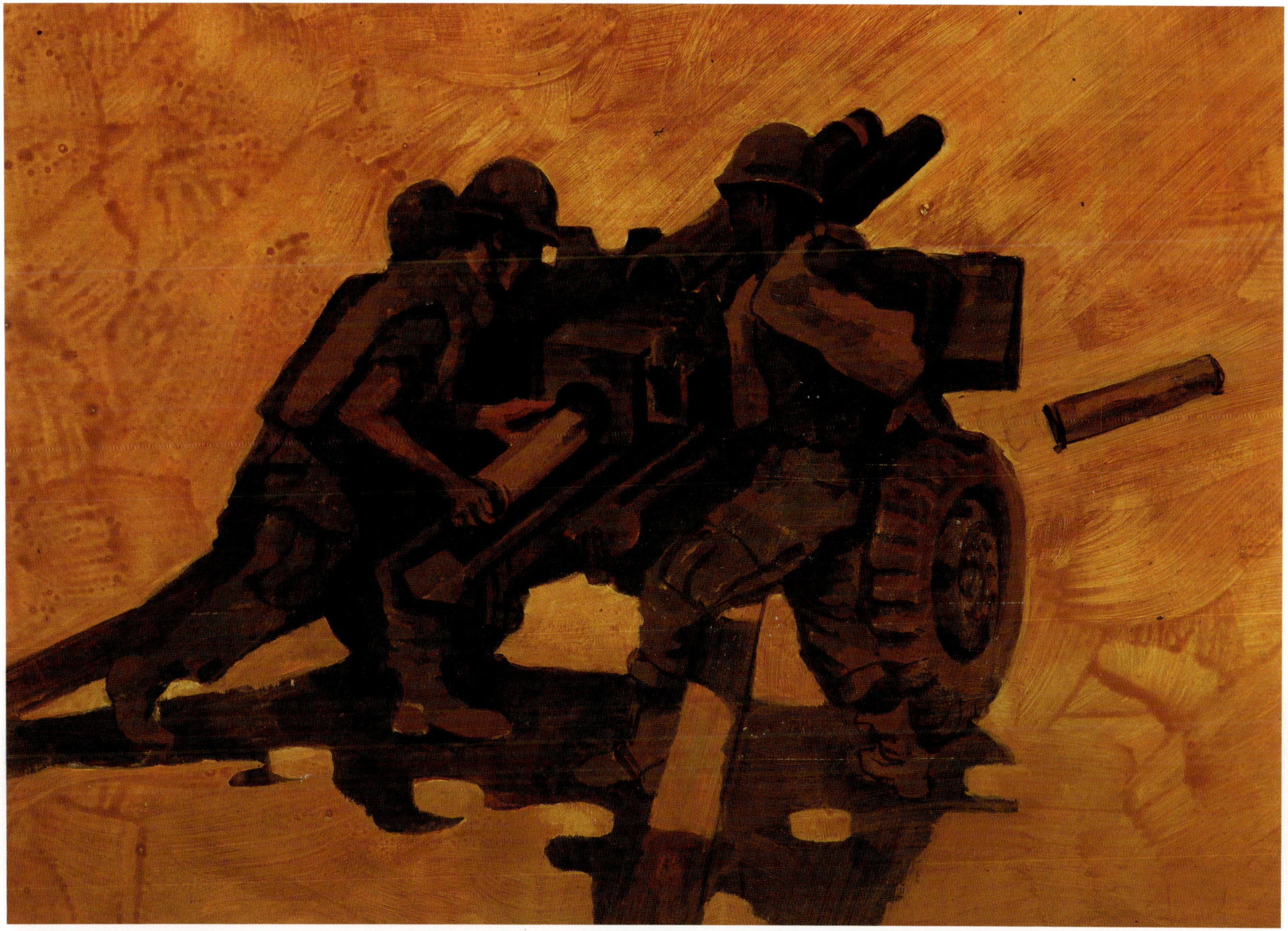

FIREBASE HARDTIMES
Vietnam
Craig L. Stewart, 1969
Acrylic on board, 13½" x 19"

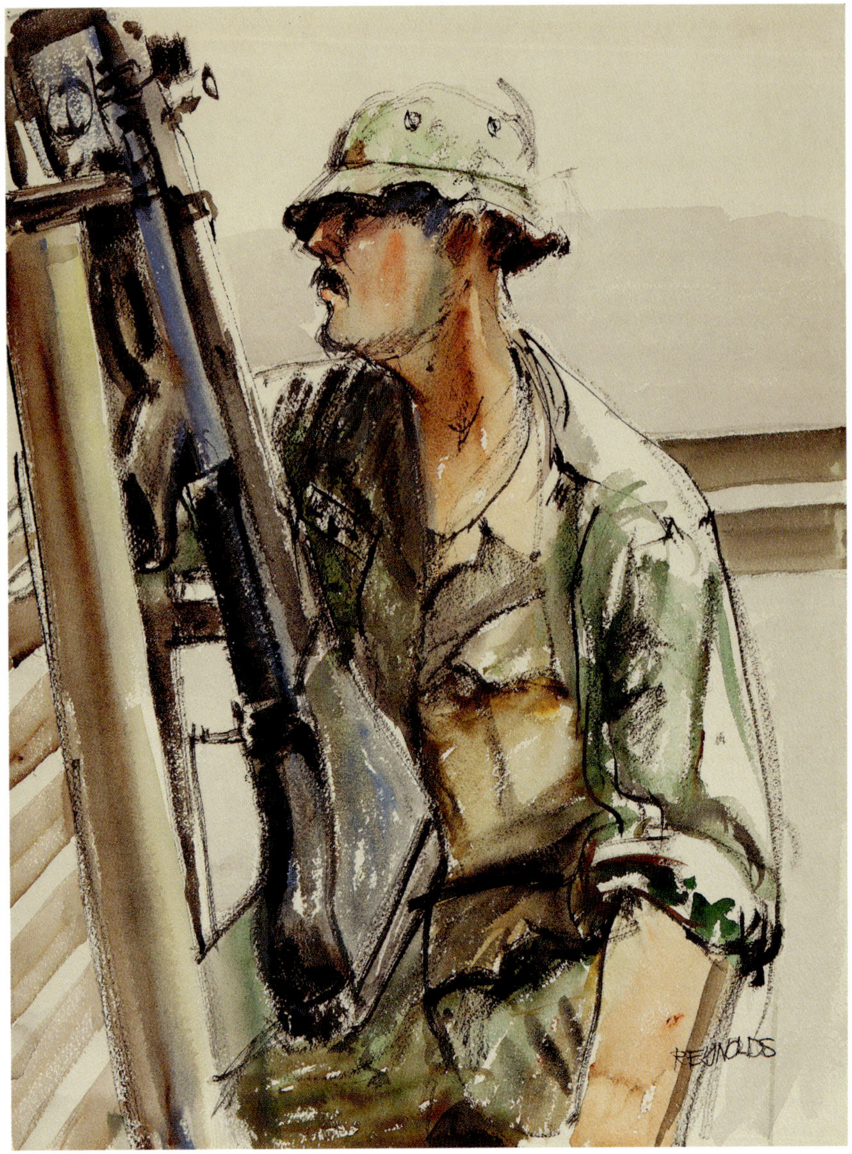

COMBAT ENGINEER
Bao Loc, Vietnam
Victory Von Reynolds, 1969
Watercolor on board, 20" x 25"

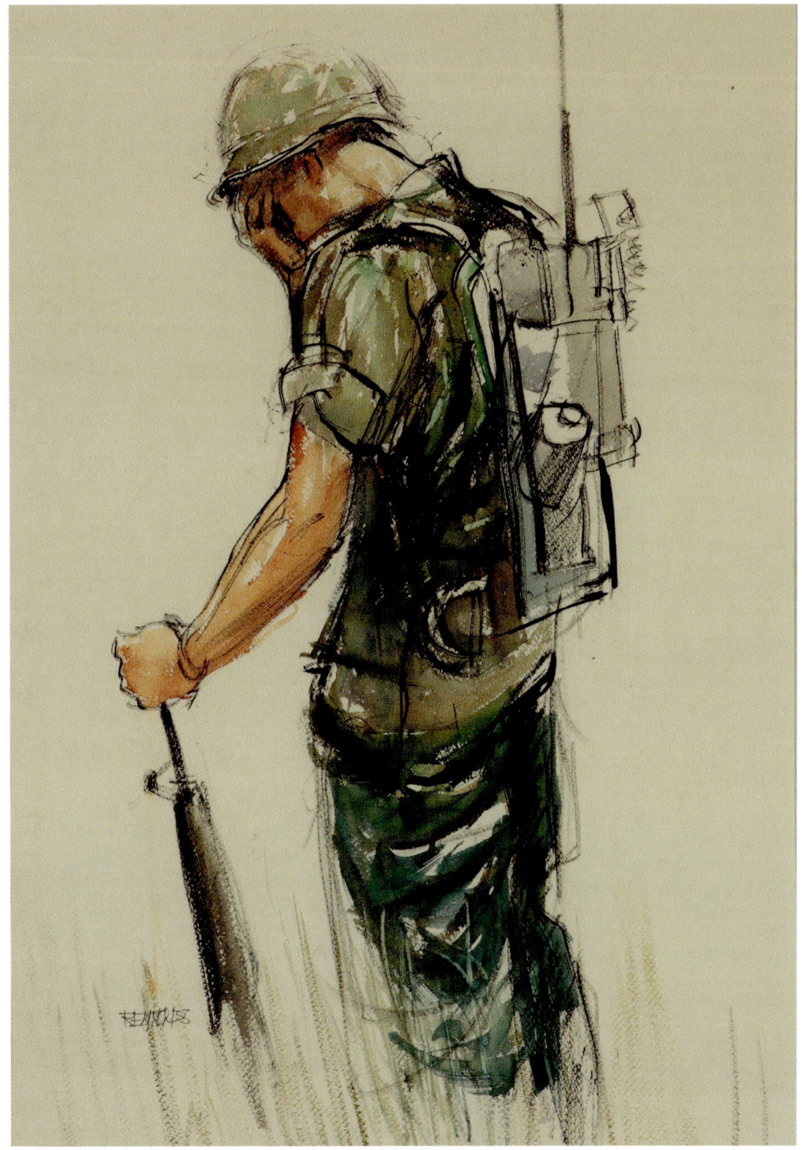

A SOLDIER
Vietnam
Victory Von Reynolds, 1969
Watercolor on paper, 28½" x 21"

Reinventing the Army
The 1970s and 1980s

The Army Art Program of the 1970s and 1980s grew out of the tradition of the Vietnam art program, and the art has its roots in that of the Vietnam War as well. With the suspension of the draft and the transition to an all-volunteer Army, the Army's emphasis shifted to attracting the individual while reshaping the Army's image. In 1980, the Army adopted the recruiting slogan "Be All You Can Be," which challenged the soldier to reach his or her full potential.

The art of this period reflects this adjustment in the Army's philosophy and signifies a change in the soldier. The focus is often on action and movement, represented by numerous images of soldiers traveling from one place to another. In a period without combat, the artists were faced with how to depict the soldier and visually answer the question "What makes a soldier a soldier?" In response to that, they often turned to the heroism of the everyday in their subject matter. S. Sgt. Roger W. Price's drawing *Combat Ready* (page 94) asks the soldier directly whether he or she has met the challenge to be all he or she can be. Art such as this piece idealizes the courage of the great stoic soldier and stresses the soldier's identity as a fighter. The modern soldier takes his or her place in a long line of classic warriors throughout the Army's history and is identified as such.

The artists also responded to this challenge by illustrating heroism through training rather than combat. Achievement and personal drive are emphasized in Sp5c. Janet Fitzgerald's *Pushing for Soldierhood* (page 81), which depicts an exhausted-looking female with an expression of intense concentration on her face as she runs. Even prior to the creation of the Be-All-You-Can-Be slogan, this soldier is clearly attempting to push her boundaries and to attain her potential as a soldier. Sgt. Adam Glenday's *Determination* (page 93) likewise portrays a soldier striving for success in a training environment. The viewer can almost hear the subject's groan as he literally struggles to grasp something just out of reach and to better himself in the process.

Another common feature of Army art from this period is a focus on quiet introspection, in which the soldier must look inward to realize his or her potential. The epitome of this is volunteer civilian artist Al Sprague's *Moment of Truth* (page 86), which portrays the instant before the combat jump into Panama. In some cases, such as Sprague's *Cruising the Panama Canal (Chinook)* (page 87), the subject of contemplation is clear. In other cases, the viewer's imagination colors the mood of the work, as in Sgt. Patrick Farrell's *The Scouts of the 24th Infantry Division* (page 83).

There is no cohesive style of the period. Some artists continued to create in brilliant colors commonly seen in the Vietnam era, such as Farrell's *Moving Out at Last Light* (page 84). Others began to work in more subdued palettes, such as Pfc. William H. Steel's *Easter at Outpost Martin* (page 65), or shunned the use of color completely, as in Steel's *Army Artist Team X* (page 188), which is the only portrayal of an entire Army Artist Team in the collection. Some artists produced in loose, expressive styles, such as volunteer civilian artist Thomas C. O'Hara's *On the Range* (page 64), and others favored tight lines and meticulous detail, as seen in Price's *Combat Ready*. This variation in approach is a symptom of the diverse training and backgrounds that each artist brought to his or her team, and the art successfully reflects the Army's process of healing from the war and reinventing its identity.

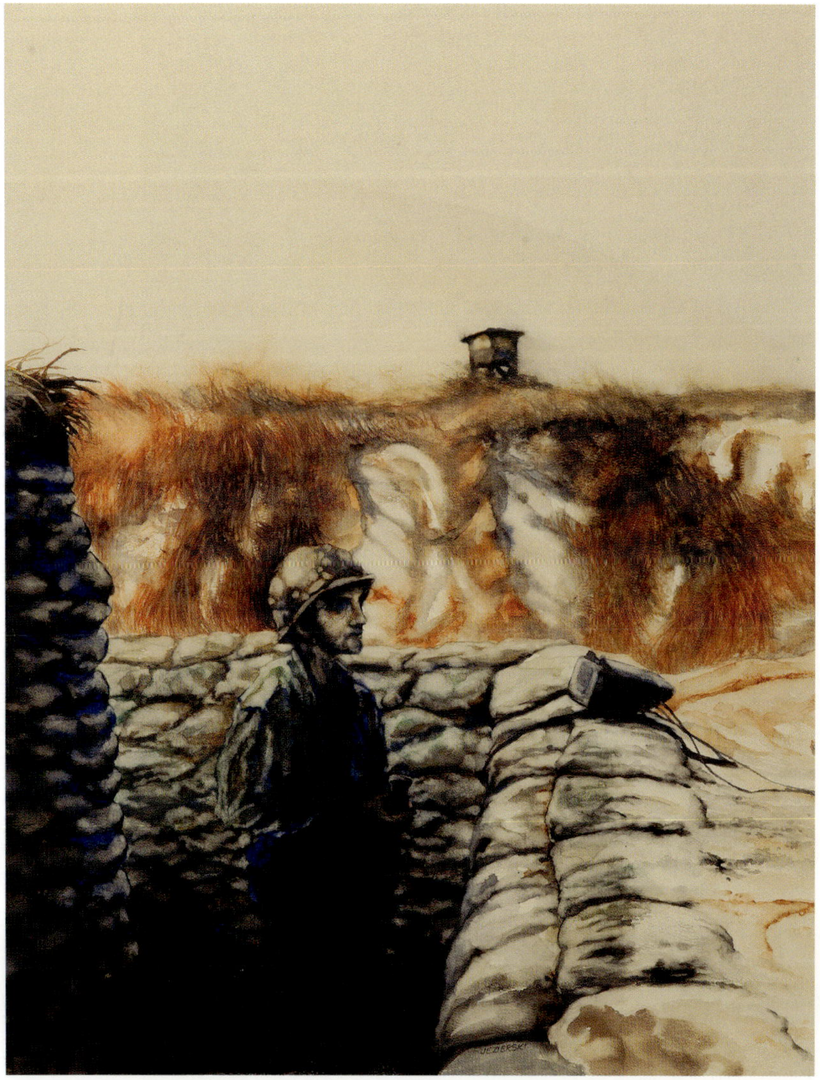

WAITING
Korea
Chester Jezierski, 1970
Watercolor on paper, 19" x 24"

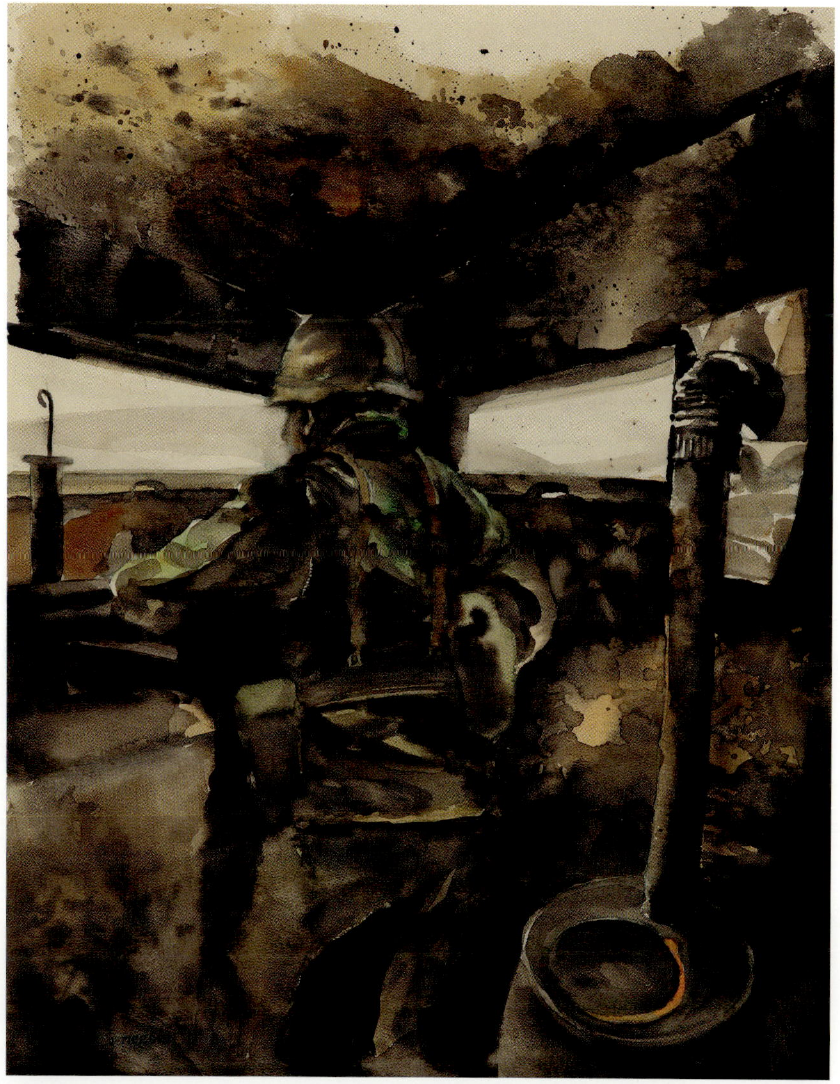

THEY WATCH
Korea
Chester Jezierski, 1970
Watercolor on paper, 19" x 24"

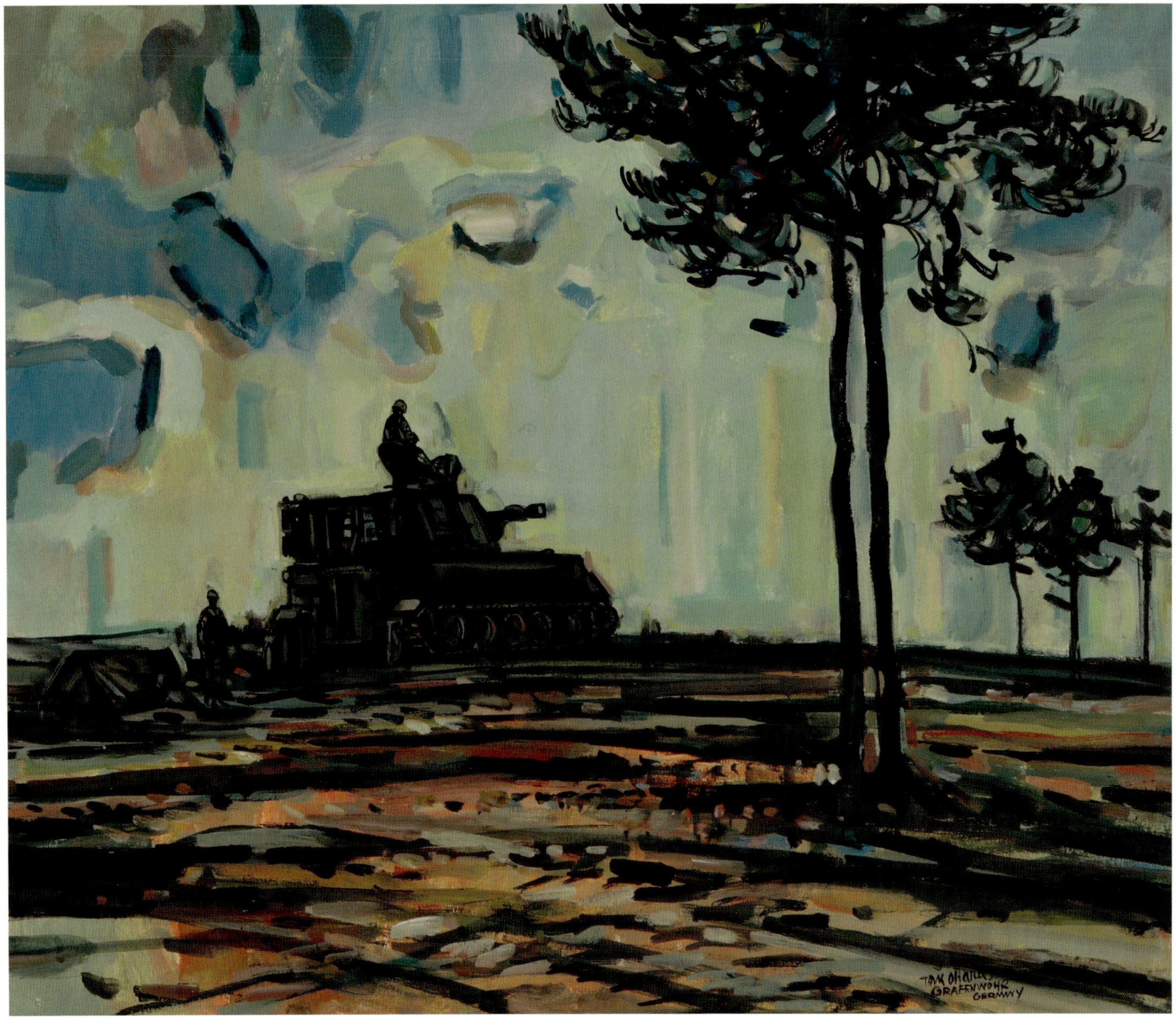

ON THE RANGE
Grafenwöhr, Germany
Thomas C. O'Hara, 1970
Acrylic on paper, 20" x 23"

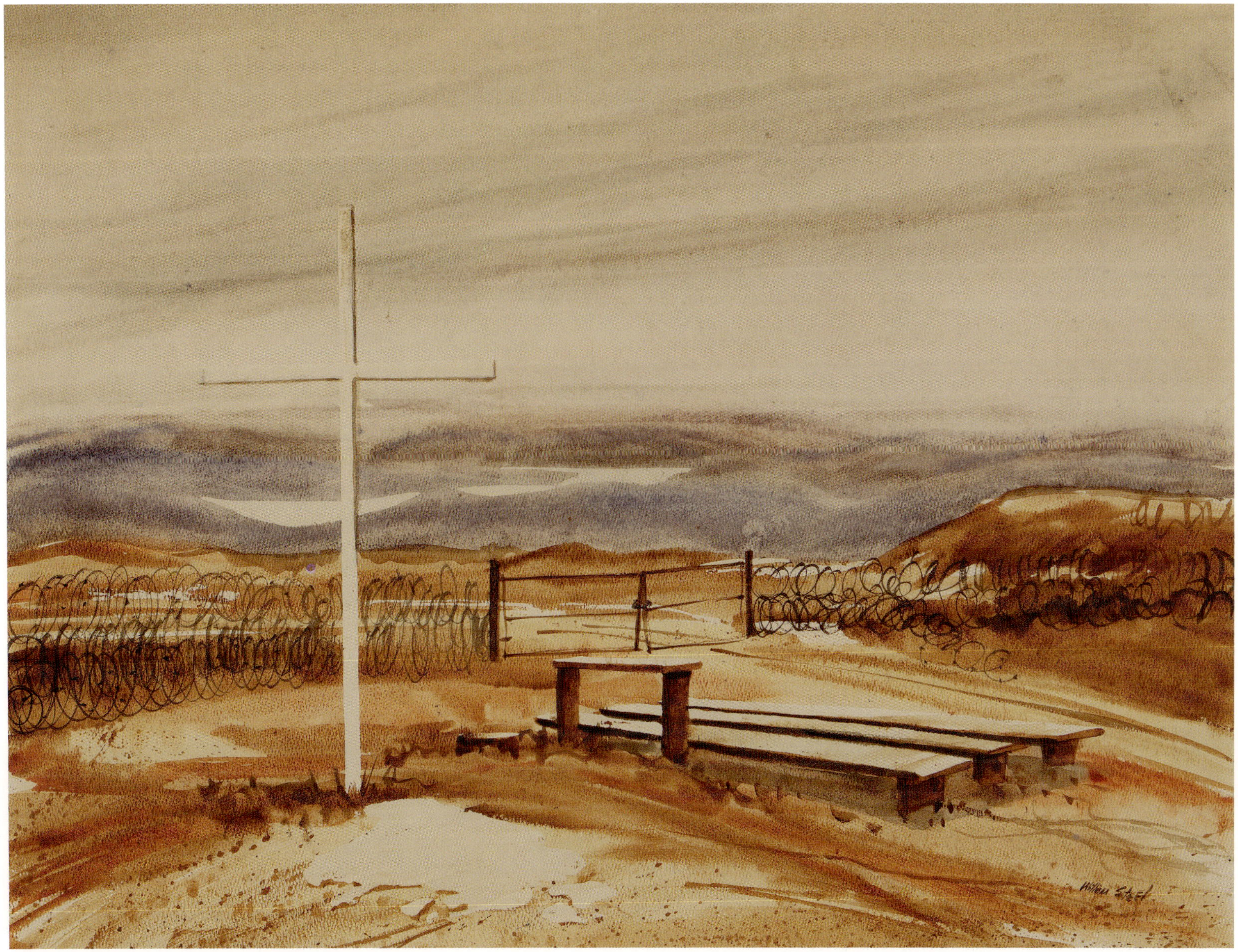

"This watercolor of a worship area located a few feet from the gate before entering the no man's land in the Z, was very moving to me. I felt it was a place where a soldier could pray for his safe return prior to his mission of entering the Z, or stop and reflect that (hopefully) he had a safe mission while in the Z, and give thanks for his or his comrades safe return. The painting is cold, as it should be, stark and windblown."

—William H. Steel

EASTER AT OUTPOST MARTIN
Korea
William H. Steel, 1970
Watercolor on paper, 18¾" x 24"

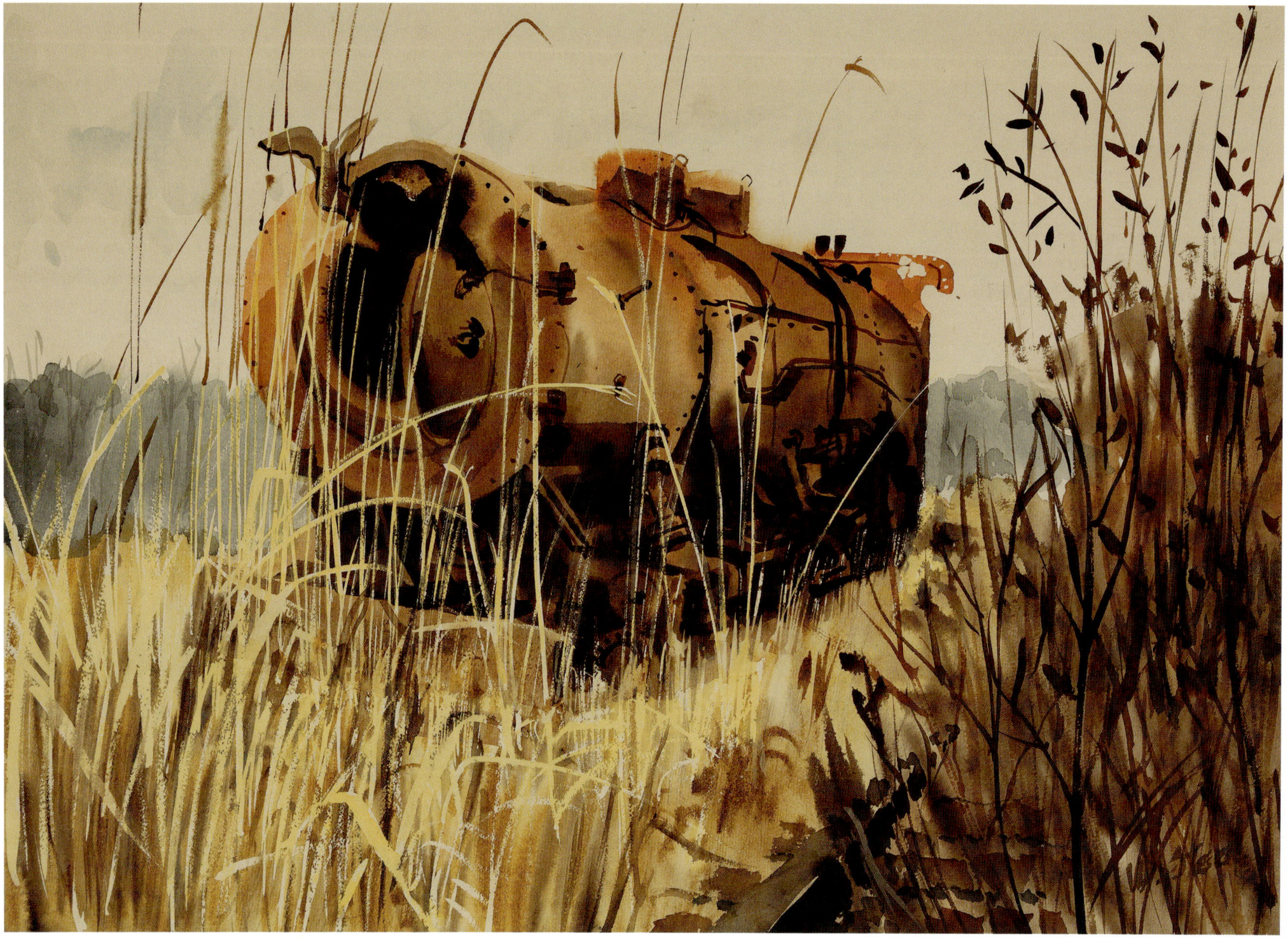

REMINDERS
Korea
William H. Steel, 1970
Watercolor on paper, 22¼" x 30½"

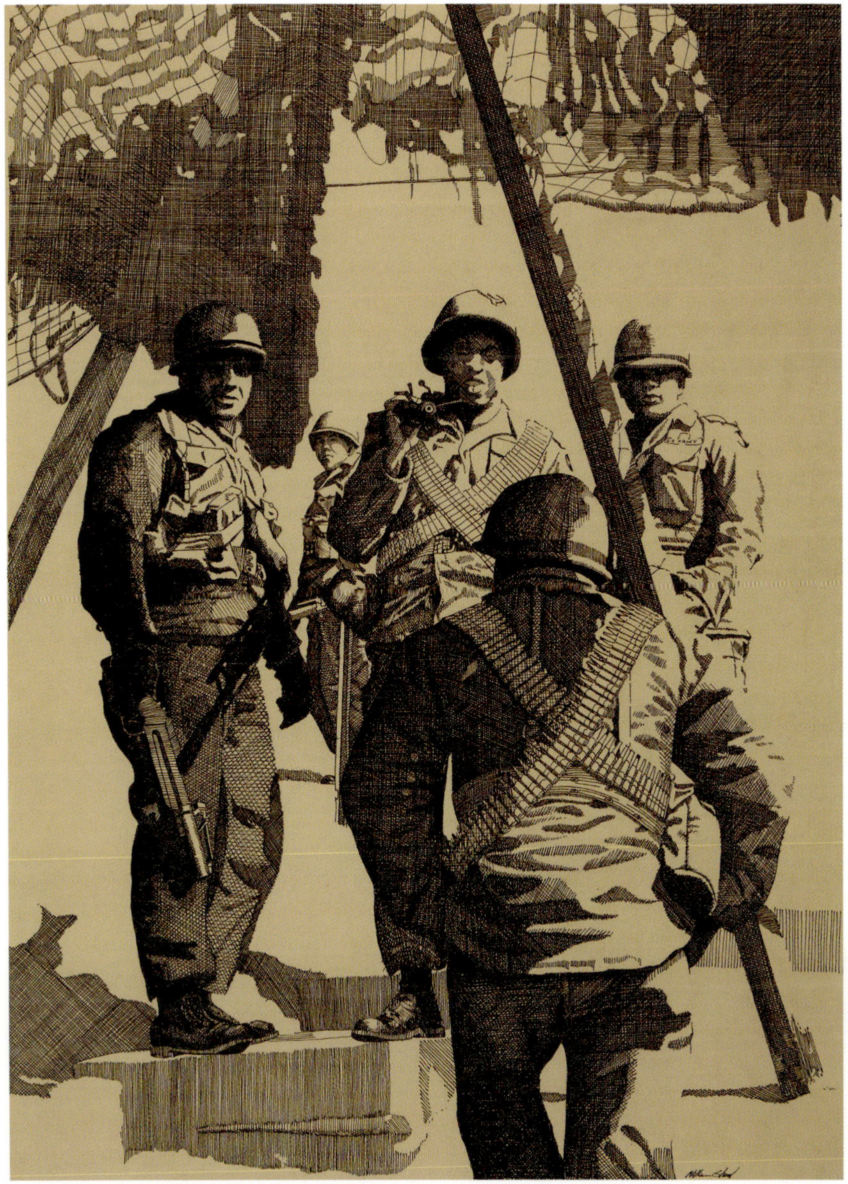

IN LIEU OF LUNCH
Korea
William H. Steel, 1970
Ink on paper, 30" x 22"

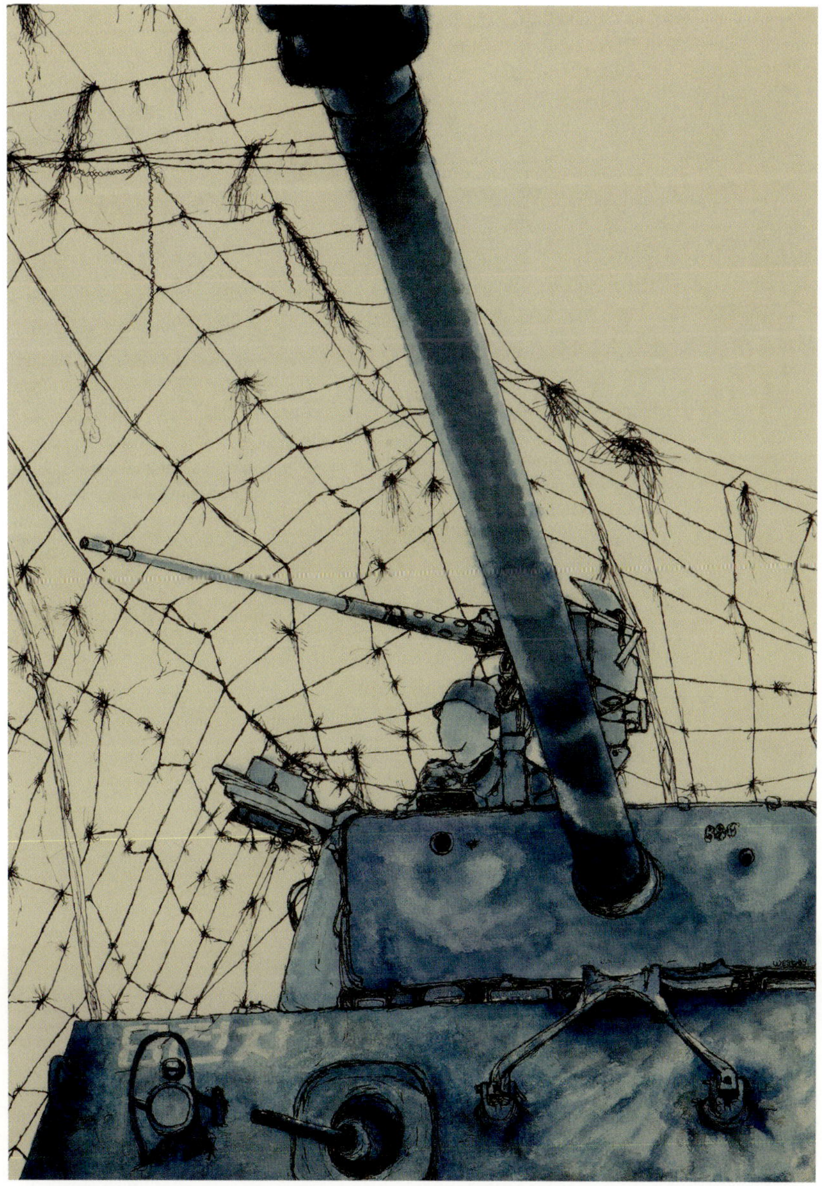

8 TANK (also TANK UNDER CAMOUFLAGE NET)
Korea
Lawrence A. Westby, 1970
Ink and watercolor on paper, 30" x 22"

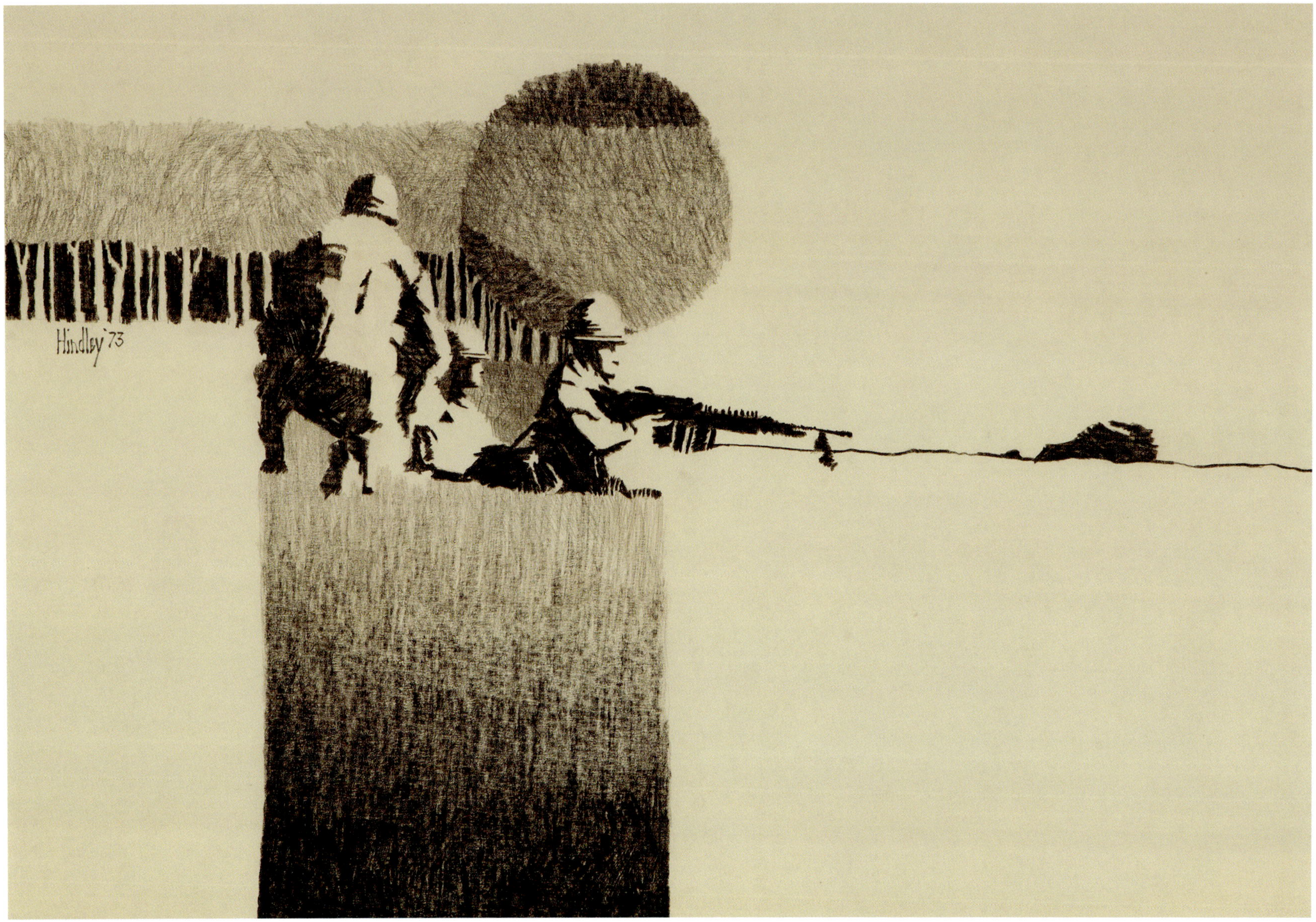

TRAINING EXERCISES
Baumholder, Germany
Joseph S. Hindley, 1972
Pencil on paper, 19" x 24"

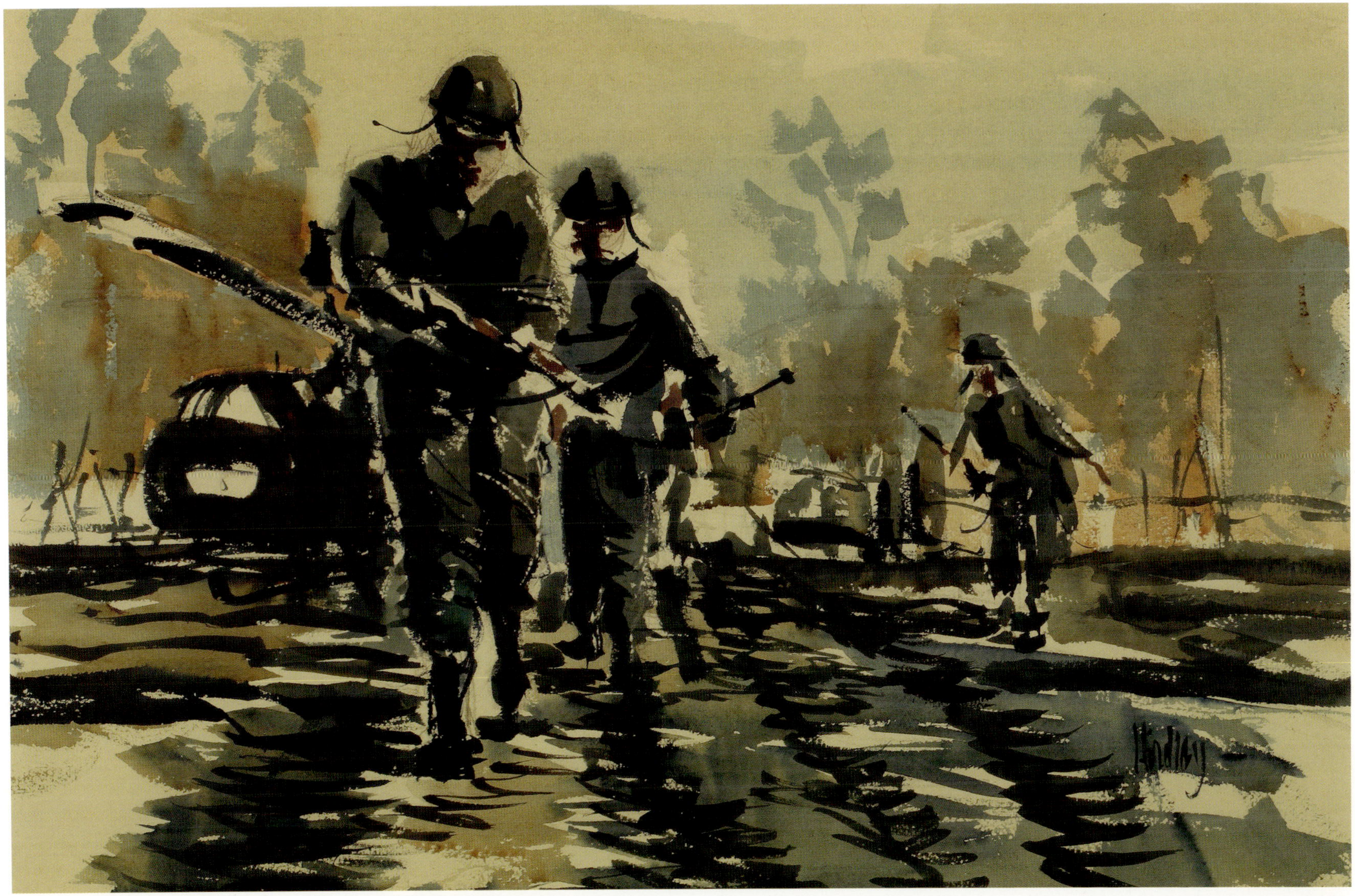

RUNNING BLUES
Fulda, Germany
Joseph S. Hindley, 1972
Watercolor on paper, 22" x 29¾"

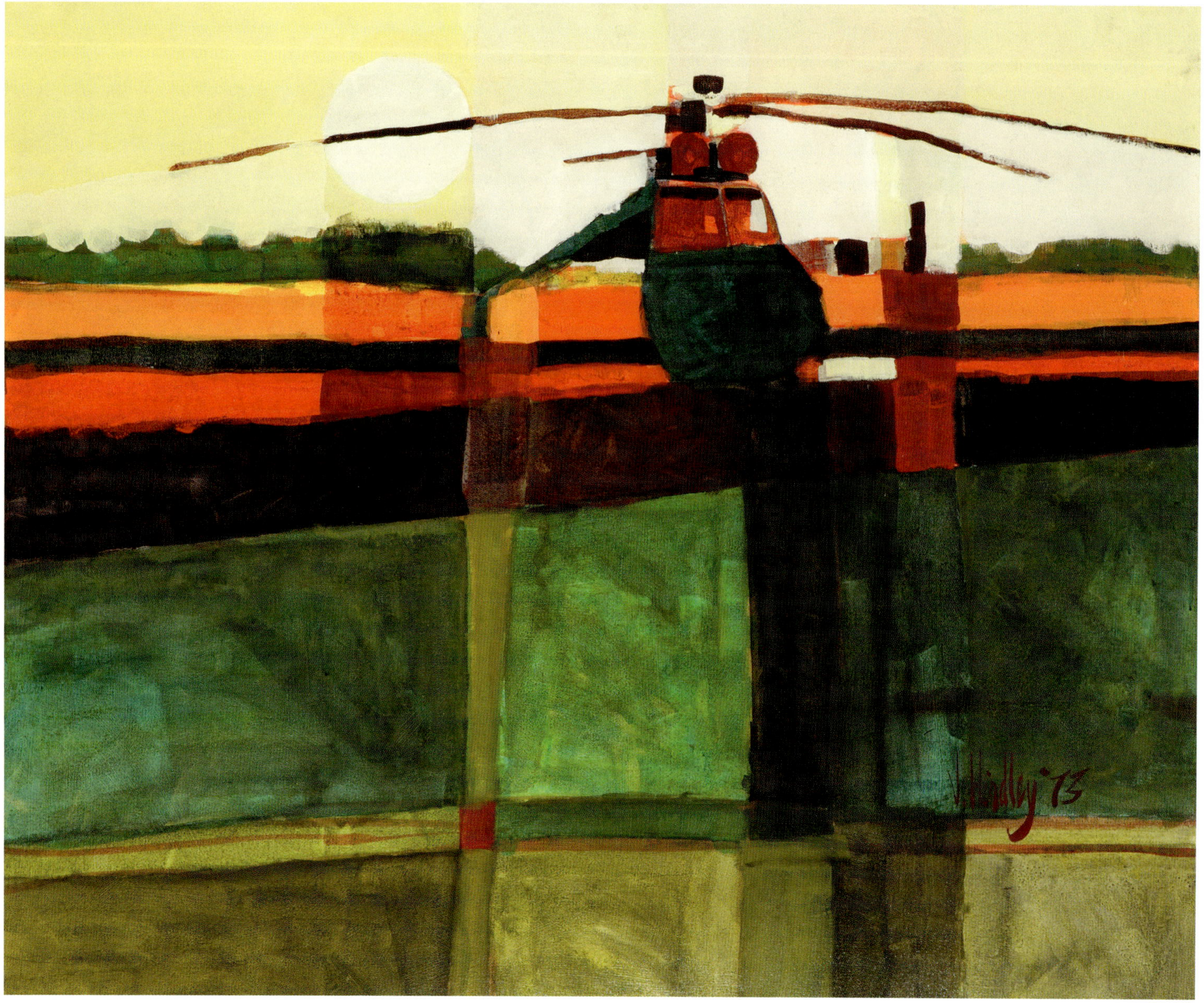

SILENT SUNDAY
Mainz, Germany
Joseph S. Hindley, 1973
Acrylic on canvas, 32" x 38"

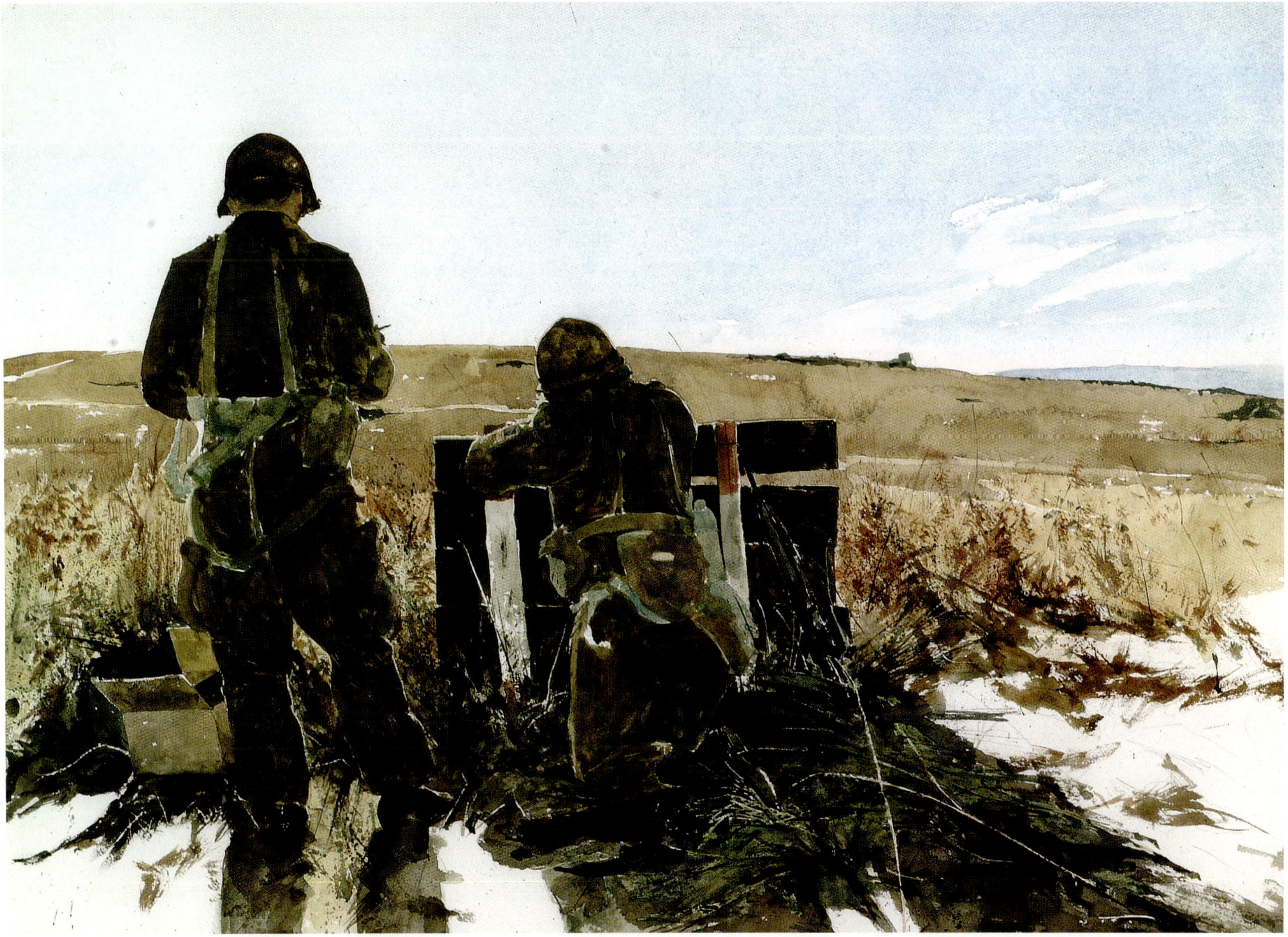

MORTAR PRACTICE AT BAUMHOLDER
Baumholder, Germany
William G. Jacobson Jr., 1972
Watercolor on paper, 22" x 29¾"

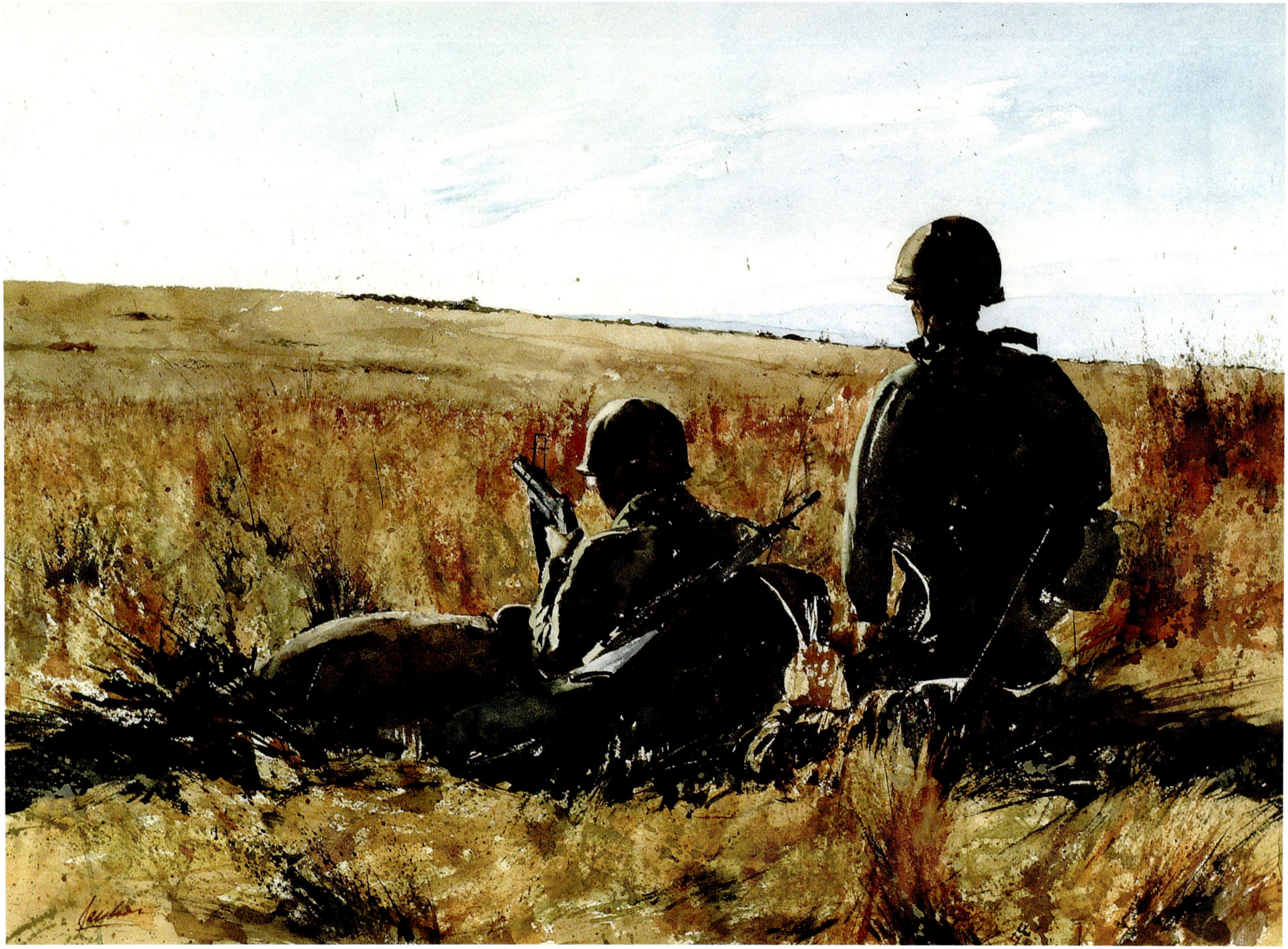

FRIDAY MORNING MORTAR PRACTICE
Baumholder, Germany
William G. Jacobson Jr., 1972
Watercolor on paper, 22¼" x 29¾"

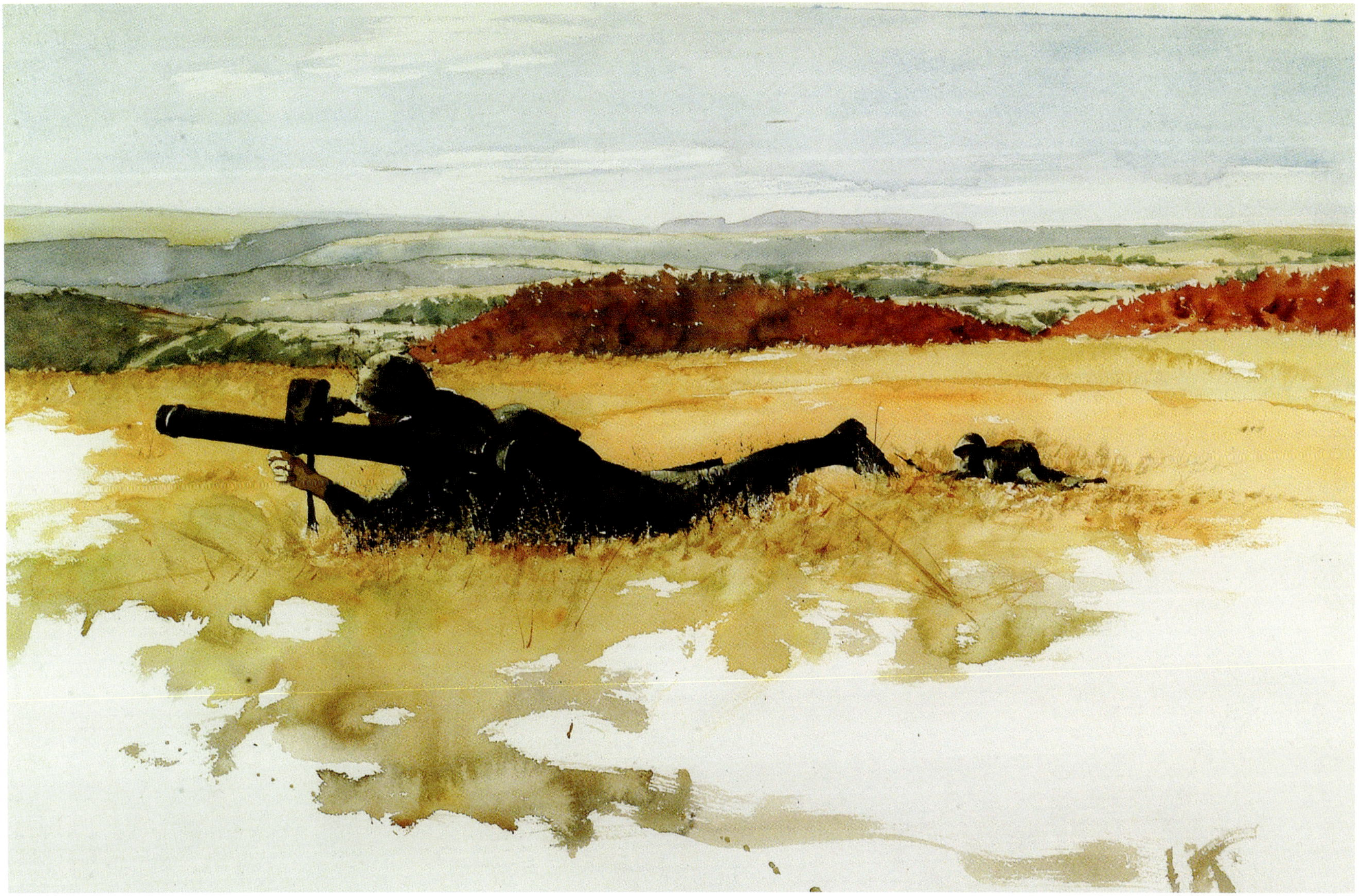

FIELD TRAINING
Baumholder, Germany
William G. Jacobson Jr., 1972
Watercolor on paper, 22" x 29¾"

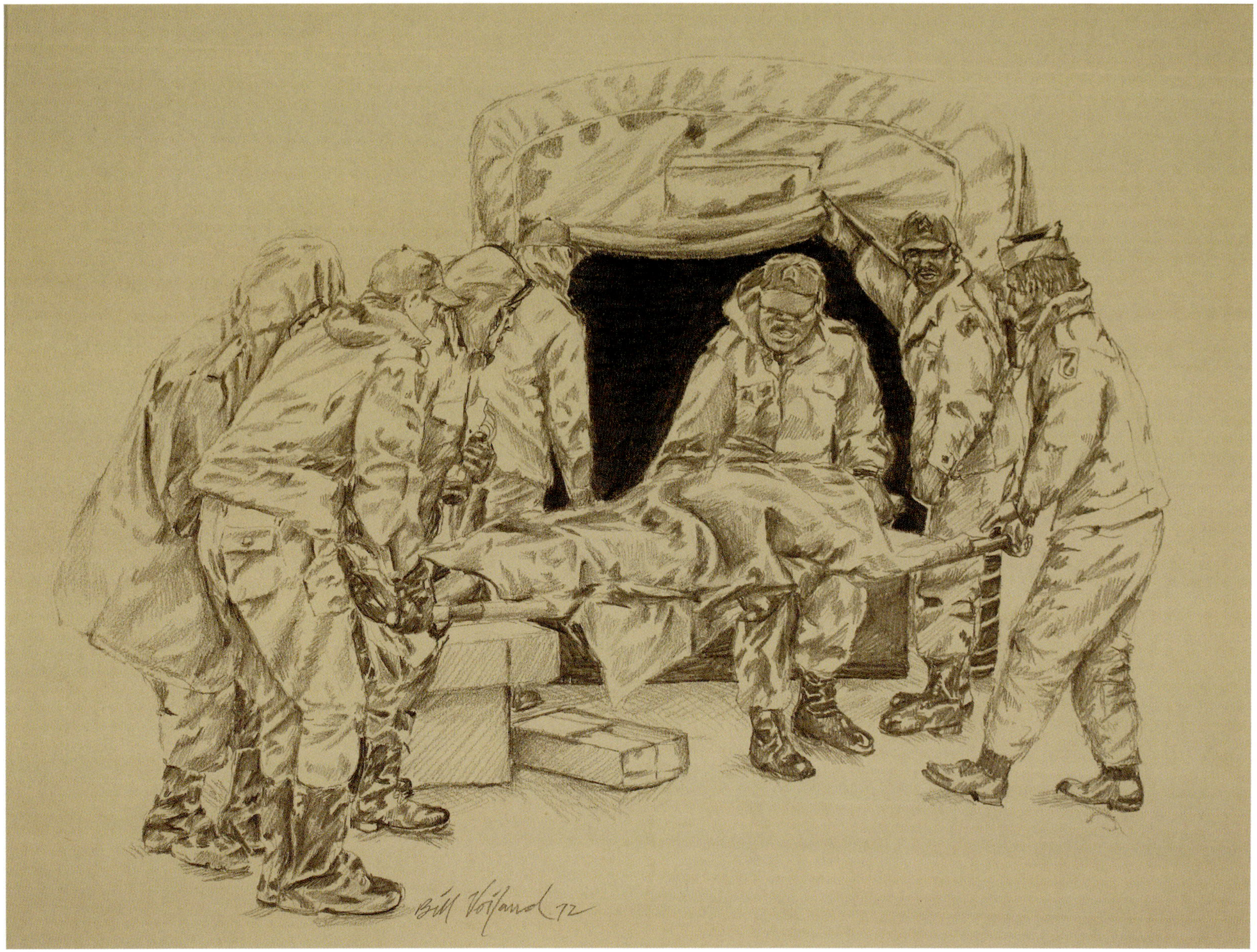

MEDICAL TRANSPORT
Germany
William F. Voiland, 1972
Pencil on illustration board, 8" x 10"

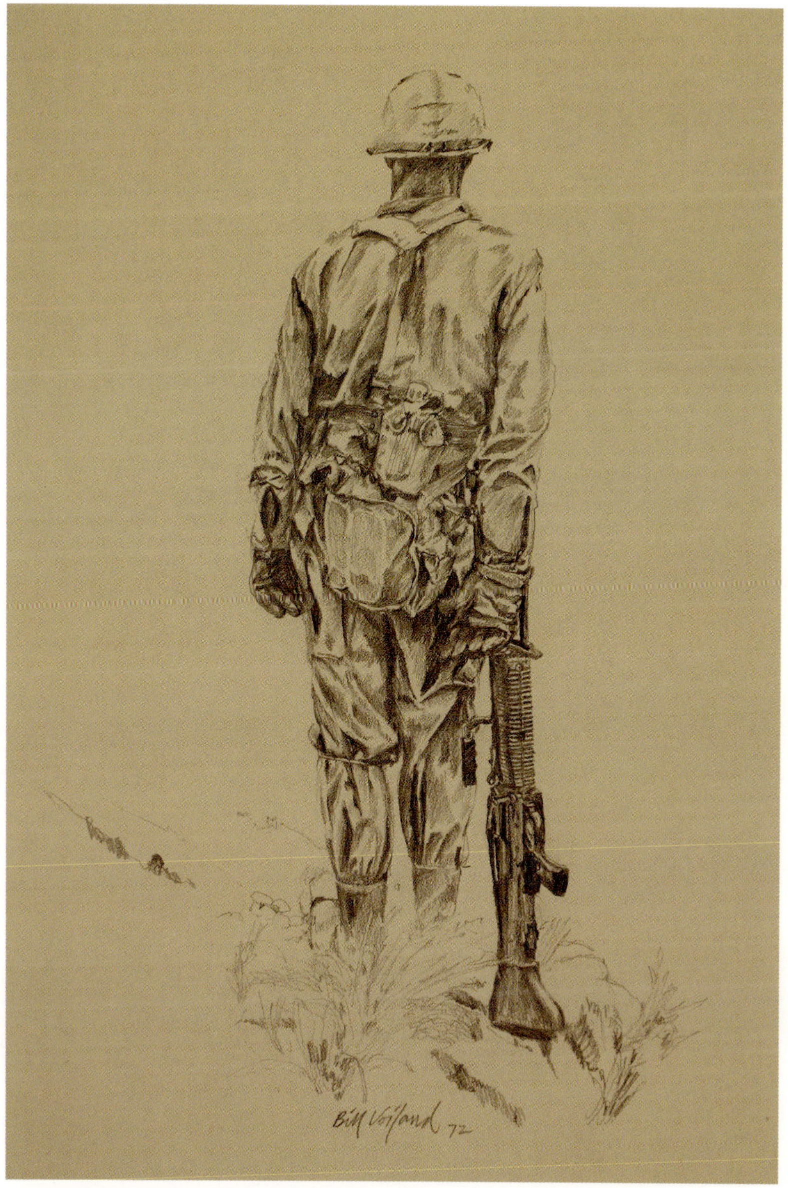

SOLDIER WITH MACHINE GUN
Germany
William F. Voiland, 1972
Pencil on illustration board, 10" x 8"

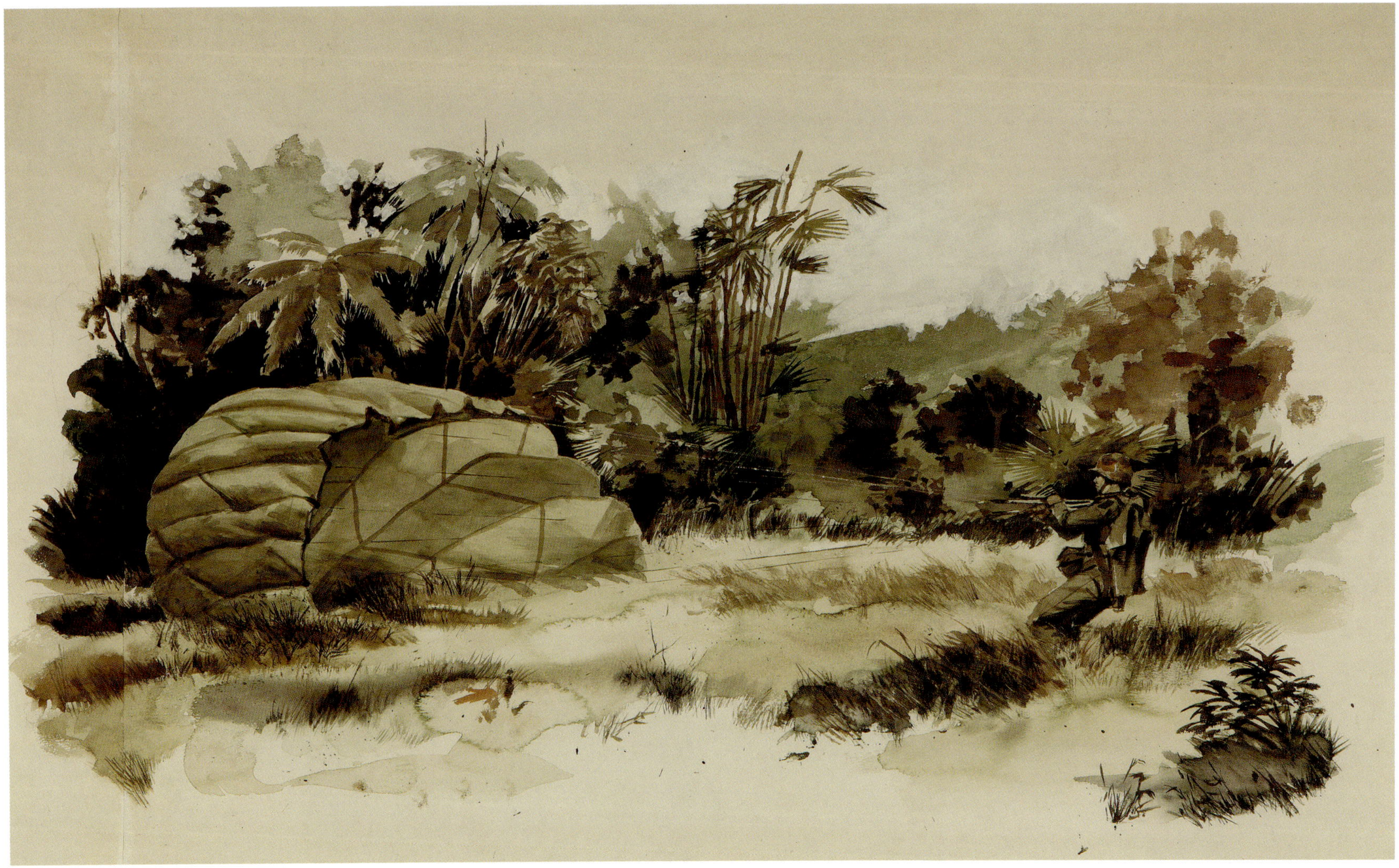

AIRBORNE TROOPER GROUNDED
Fort Cimarrón, Panama
Steven M. Hyatt, 1974
Watercolor on paper, 22¾" x 33¼"

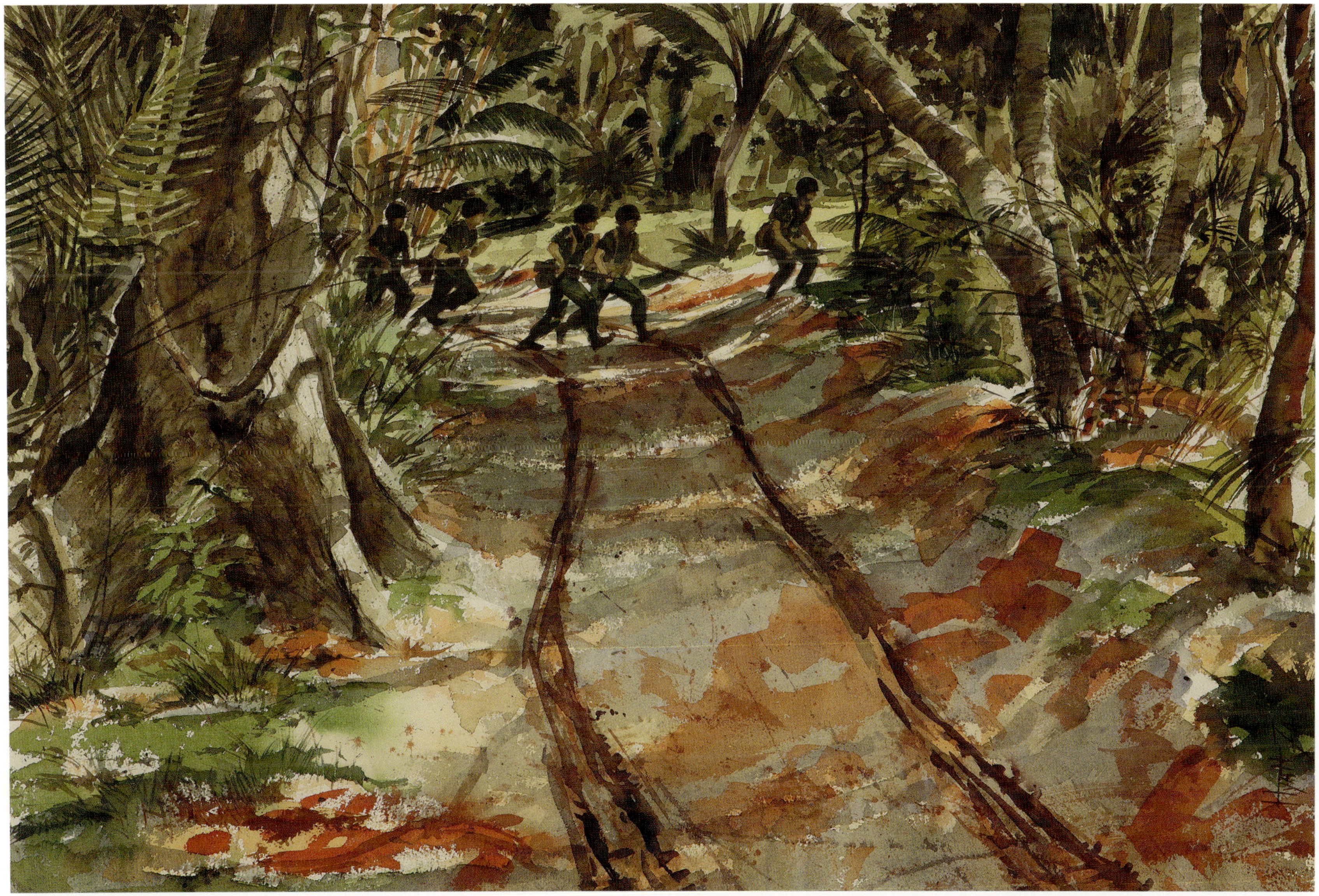

ROAD CROSSING
Fort Gulick, Canal Zone, Panama
Steven M. Hyatt, 1974
Watercolor on paper, 22½" x 30¼"

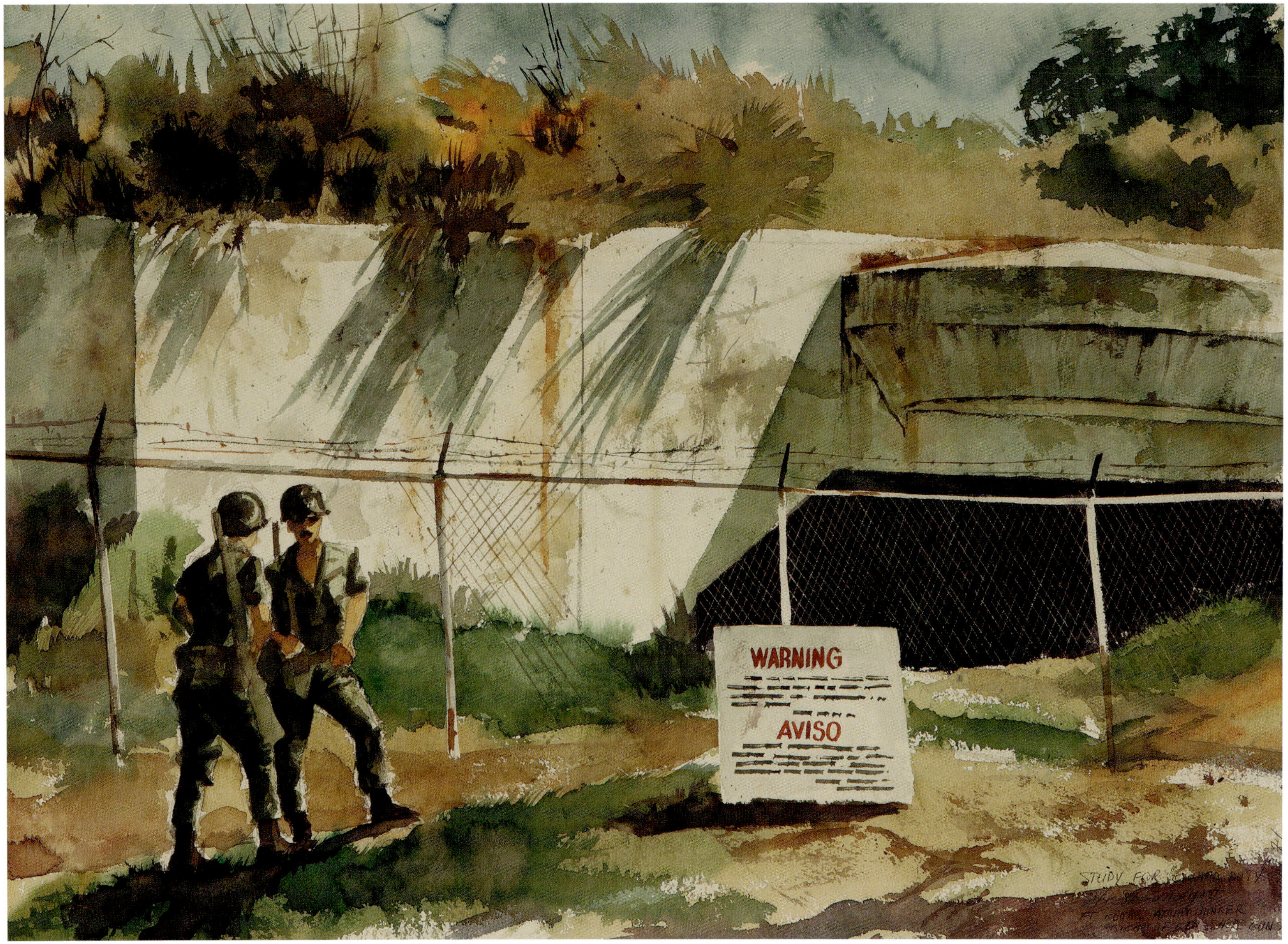

STUDY FOR "GUARD DUTY"
Fort Kobbe, Canal Zone, Panama
Steven M. Hyatt, 1974
Watercolor on paper, 22¾" x 30"

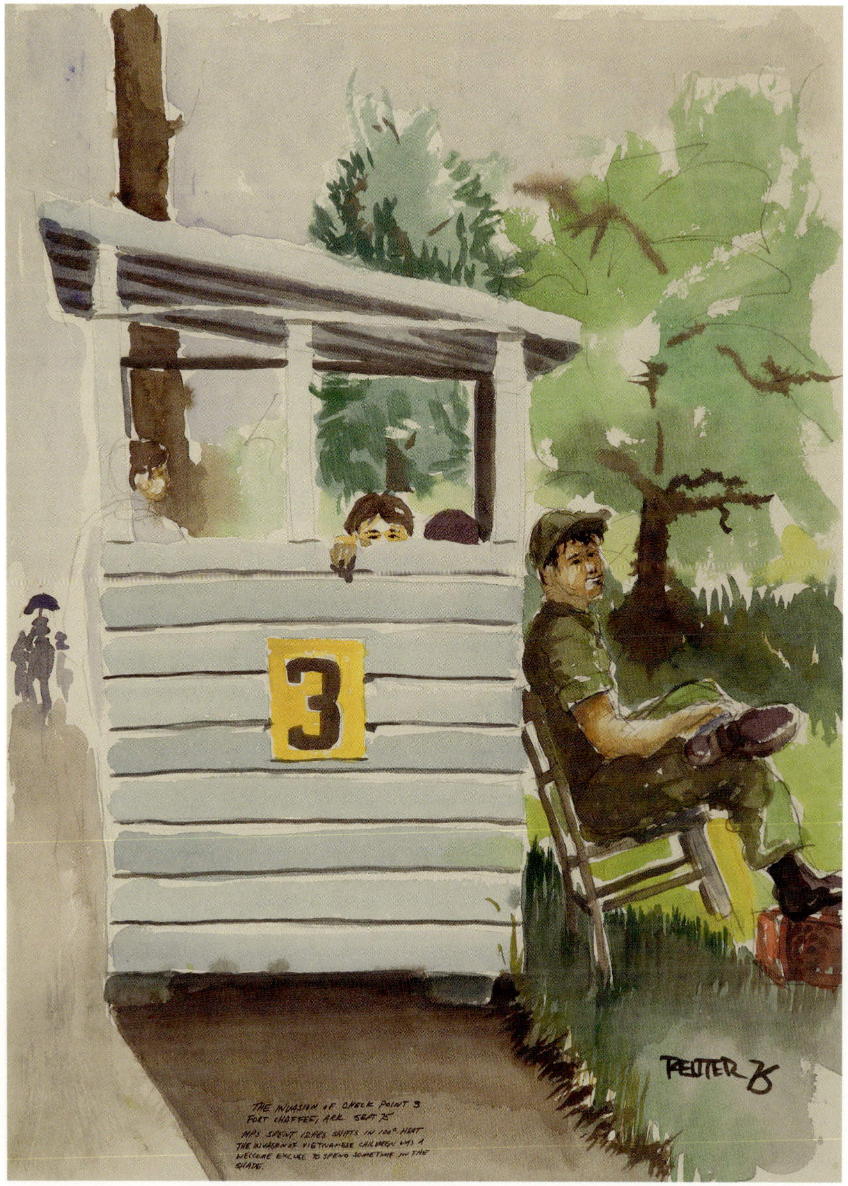

"Most of my time was spent just experiencing life at Chaffee. I ate in the Vietnamese mess halls, talked to them, and observed them and the various soldiers and civilians here."
—Raymond R. Reuter

INVASION OF CHECK POINT #3
Fort Chaffee, Arkansas
Raymond R. Reuter, 1975
Watercolor on paper, 30" x 22"

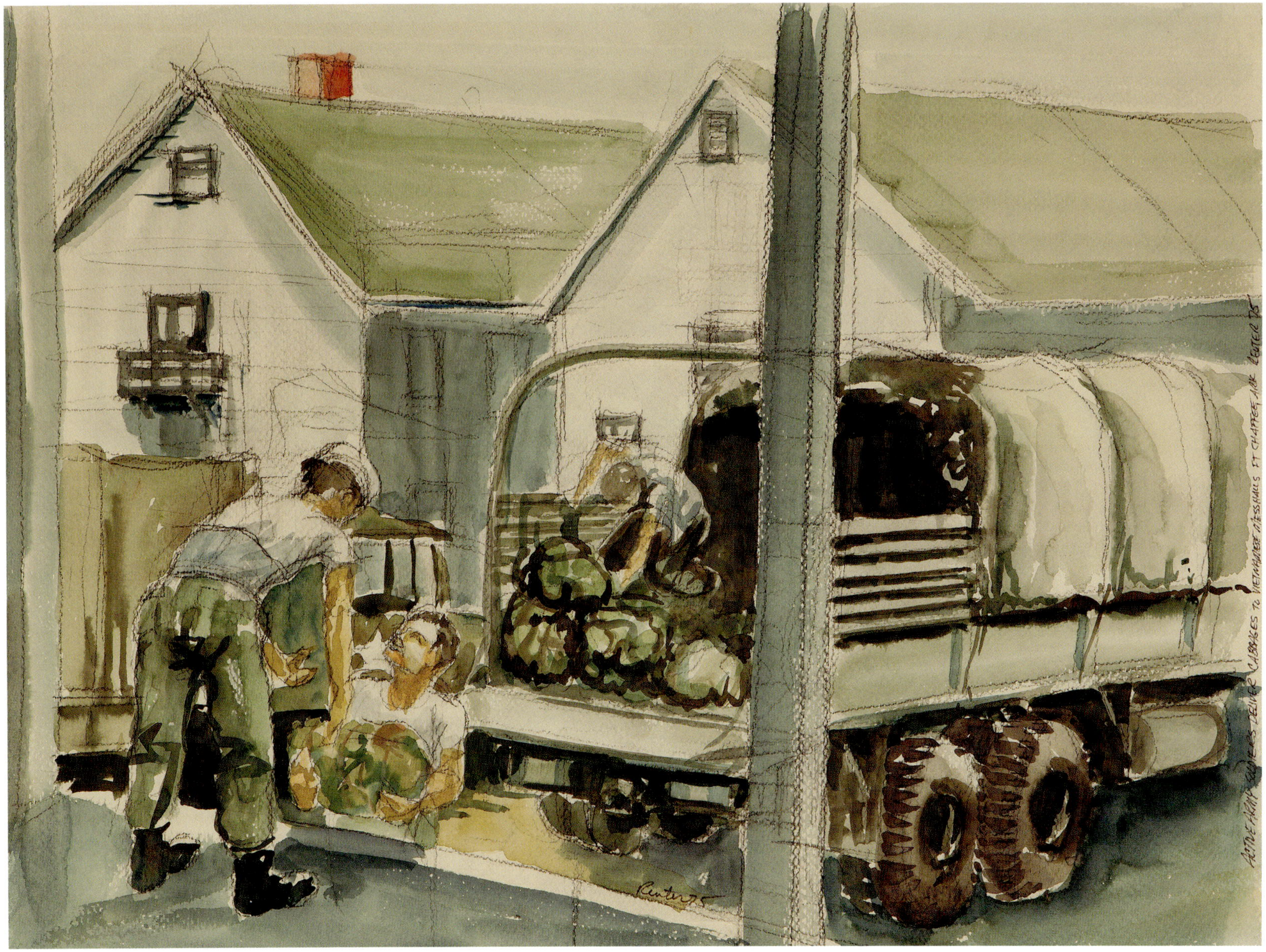

FOOD DELIVERY
Fort Chaffee, Arkansas
Raymond R. Reuter, 1975
Watercolor on paper, 18" x 24"

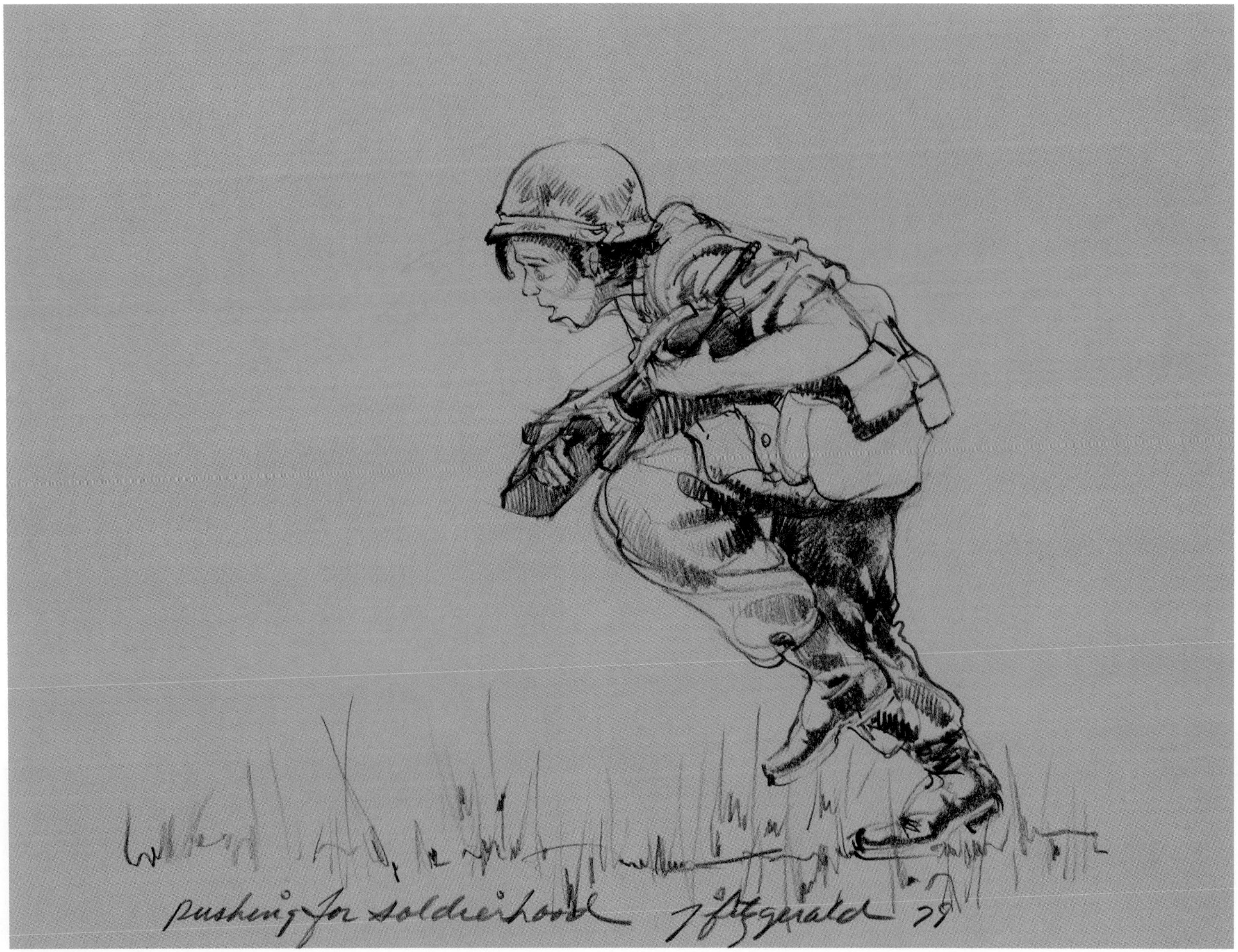

PUSHING FOR SOLDIERHOOD
Fort Jackson, South Carolina
Janet Fitzgerald, 1979
Conte crayon on paper, 9" x 12"

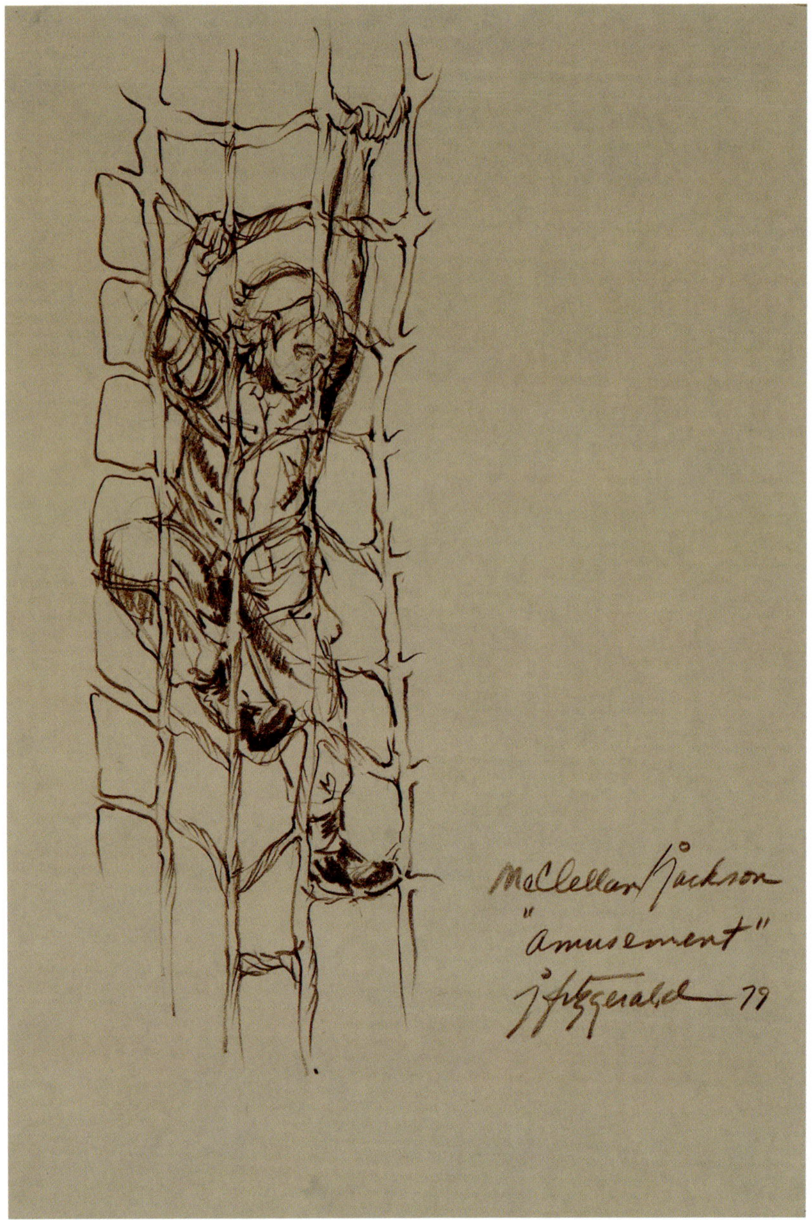

AMUSEMENT
Fort McClellan, Alabama
Janet Fitzgerald, 1979
Conte crayon on paper, 12" x 9"

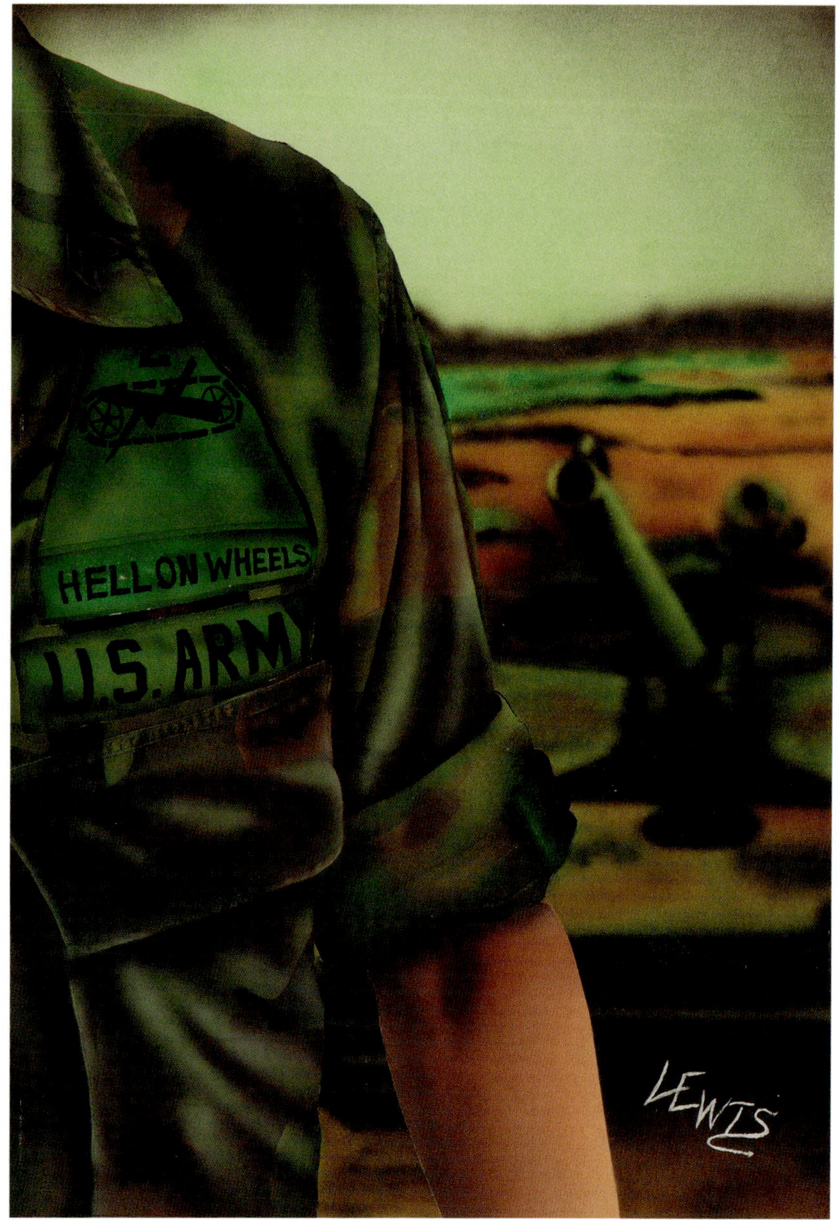

HELL ON WHEELS
Germany
Gary Lewis, 1987
Airbrush on paper, 20" x 14½"

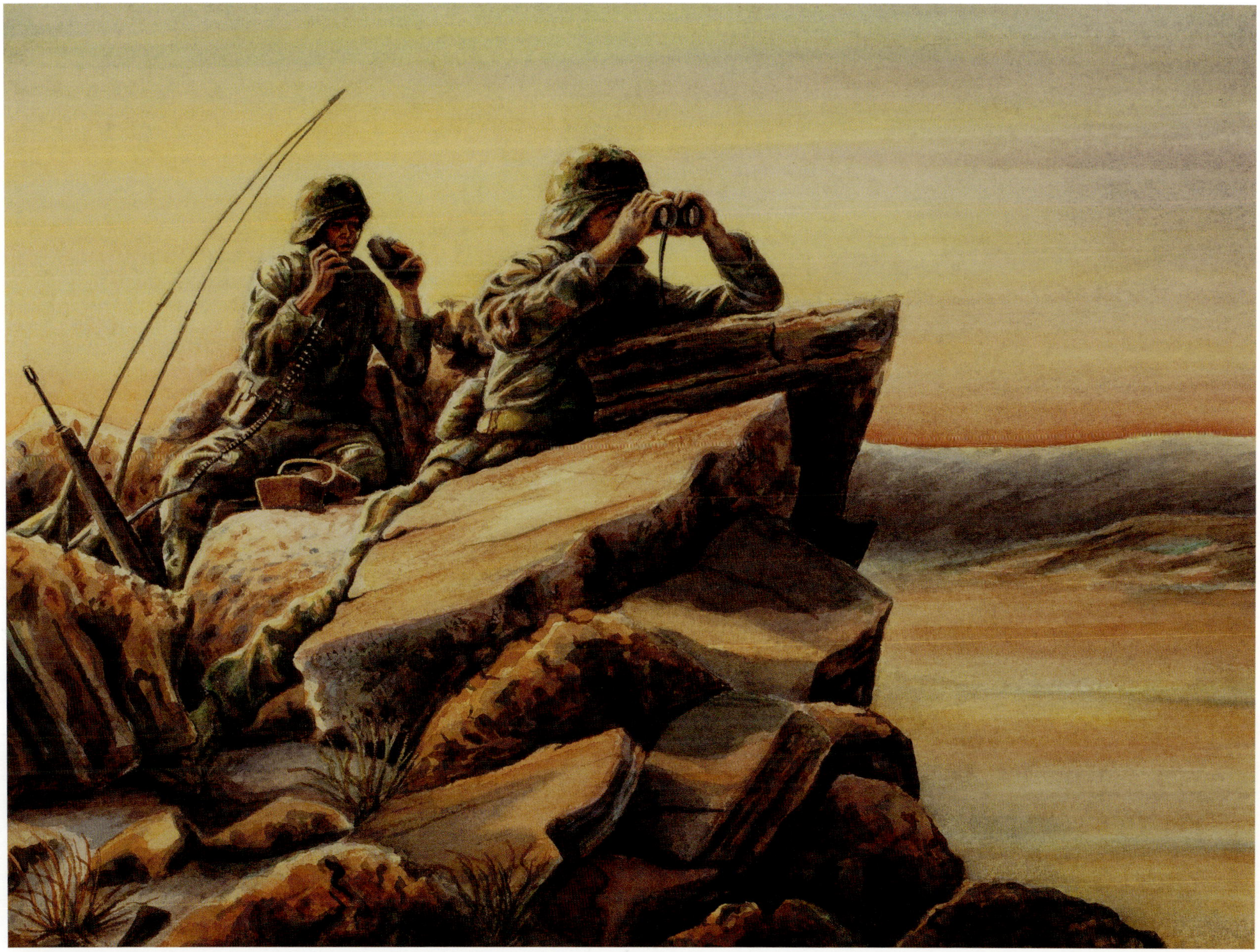

THE SCOUTS OF THE 24TH INFANTRY DIVISION
Fort Irwin, California
Patrick Farrell, 1988
Watercolor on paper, 11" x 14¼"

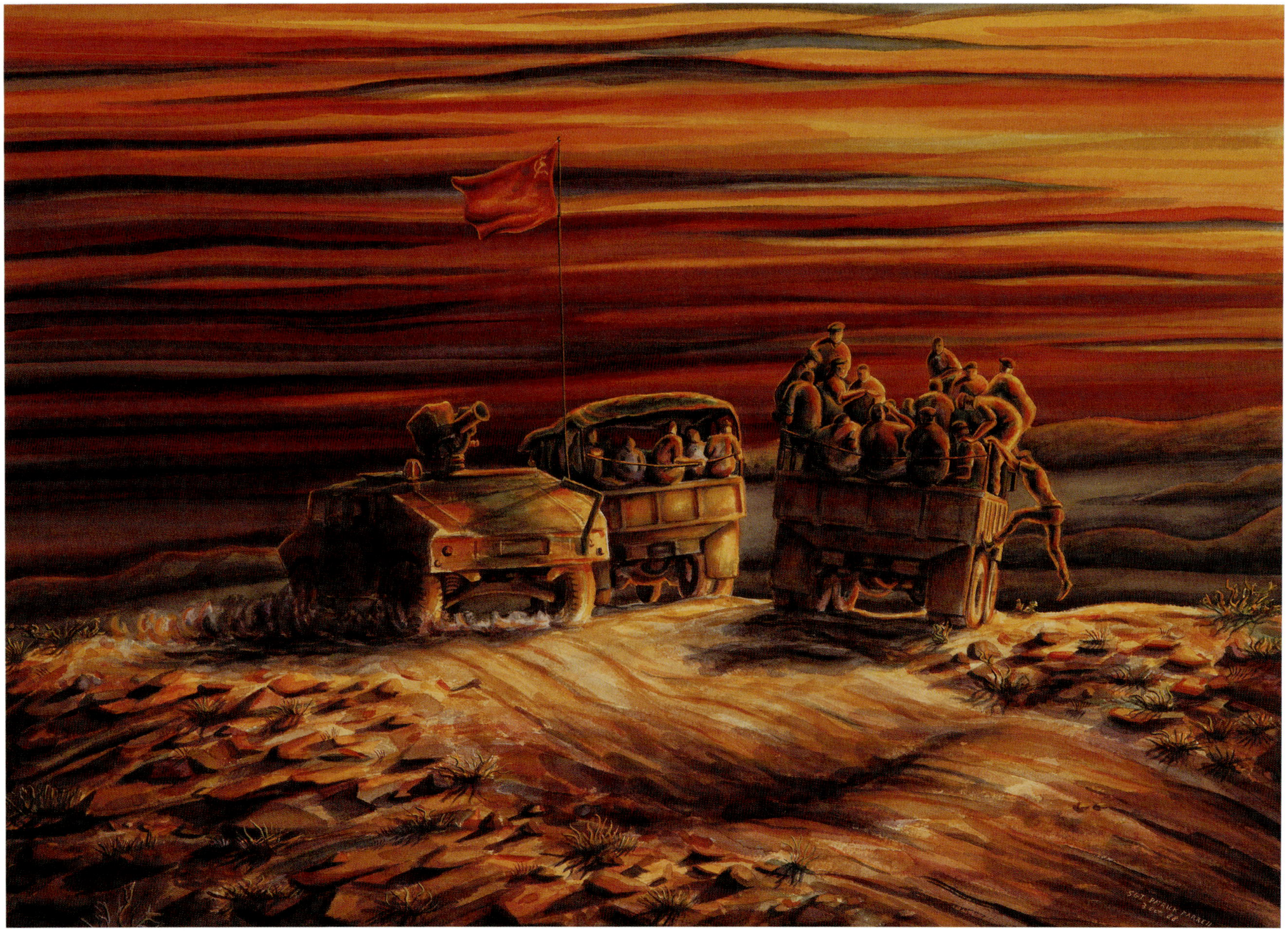

MOVING OUT AT LAST LIGHT
Fort Irwin, California
Patrick Farrell, 1988
Watercolor on paper, 21½" x 30½"

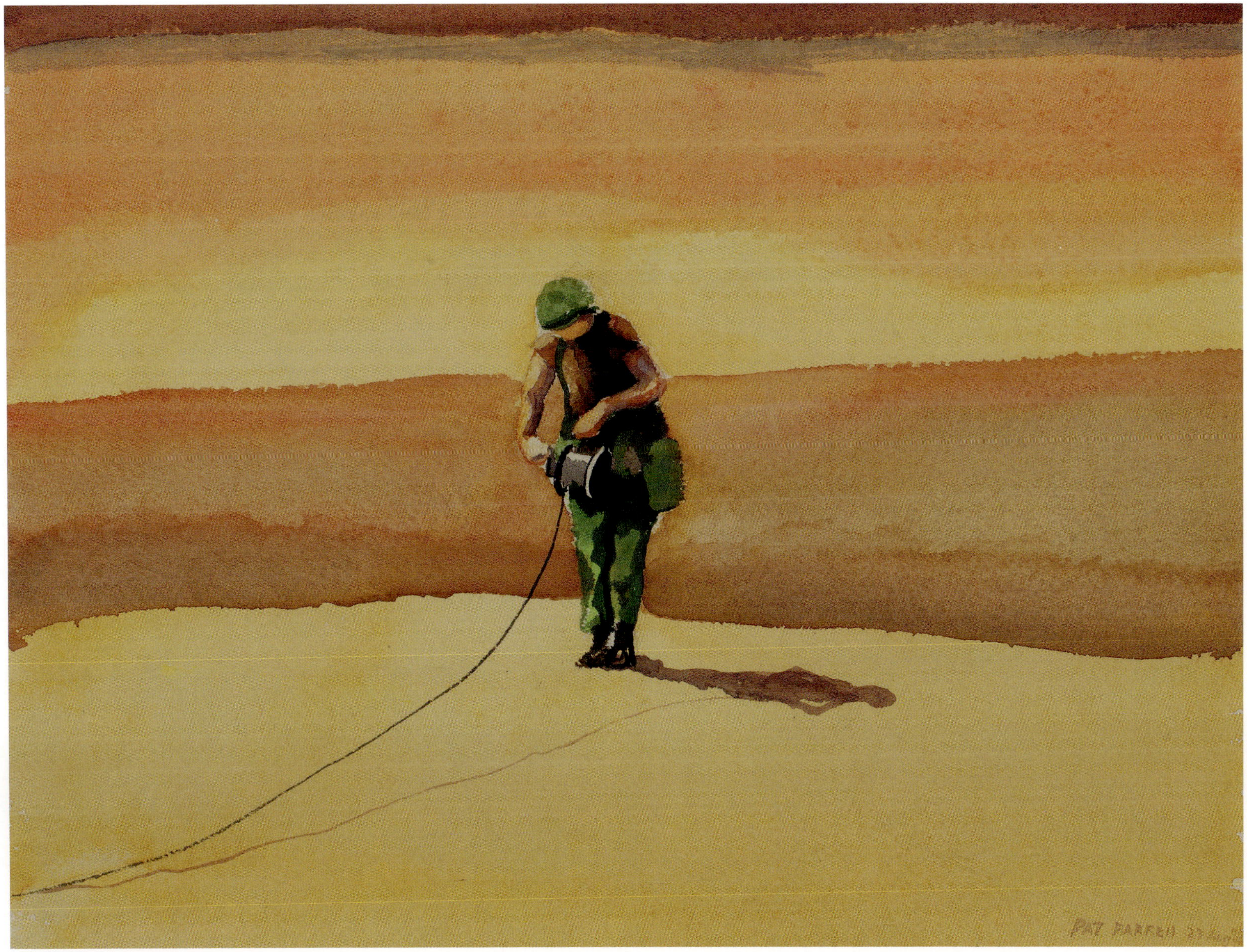

LAYING THE WIRE
Fort Irwin, California
Patrick Farrell, 1988
Watercolor on paper, 11¾" x 15"

"As I watched each soldier approach the open door, I could feel the tension and fear—but that did not stop him from launching himself into the air. The excitement and determination was so palpable that I wished I had on gear so that I could jump, too."

—Al Sprague

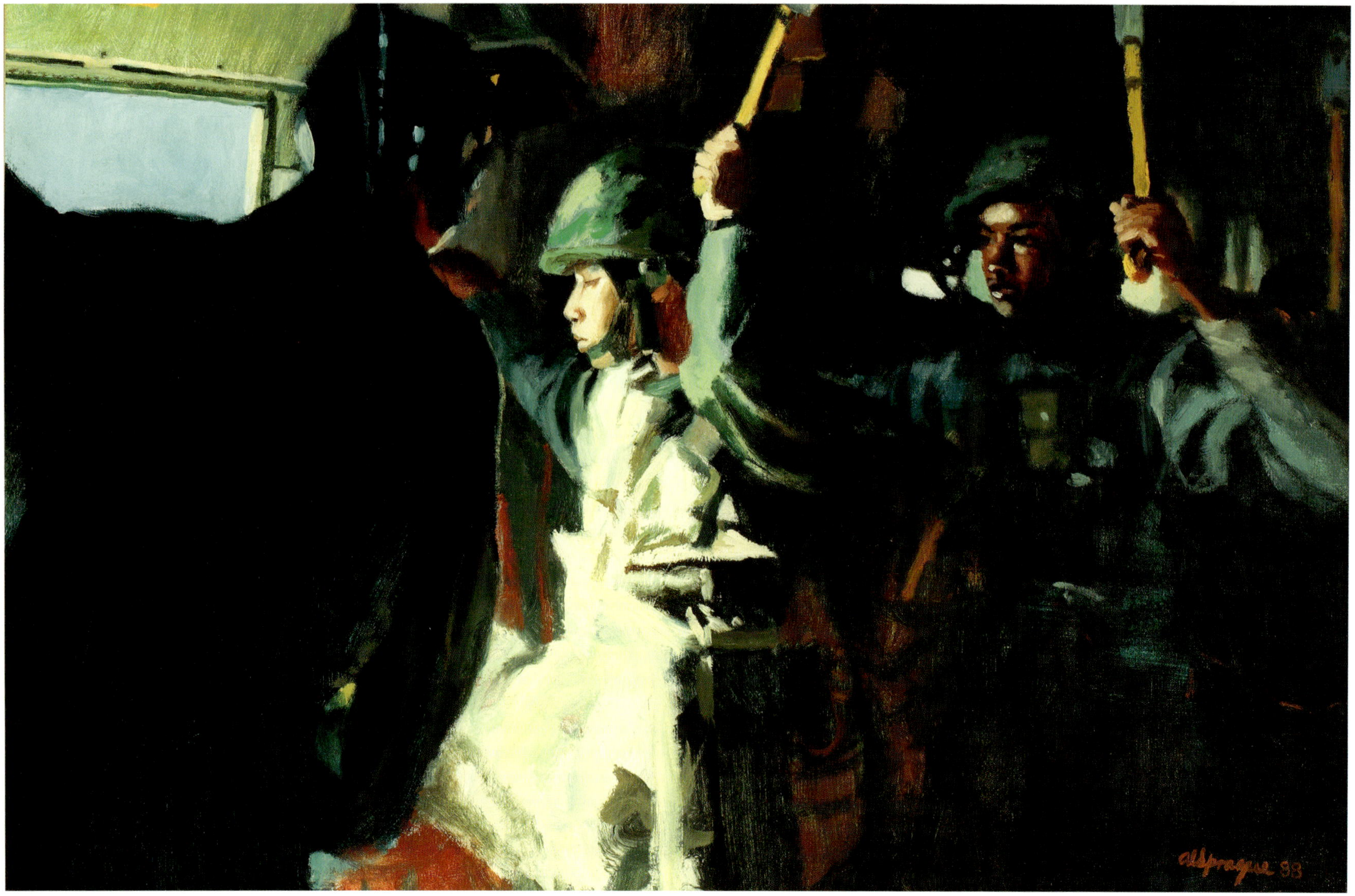

MOMENT OF TRUTH
Panama
Al Sprague, 1988
Oil on canvas, 20½" x 30½"

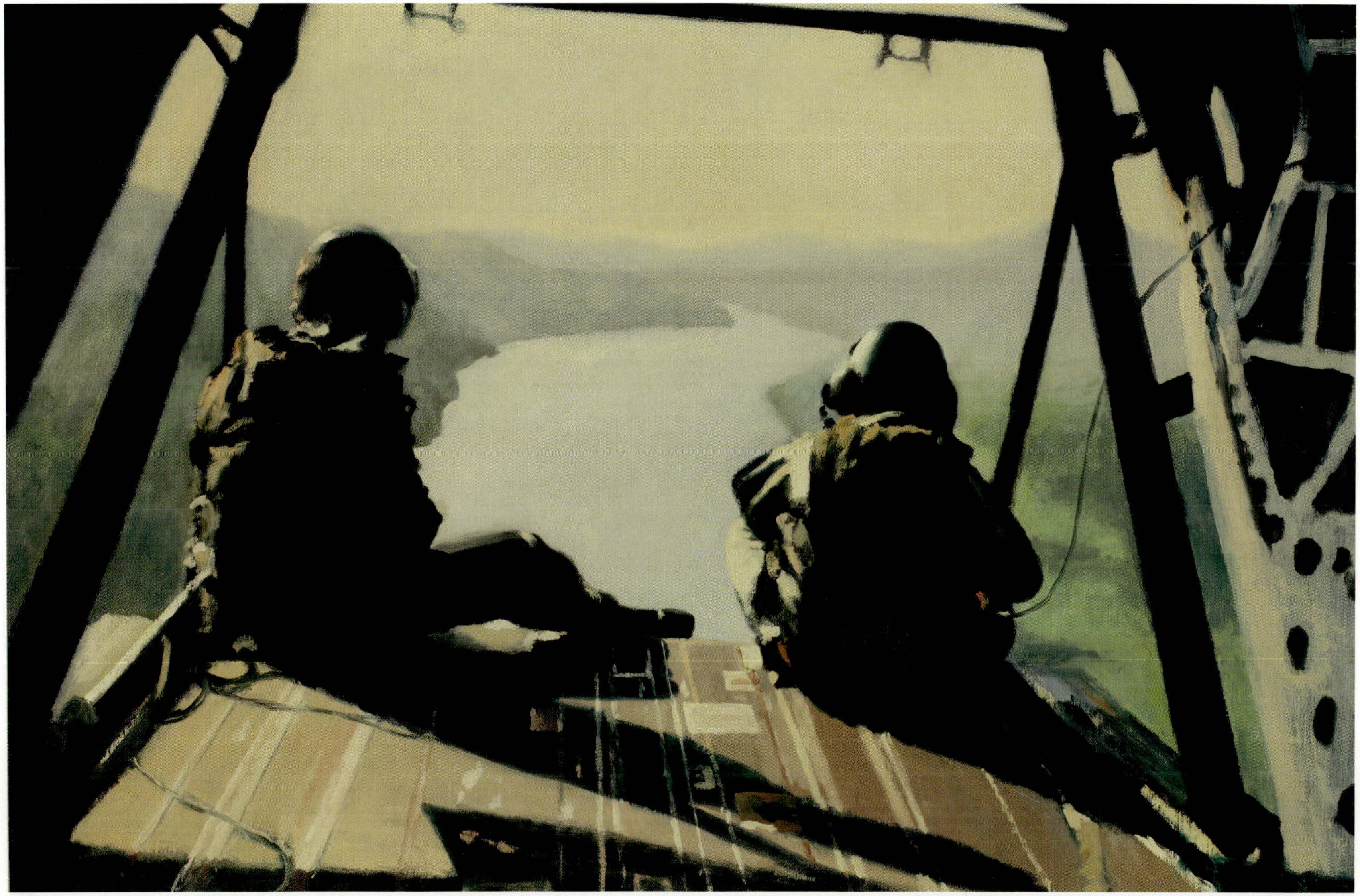

CRUISING THE PANAMA CANAL (CHINOOK)
Panama, Operation JUST CAUSE
Al Sprague, 1990
Oil on canvas, 24" x 36"

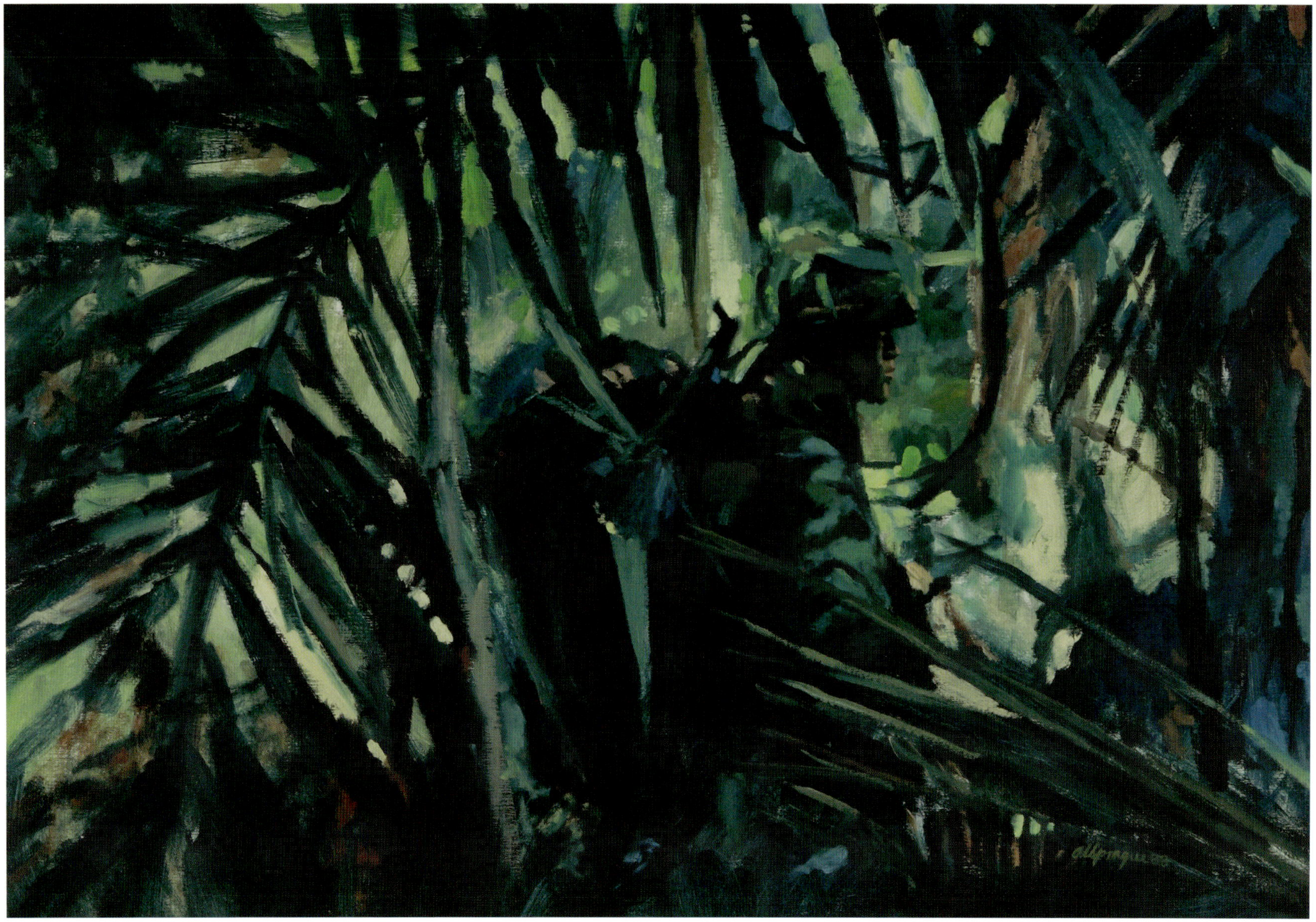

JUNGLE NOON
Panama
Al Sprague, 1988
Oil on canvas, 28" x 40"

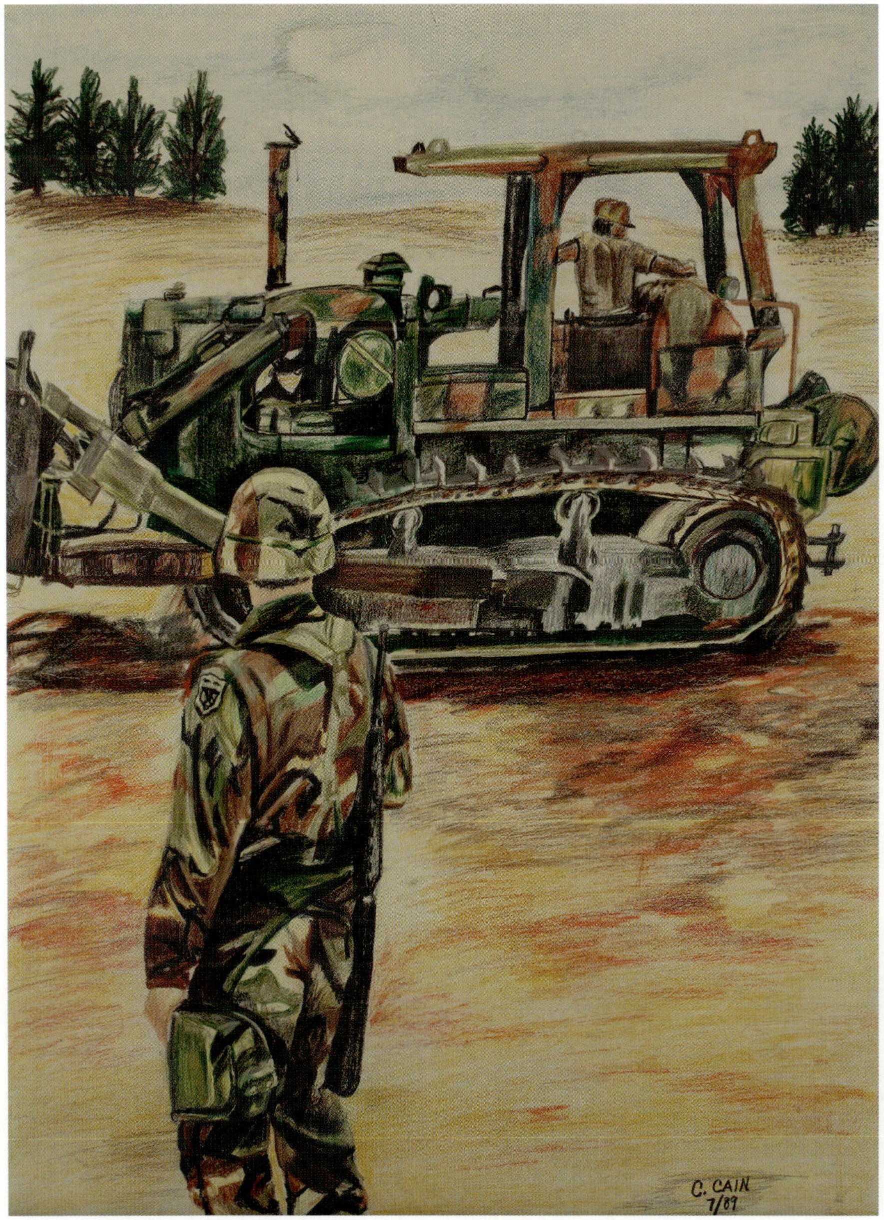

ENGINEERS PREPARING THE SITE
Fort McCoy, Wisconsin
Mary Colleen Cain, 1989
Pastel on paper, 24" x 18"

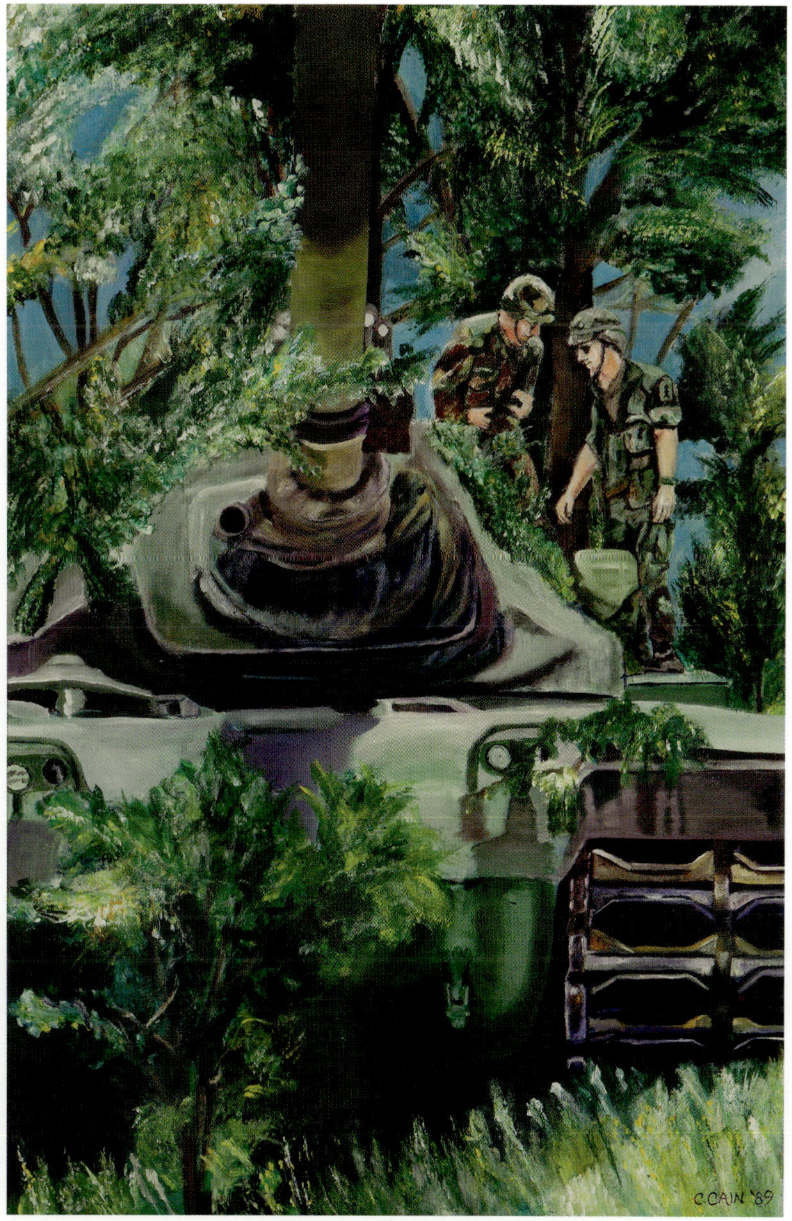

THE ART OF CAMOUFLAGE
Fort McCoy, Wisconsin
Mary Colleen Cain, 1989
Acrylic on canvas, 36" x 24"

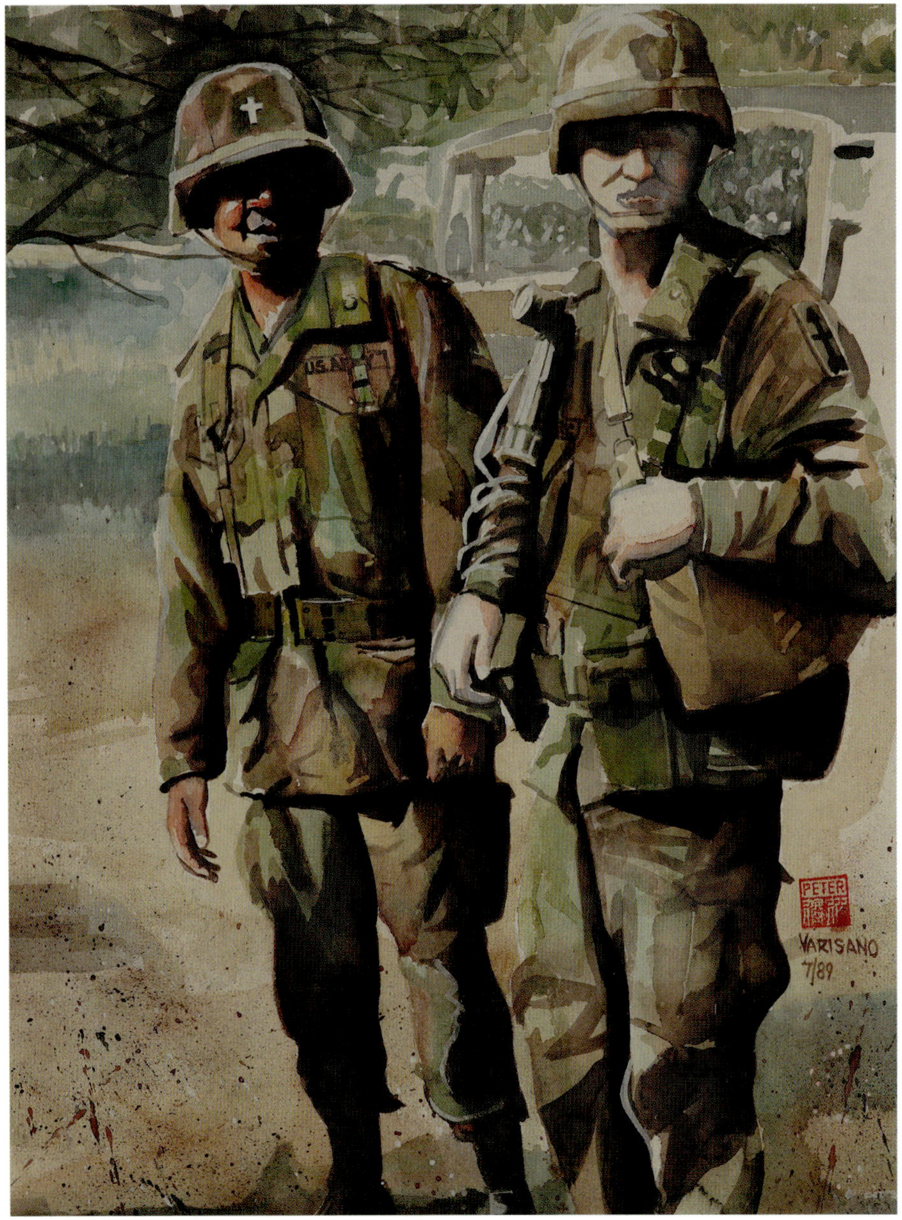

THE CHAPLAIN #1
Fort McCoy, Wisconsin
Peter G. Varisano, 1989
Watercolor on paper, 24" x 18"

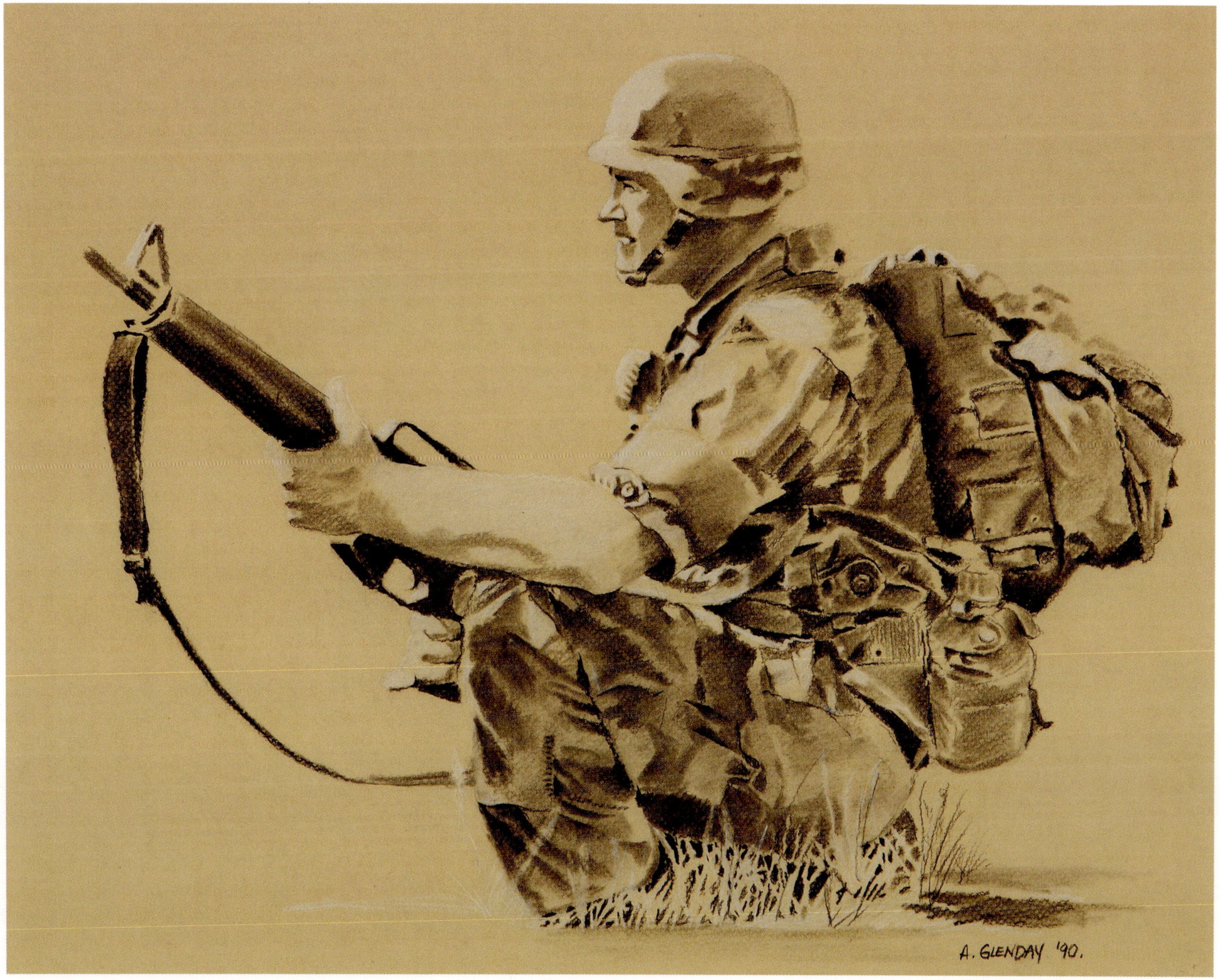

"It was my desire to allow the public to see and get a better feel for today's Army through my artwork."
—Adam Glenday

SUNLIT SOLDIER-FIRE TEAM MANEUVERS
Erlangen, Germany
Adam Glenday, 1990
Chalk and pastel on paper, 25½" x 19½"

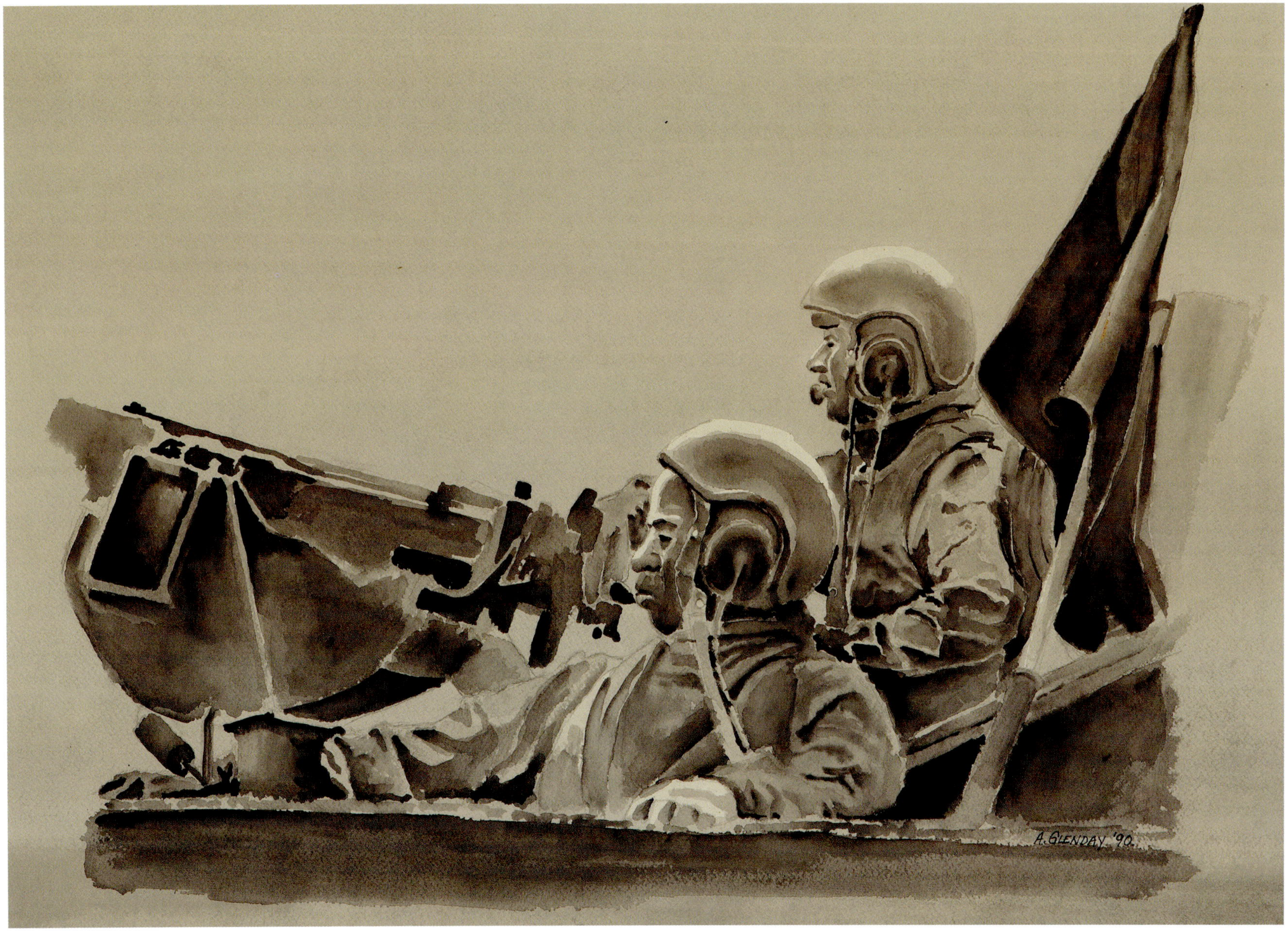

THE CREW
Grafenwöhr, Germany
Adam Glenday, 1990
Watercolor on paper, 18" x 24"

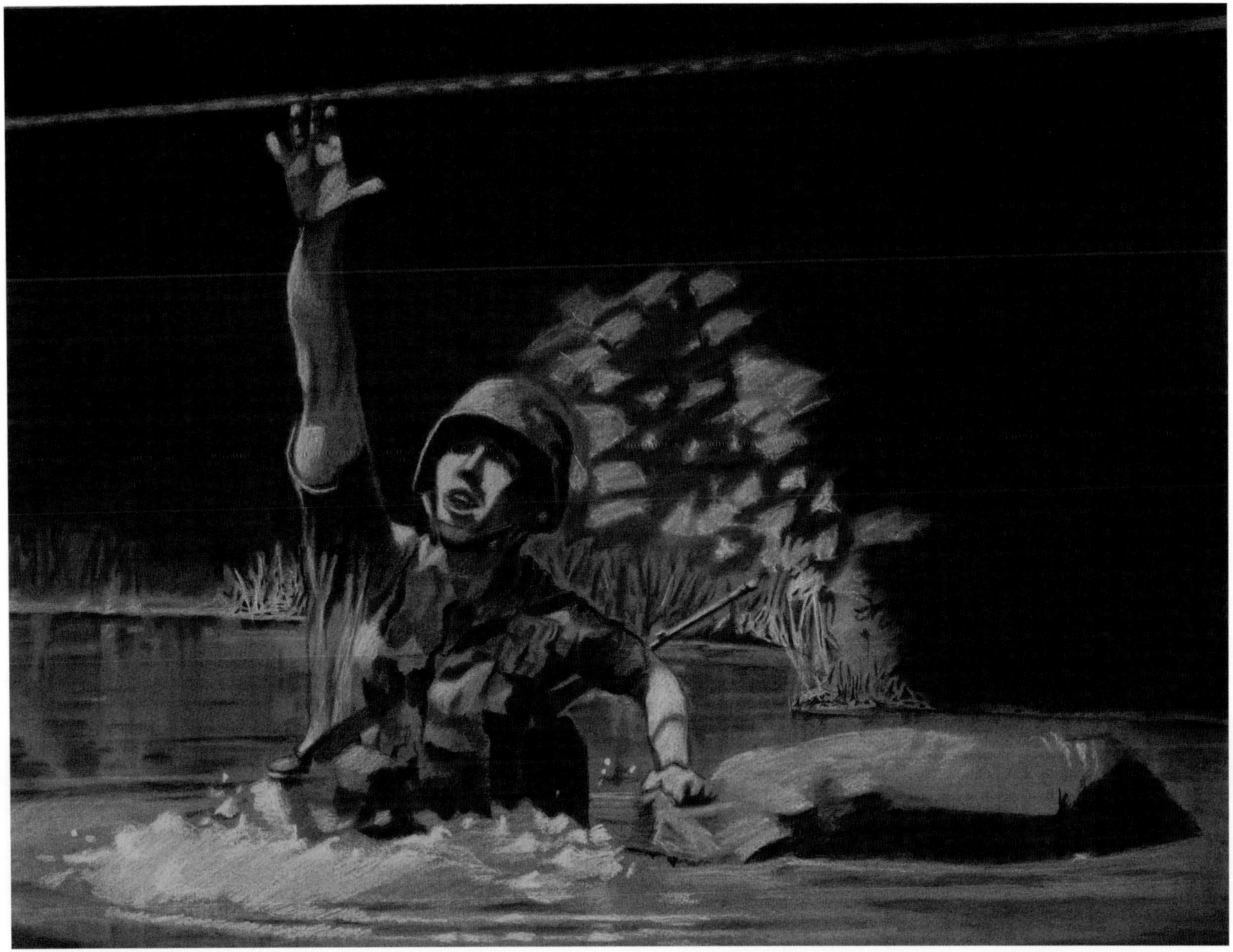

DETERMINATION
Wildflecken, Germany
Adam Glenday, 1990
Watercolor on paper, 19¾" x 25½"

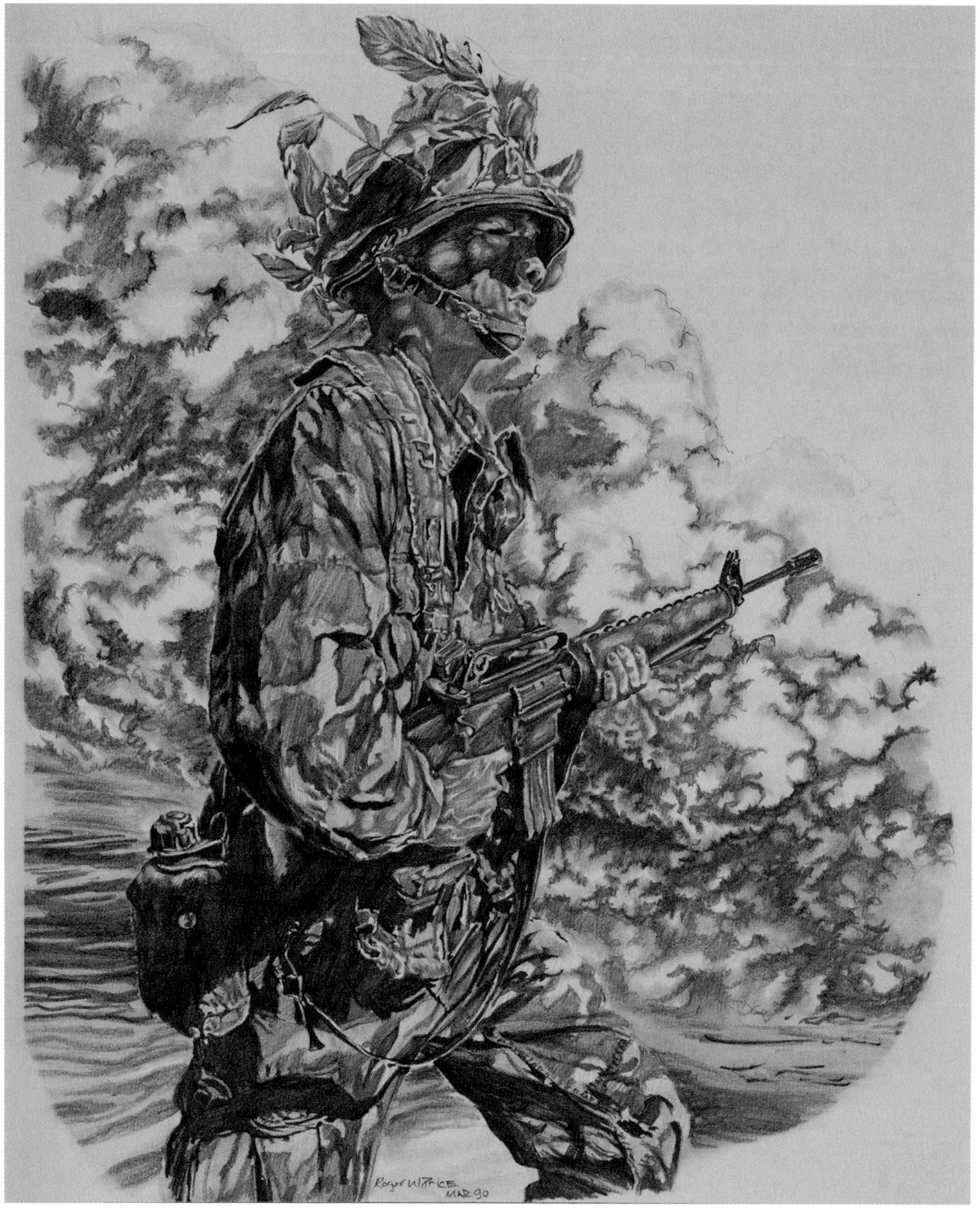

COMBAT READY
Germany
Roger W. Price, 1990
Pencil on paper, 10" x 8½"

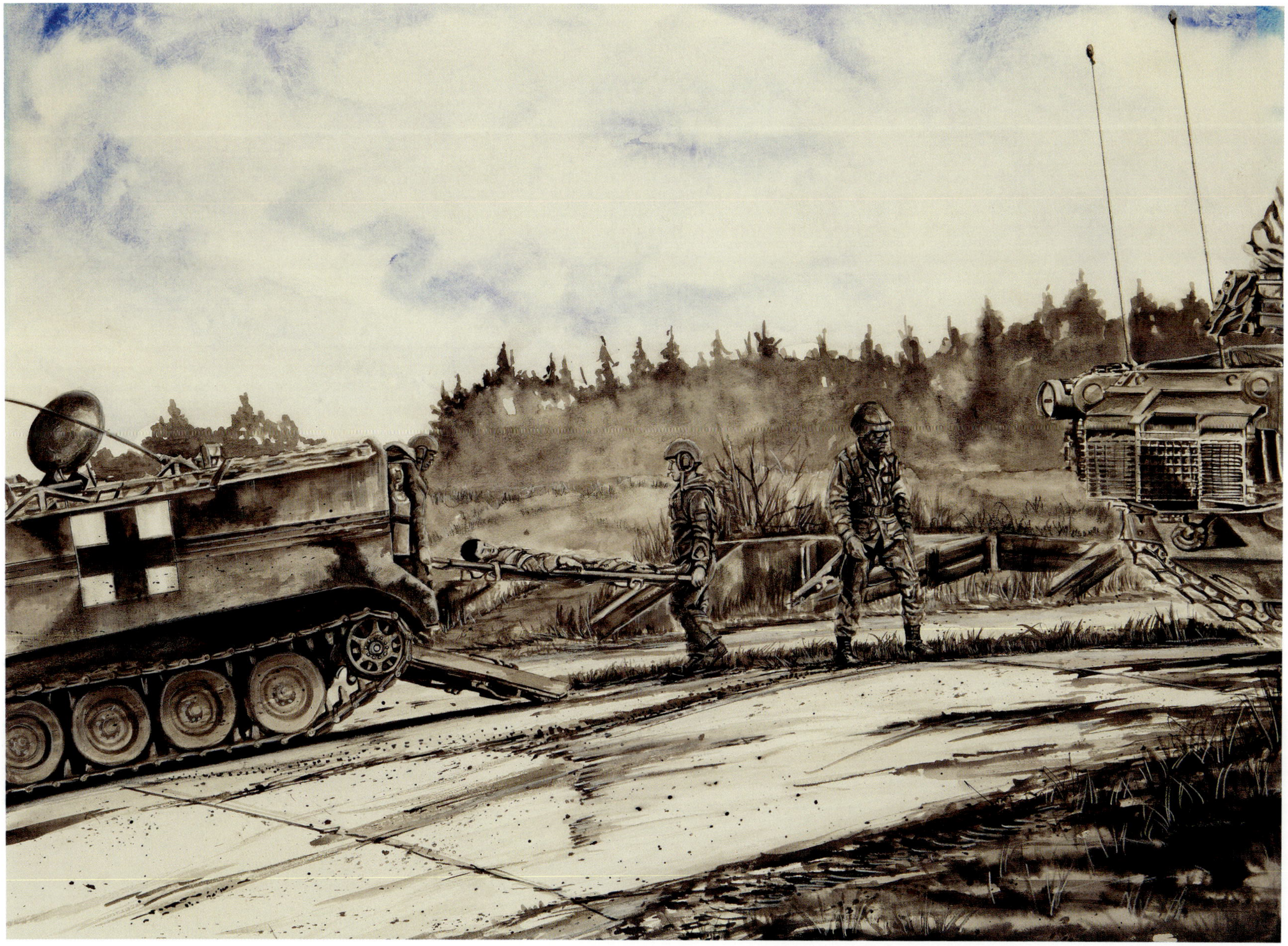

IN SUPPORT
Grafenwöhr, Germany
Roger W. Price, 1990
Watercolor on paper, 18" x 24"

Desert Shield and Desert Storm

In viewing the artwork associated with Operations Desert Shield and Desert Storm, one would conclude that the artists were either willfully or unconsciously making a break with their predecessors who documented the American soldier in the jungles of Vietnam. Overall, the body of work is bright and bathed in desert sunlight against a very open and flat landscape that is only troubled by the distant black plumes of burning oil wells on the horizon. The mood is decidedly one of rejoicing, which is associated with the U.S. victory in the now famous 100-hour ground war.

It is important to note that the Army sent two teams of artists to cover the operations. The first team, composed of Sfc. Sieger Hartgers and Sfc. Peter G. Varisano, was sent to chronicle Operation Desert Shield (August 1990–January 1991), which was the deployment and build-up phase. This team produced art featuring soldiers who appear relaxed, youthful, and clean but are imbued with a sense of determination. The ever-resilient GI is depicted resting, playing chess, attending religious services, and taking care of the wash as in *Soldier Doing the Laundry* by Hartgers (page 102). These light moments are reminiscent of camp scenes found in the art of the world wars. The soldiers' faces are recognizable, well defined, and some even dare a smile. The chosen mediums aid the light feeling but are also handled in a very traditional manor. Watercolor and pencil sketches replace the heavy oils and acrylics that contributed so much to the atmospheric humidity of the art of the Vietnam War.

In contrast to art of previous periods, this art reveals a marked increase in female personnel representing typical soldiers. Varisano's watercolor of a passing female soldier entitled *24th Infantry Division Soldier* (page 106) shows the woman as an equal with her armed peers. She is not identified in the title as a female soldier, but rather as a standard example of a soldier from the division, marking a pivotal change in the portrayal of the American soldier. One aspect of the unique nature and anxiety of the conflict is captured in the portraits of soldiers wearing nuclear, biological, and chemical (NBC) protective gear, as seen in Varisano's *Watch & Listen for Threats* (page 104), which recalls the insect-like appearance of soldiers of World War I.

The second team, reservists Capt. Mario H. Acevedo and Lt. Col. Frank M. Thomas, deployed after the ground war had concluded and based its impressions on first-person interviews, photographs, and visits to the ravaged terrain where the action took place. This secondhand experience contributed to the commercial arts style of many of the combat action images of Operation DESERT STORM. These treatments are markedly devoid of the themes of death or suffering because the artists did not feel the emotions and react to the events firsthand.

This team's subject matter and compositions reference themes from World War II that underscore decisive military might, such as tanks and helicopters depicted in dynamic action and victorious engagements. In contrast, the enemy is portrayed only through destroyed equipment. The technology of war is carefully rendered as in the edgy action in *The Cavalry Advances* by Acevedo (page 111), contrasted to the painterly *Tent City* by Hartgers (page 101), with its muted strokes that capture a scene of canvas and ropes universal to the Army of the twentieth century.

Victory, which was defined as removing Iraqi forces from Kuwait, was covered by yet a third group of soldier-artists who documented homecoming activities as jubilant forces returned to the United States. In the midst of this celebratory atmosphere, the artists also focus on intimate moments, such as the reuniting of family members seen in S. Sgt. Emilio E. Arias' *Mother With Child* (page 117). This triumphant mood is prevalent throughout all of the art of the Persian Gulf war, as victory was imminent and troop morale was high.

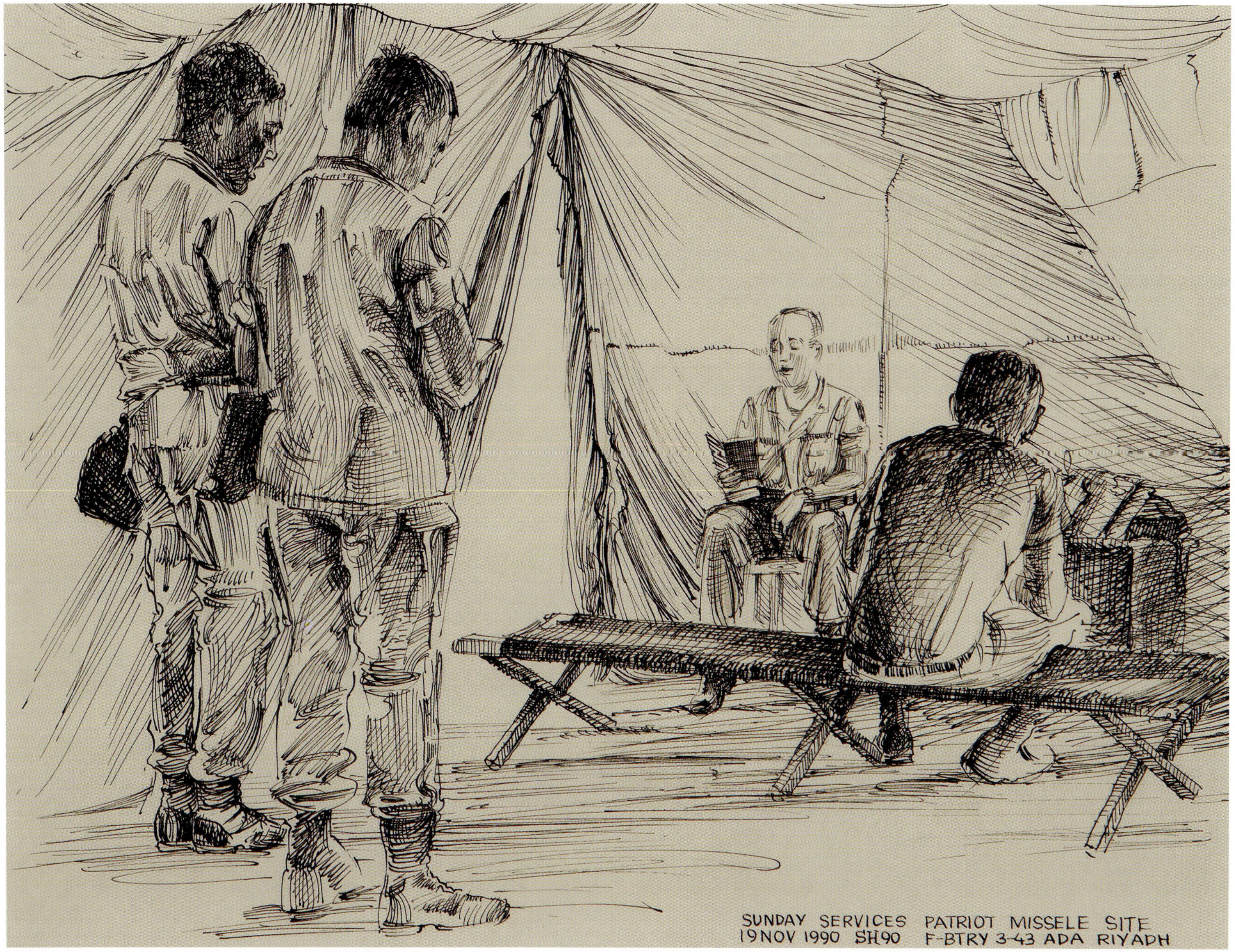

"Machinery comes and goes. I show the human side. Art from 600 years ago still relates to us now because it shows the human story; that's what gives us value. . . . We can put feelings into our paintings. It's what writers have, what musicians have, what artists have."

—Sieger Hartgers

SUNDAY SERVICES, PATRIOT MISSILE SITE
Saudi Arabia, Operation DESERT SHIELD
Sieger Hartgers, 1990
Ink on paper, 13" x 12½"

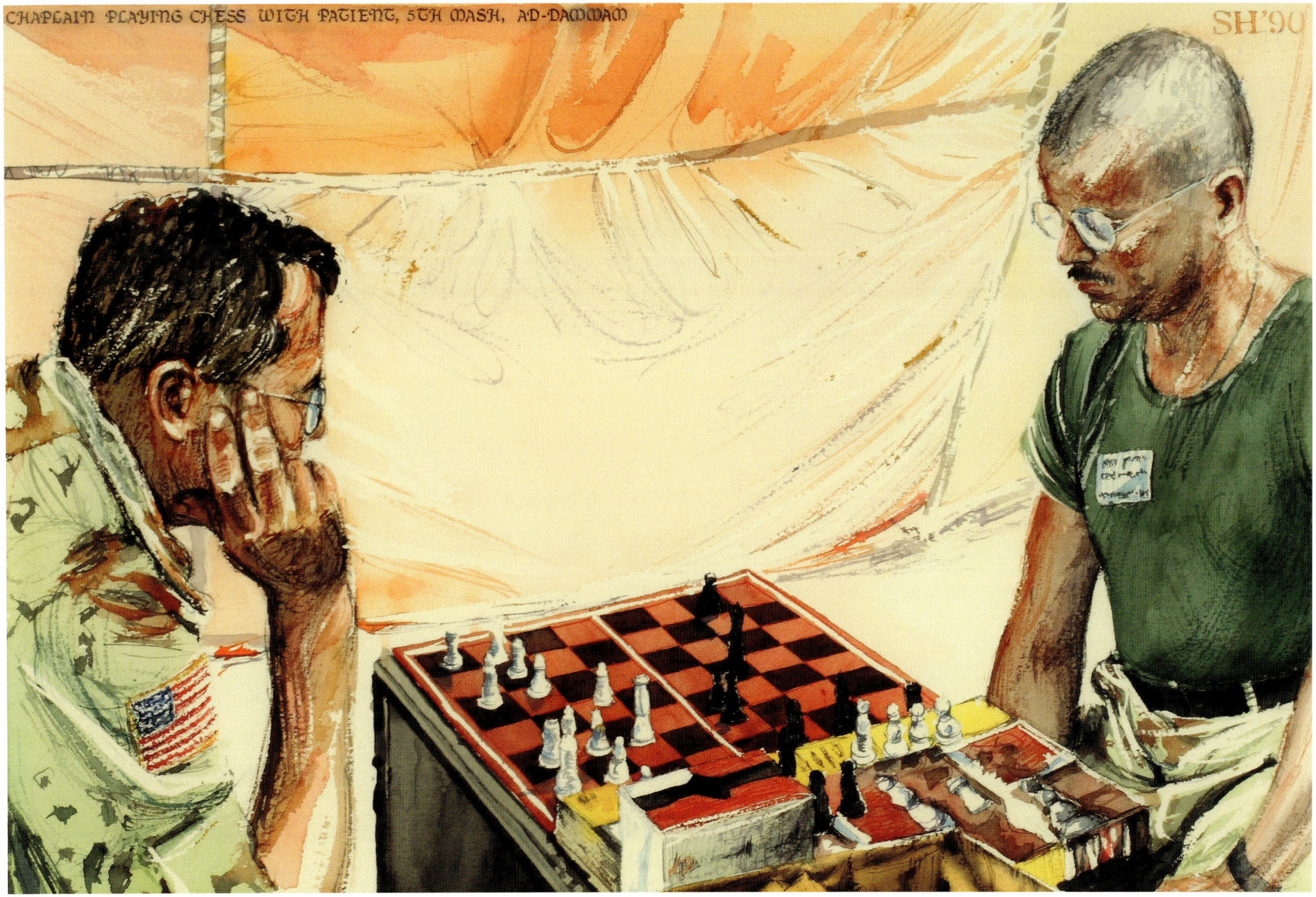

CHAPLAIN PLAYING CHESS WITH PATIENT, 5TH MASH, AD-DAMMAM
Saudi Arabia, Operation Desert Shield
Sieger Hartgers, 1990
Watercolor on paper, 22¾" x 30"

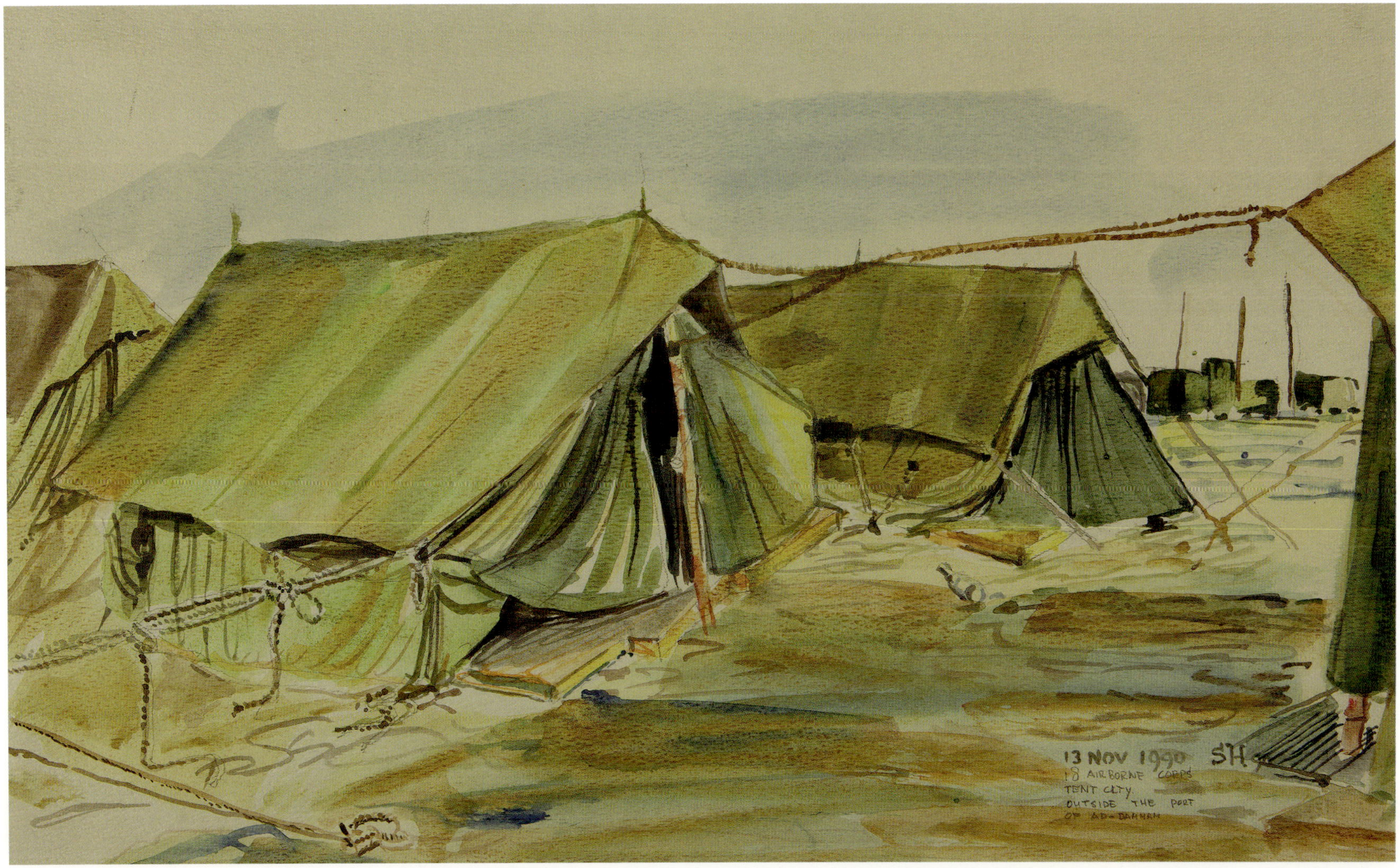

TENT CITY
Ad Dammam, Saudi Arabia, Operation Desert Shield
Sieger Hartgers, 1990
Watercolor on paper, 12" x 18"

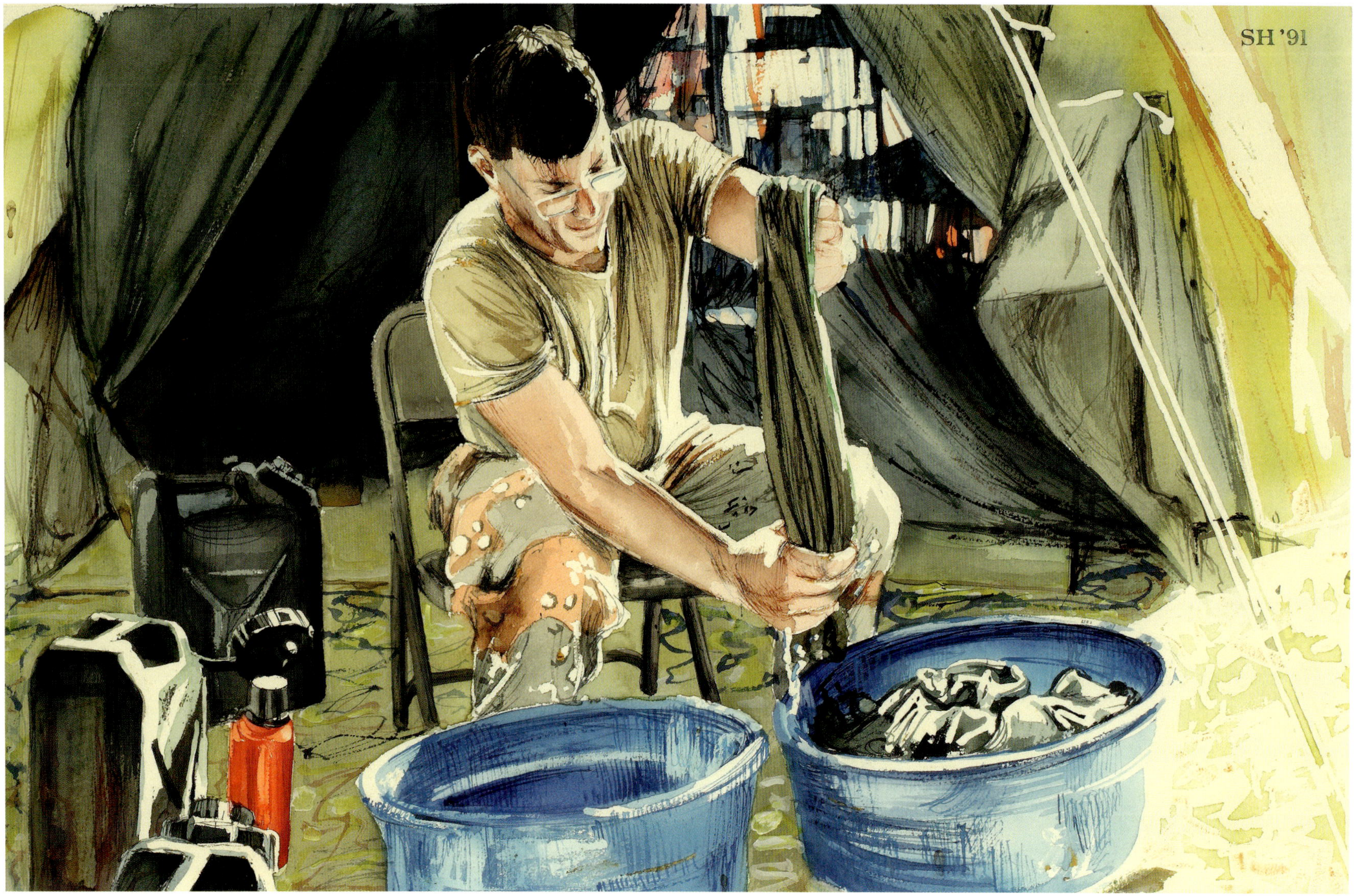

SOLDIER DOING THE LAUNDRY
North of Ad Dammam, Saudi Arabia, Operation Desert Shield
Sieger Hartgers, 1991
Watercolor on paper, 24" x 36"

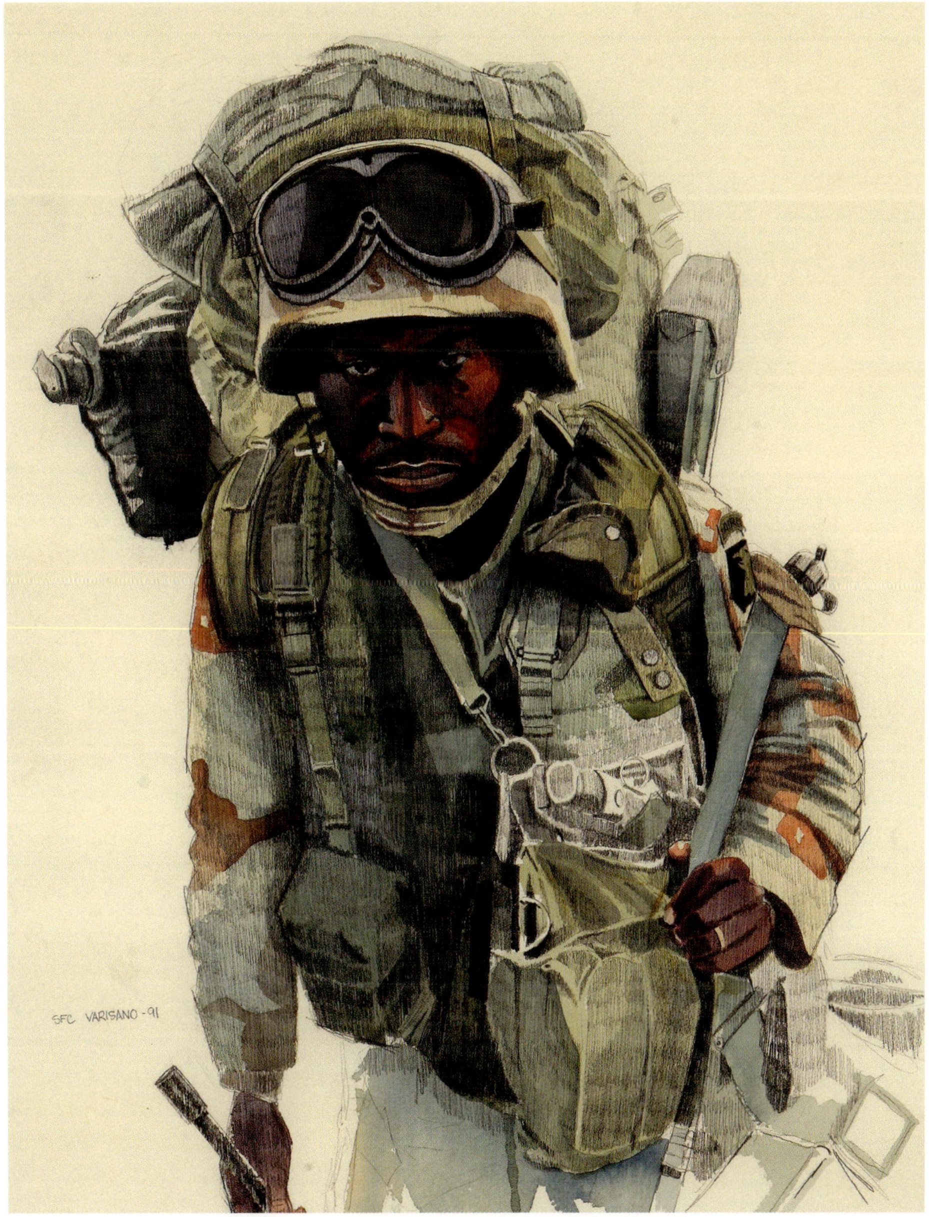

"This soldier's look moved me because of the confrontational stare and the amount of equipment he was carrying. . . . He was 'rucking up' to go forward and I thought, when I saw this image, that there is no way we are going to lose this war if we have soldiers like this."
—Peter G. Varisano

DON'T MESS WITH THE 101ST
Saudi Arabia, Operation Desert Shield
Peter G. Varisano, 1991
Pencil and watercolor on paper, 30½" x 22½"

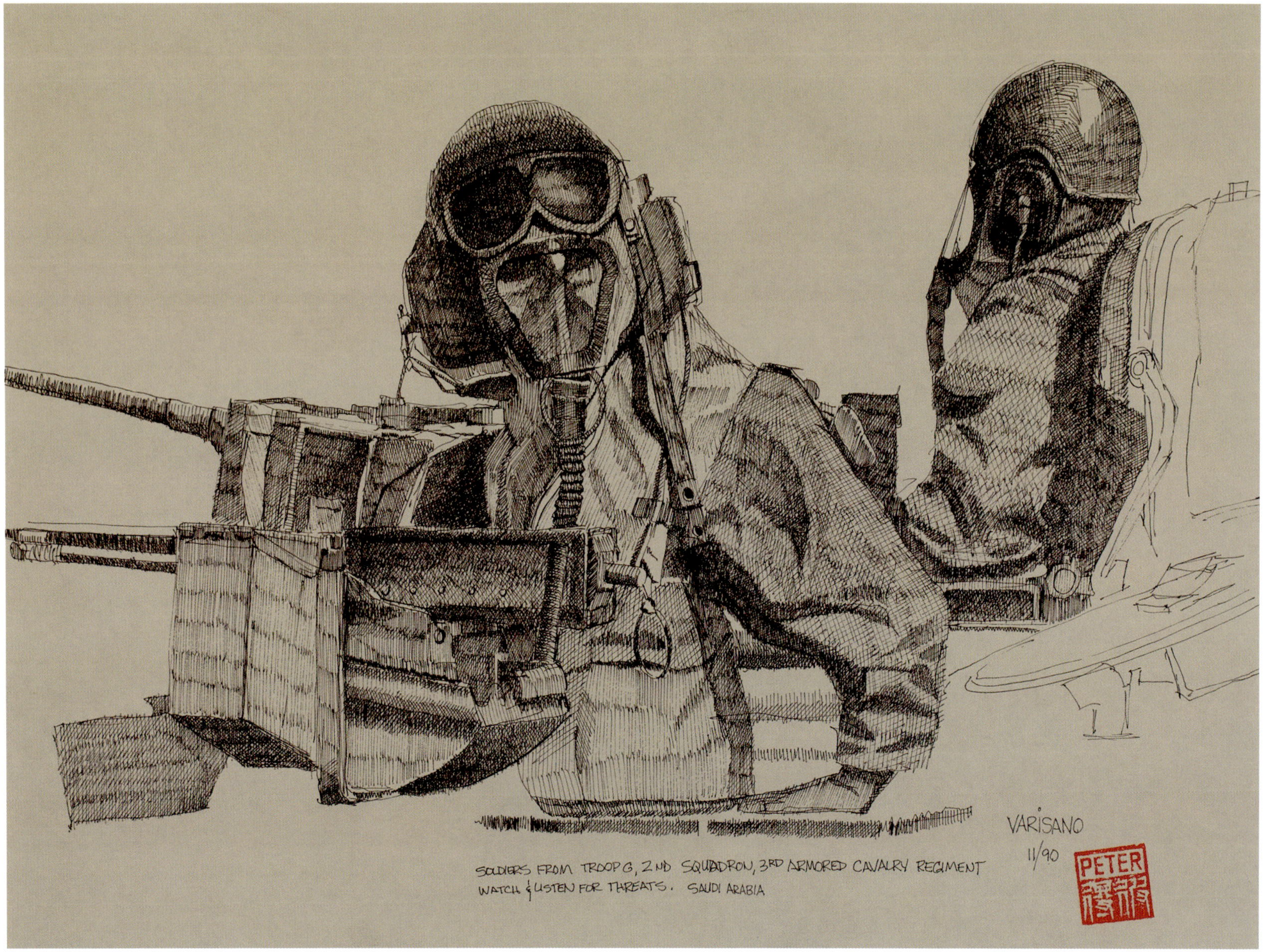

WATCH & LISTEN FOR THREATS
Saudi Arabia, Operation Desert Shield
Peter G. Varisano, 1990
Ink on paper, 14" x 17"

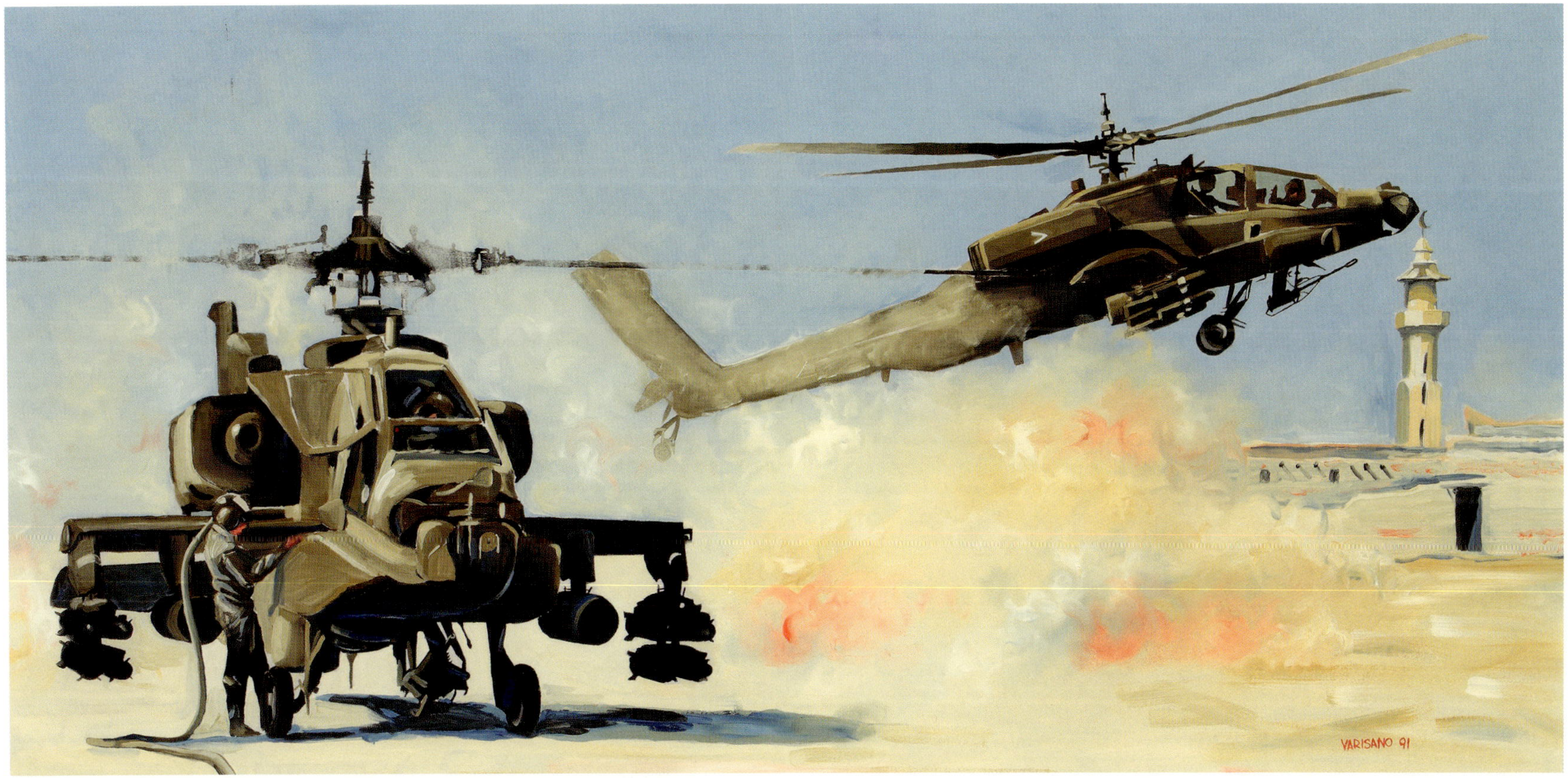

DESERT STORM 101ST STYLE
Saudi Arabia, Operation Desert Storm
Peter G. Varisano, 1991
Oil on canvas, 24" x 48"

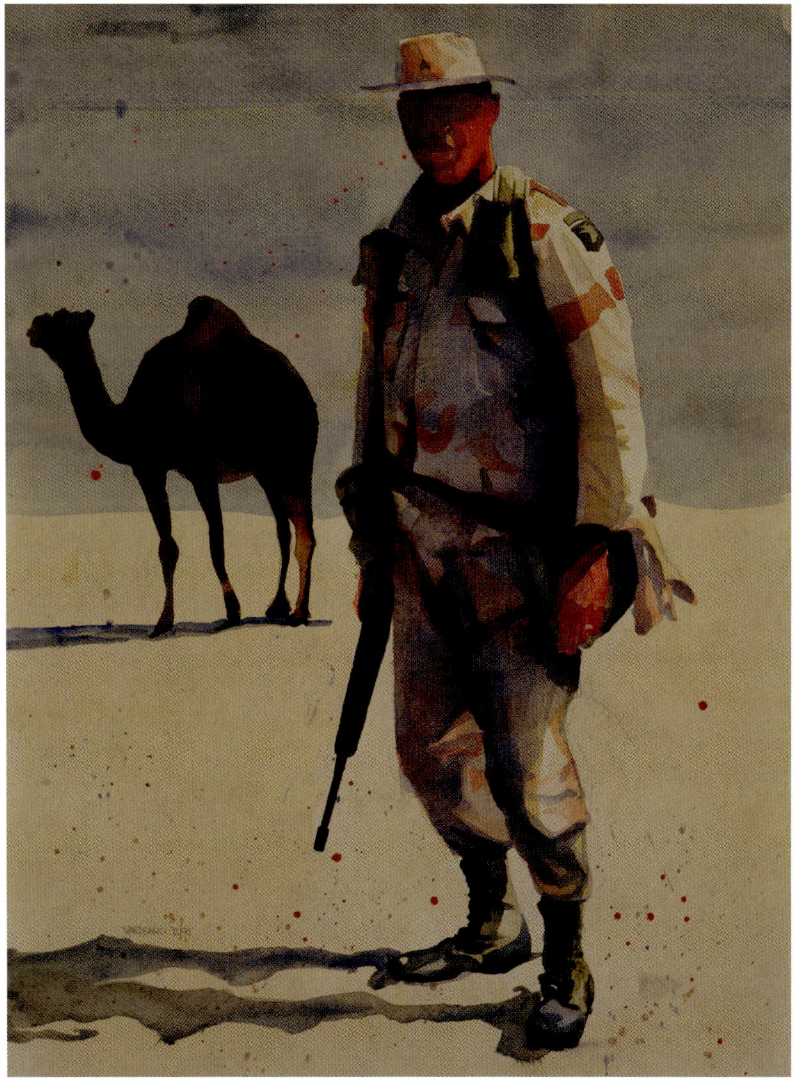

THE CAMEL AND THE EAGLE
Saudi Arabia, Operation Desert Shield
Peter G. Varisano, 1991
Watercolor on paper, 24" x 18½"

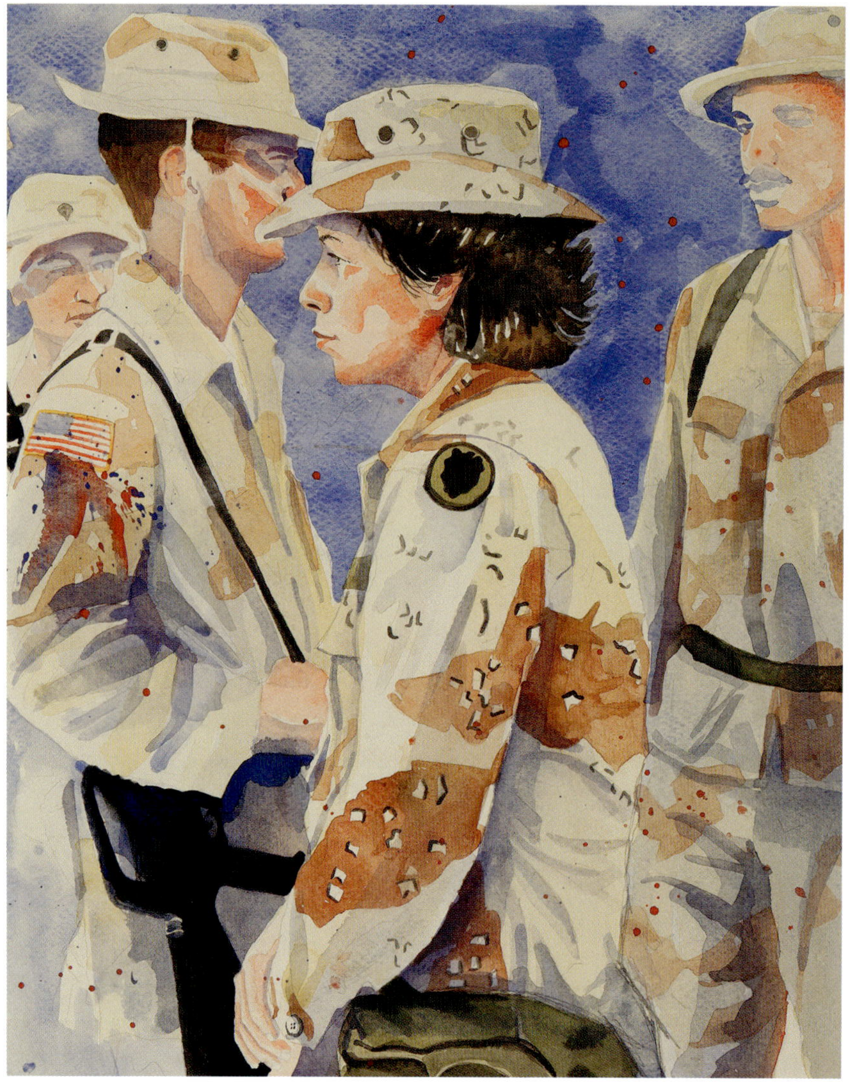

24TH INFANTRY DIVISION SOLDIER
Saudi Arabia, Operation Desert Shield
Peter G. Varisano, 1991
Watercolor on paper, 24" x 18"

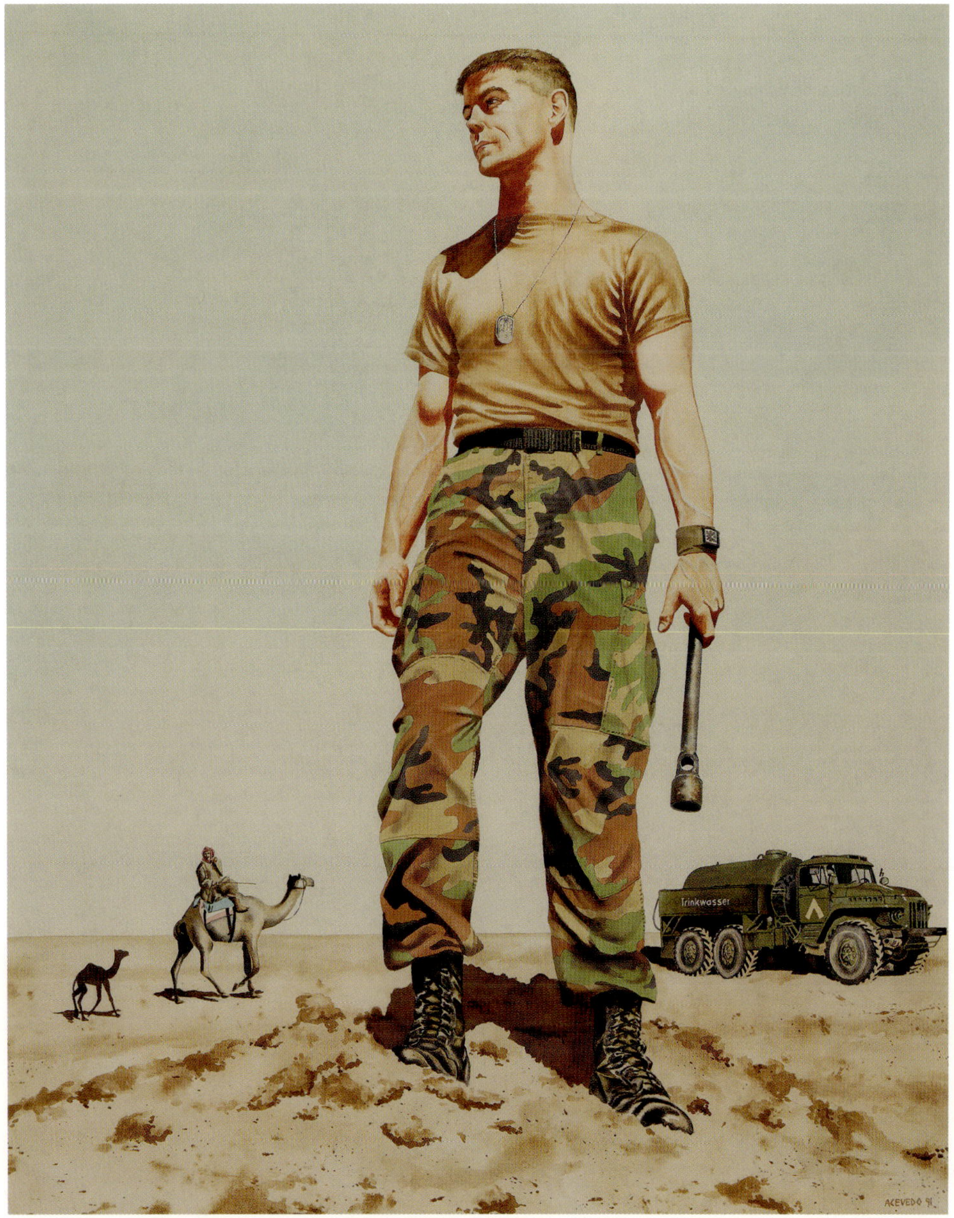

"Most of my paintings are focused on a soldier. The perception was it was a very technical, automated war. I wanted to make sure people understood the burden was on the soldiers."
—Mario H. Acevedo

TRUCK DRIVER IN THE VII CORPS
Southwest Asia, Operation Desert Storm
Mario H. Acevedo, 1991
Watercolor on paper, 51¼" x 40¾"

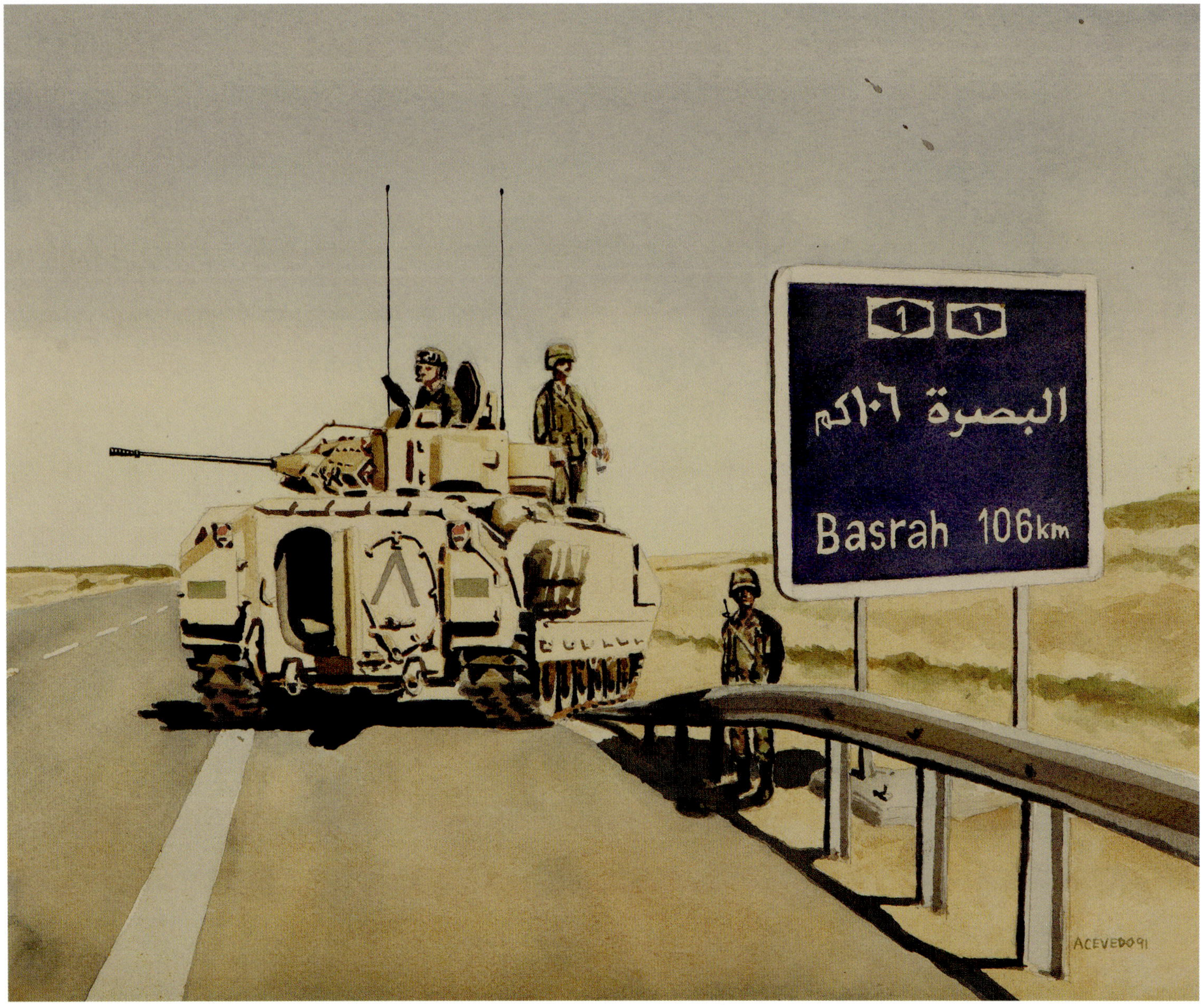

ROAD TO BASRAH
Southwest Asia, Operation Desert Storm
Mario H. Acevedo, 1991
Watercolor on paper, 10" x 12"

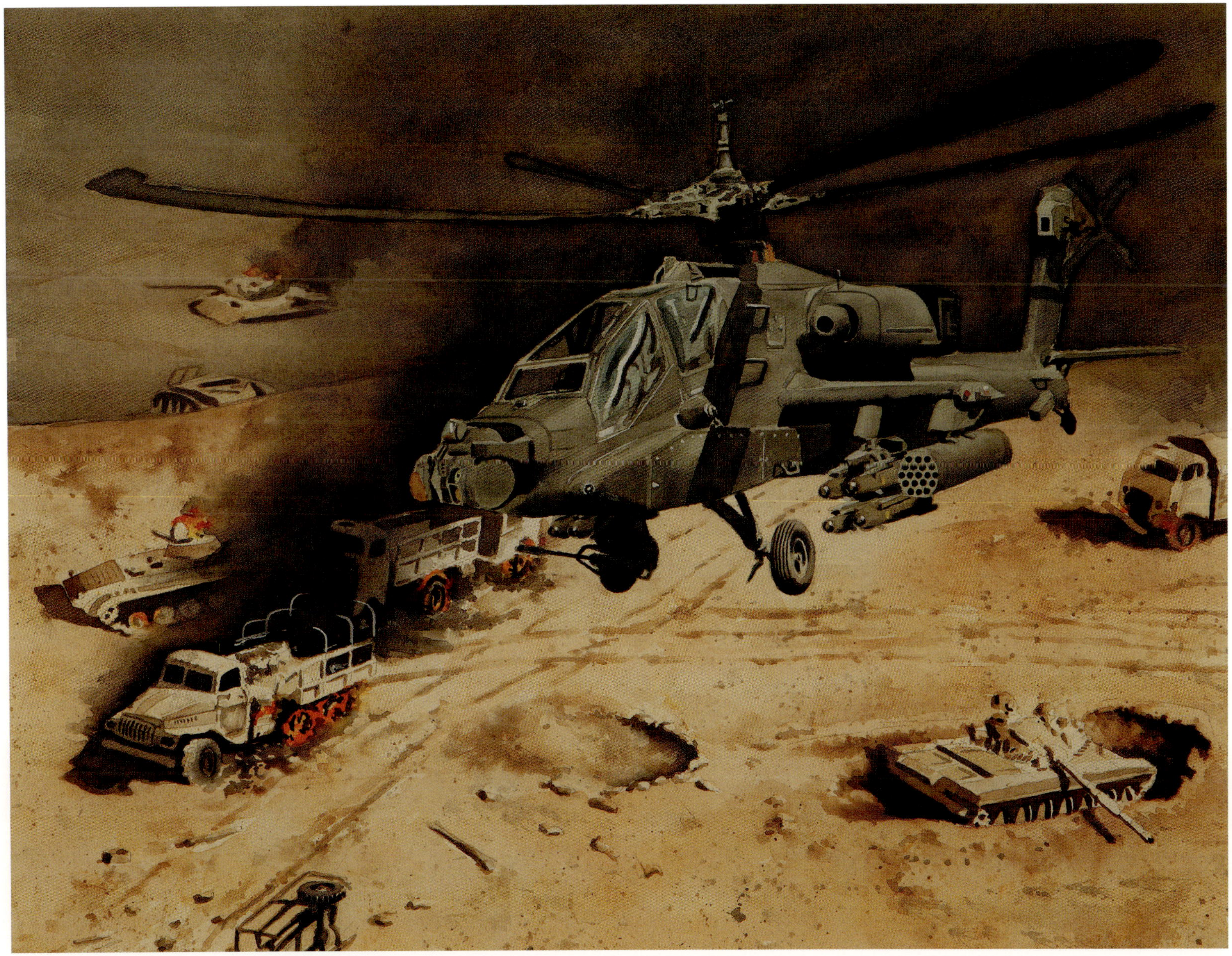

AH–64 PURSUES THE RETREATING IRAQI ARMY
Saudi Arabia, Operation Desert Storm
Mario H. Acevedo, 1995
Watercolor on paper, 16" x 18"

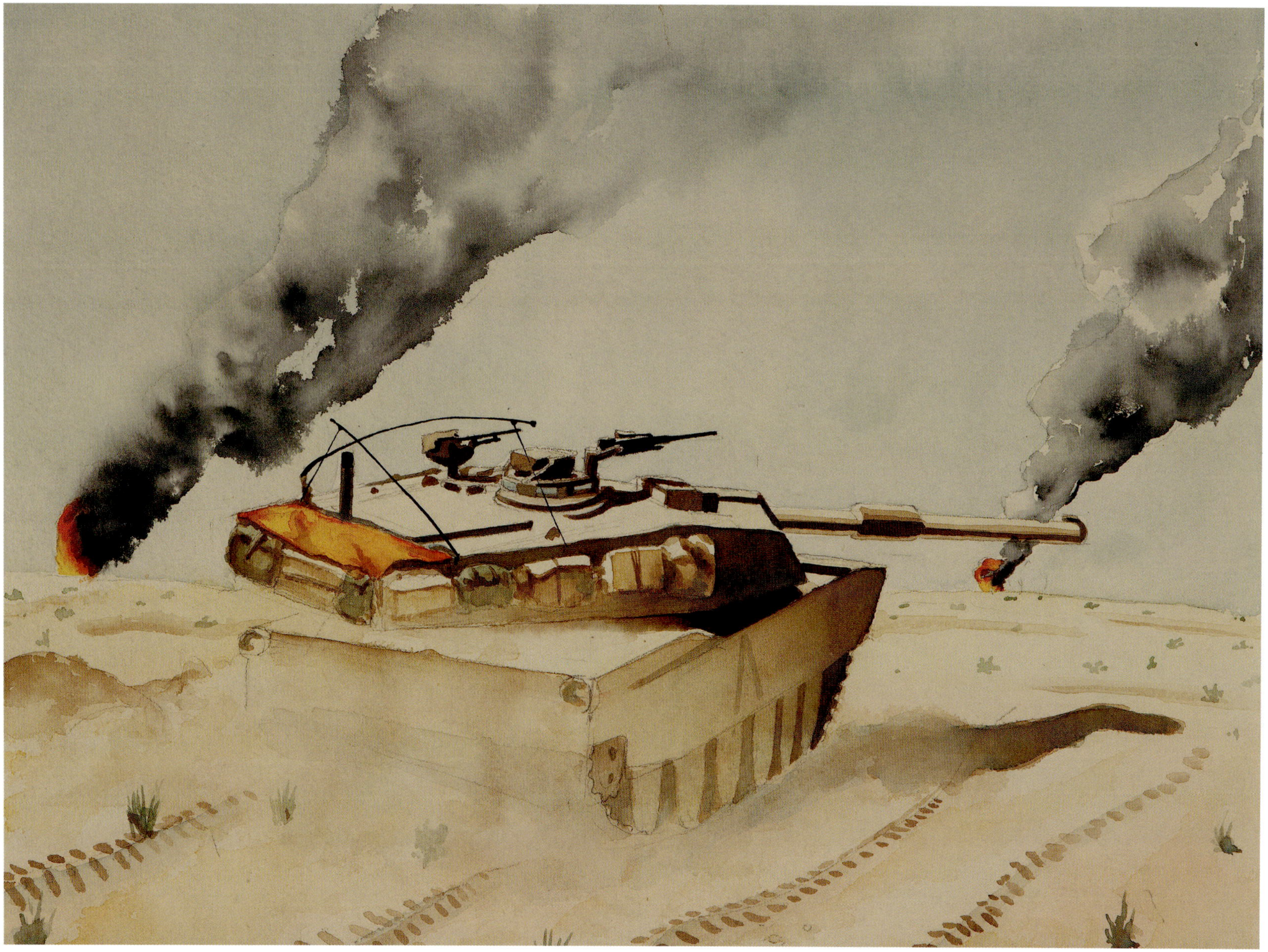

STUDY OF THE 1ST ARMORED DIVISION IN COMBAT, FEB '91
Southwest Asia, Operation DESERT STORM
Mario H. Acevedo, 1991
Watercolor on paper, 12" x 16"

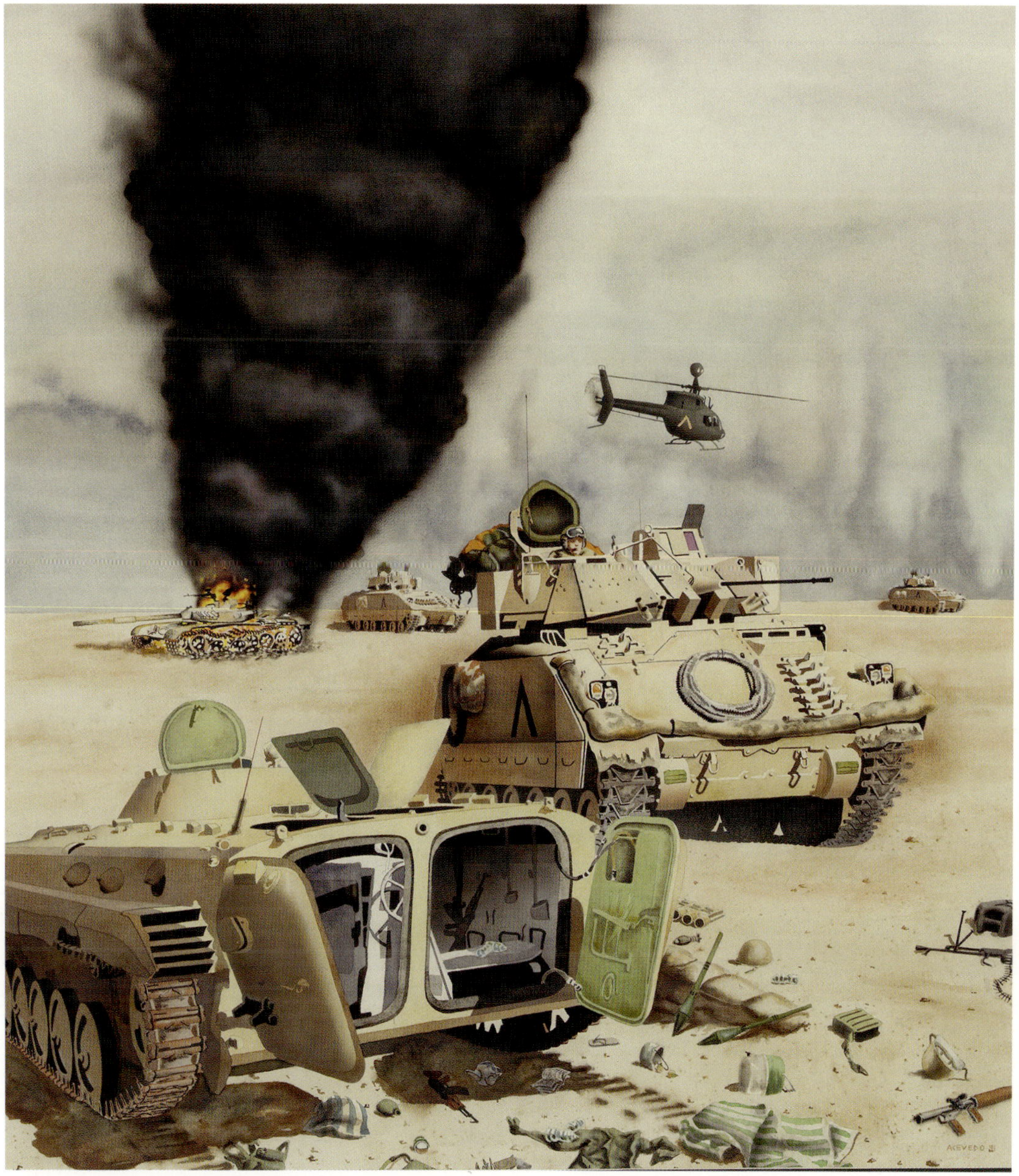

THE CAVALRY ADVANCES
Southwest Asia, Operation Desert Storm
Mario H. Acevedo, 1991
Watercolor on paper, 42" x 36¾"

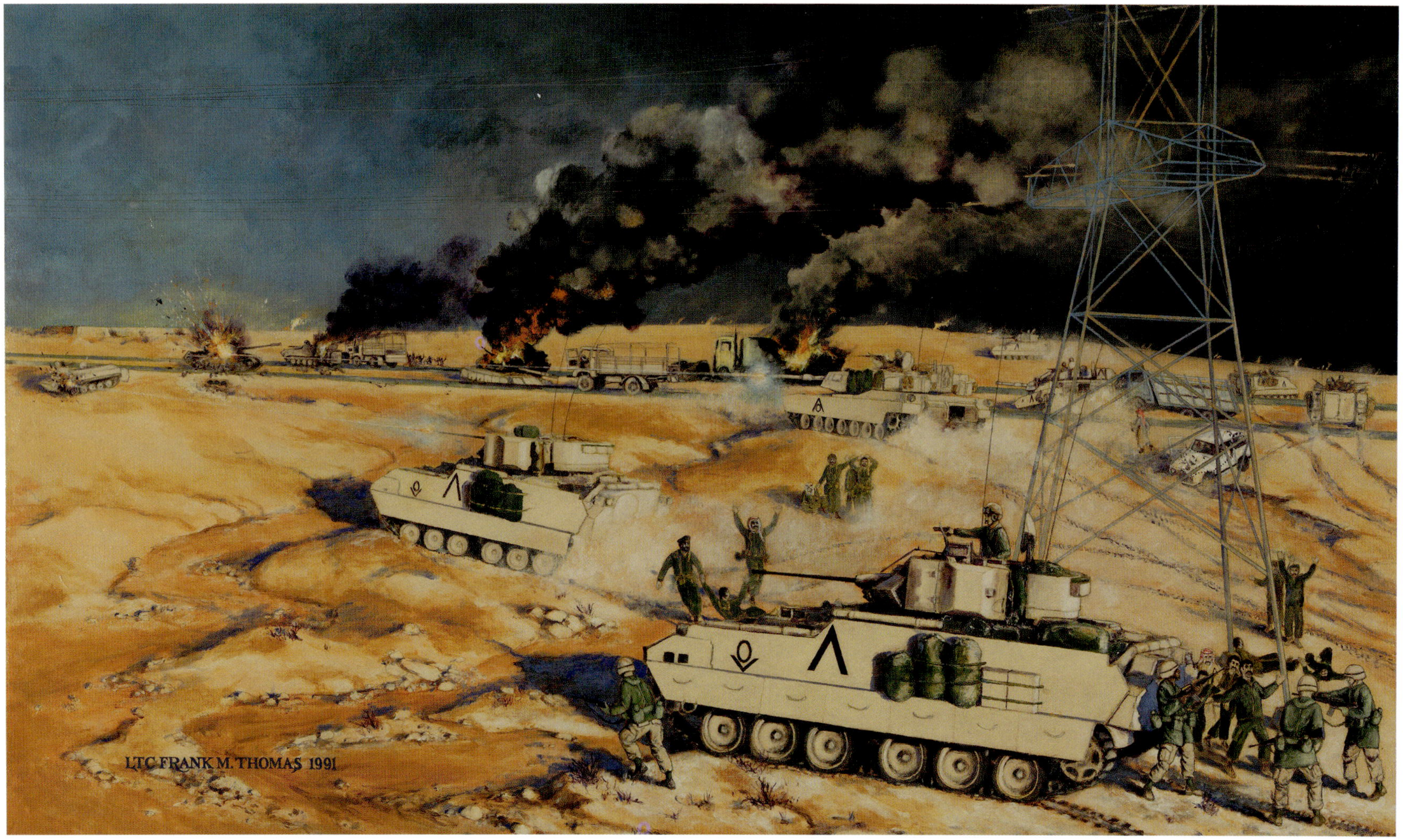

CUTTING THE BASRA HIGHWAY
Southwest Asia, Operation Desert Storm
Frank M. Thomas, 1991
Acrylic on canvas, 36" x 60"

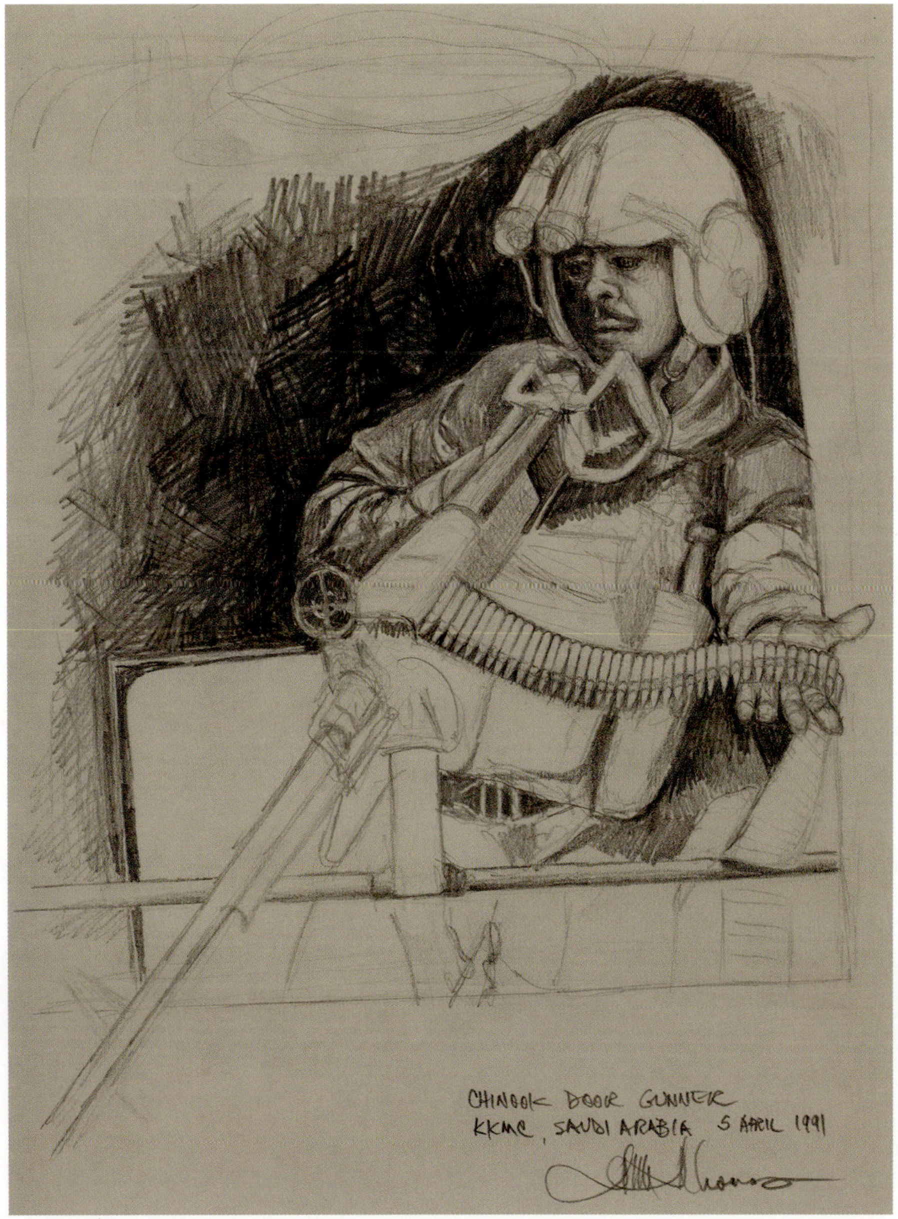

CHINOOK DOOR GUNNER
King Khalid Military City, Saudi Arabia, Operation DESERT STORM
Frank M. Thomas, 1991
Pencil on paper, 10$\frac{5}{8}$" x 8$\frac{1}{4}$"

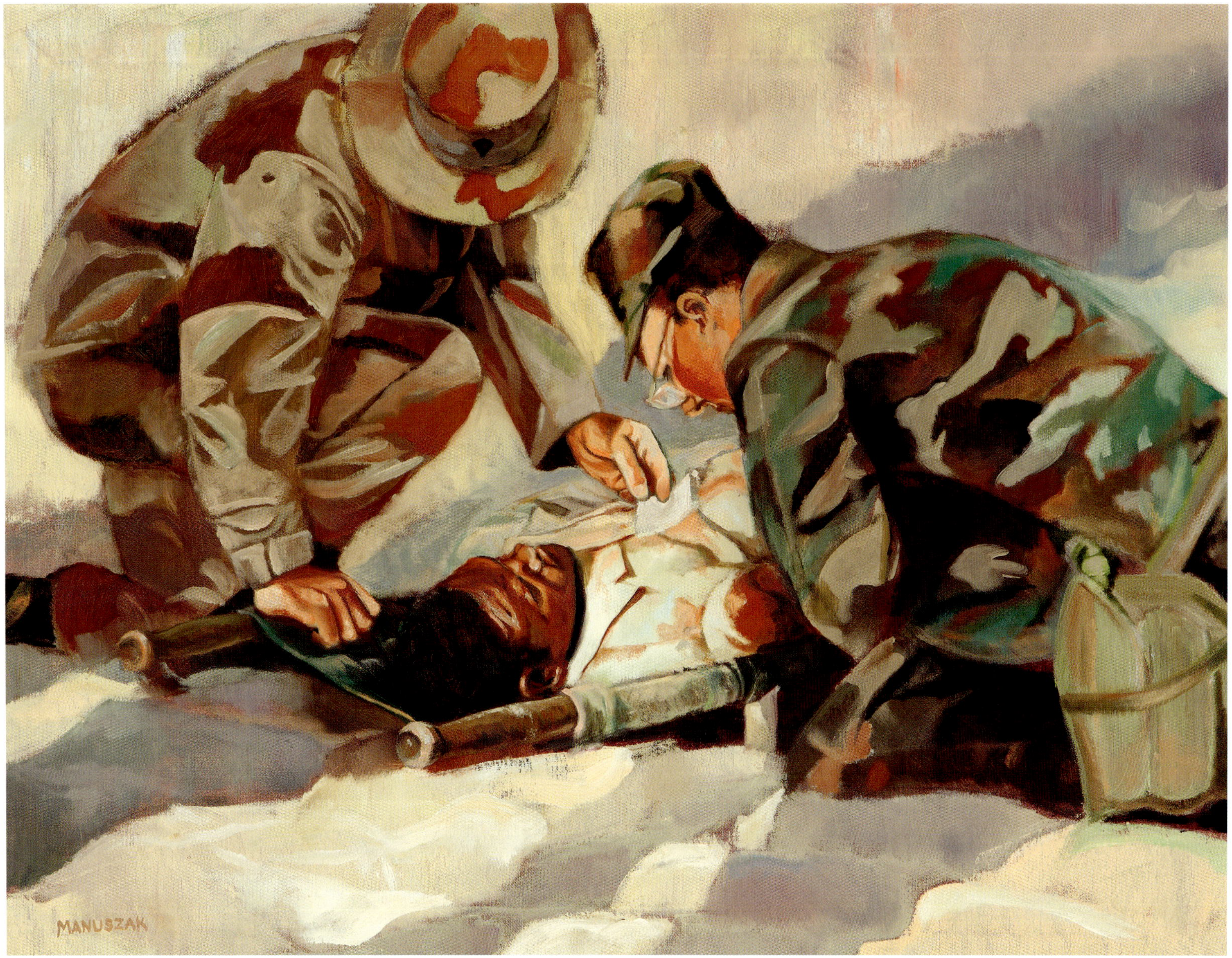

MEDICAL PERSONNEL PERFORMING TRIAGE
Dhahran, Saudi Arabia, Operation Desert Storm
Jeffrey T. Manuszak, 1996
Oil on canvas, 22" x 28"

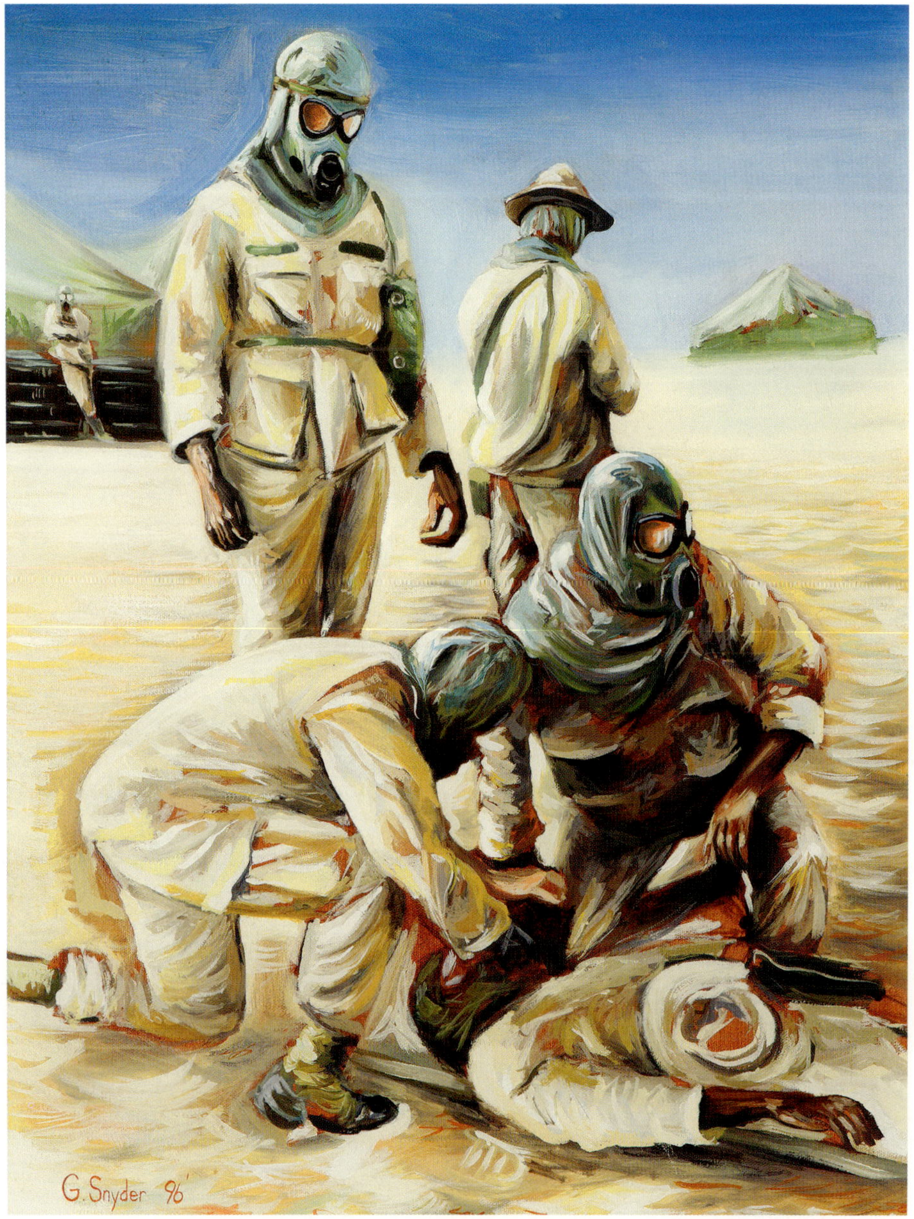

31ST COMBAT SUPPORT HOSPITAL
Saudi Arabia, Operation Desert Storm
Carl E. "Gene" Snyder, 1996
Oil on canvas, 24" x 18"

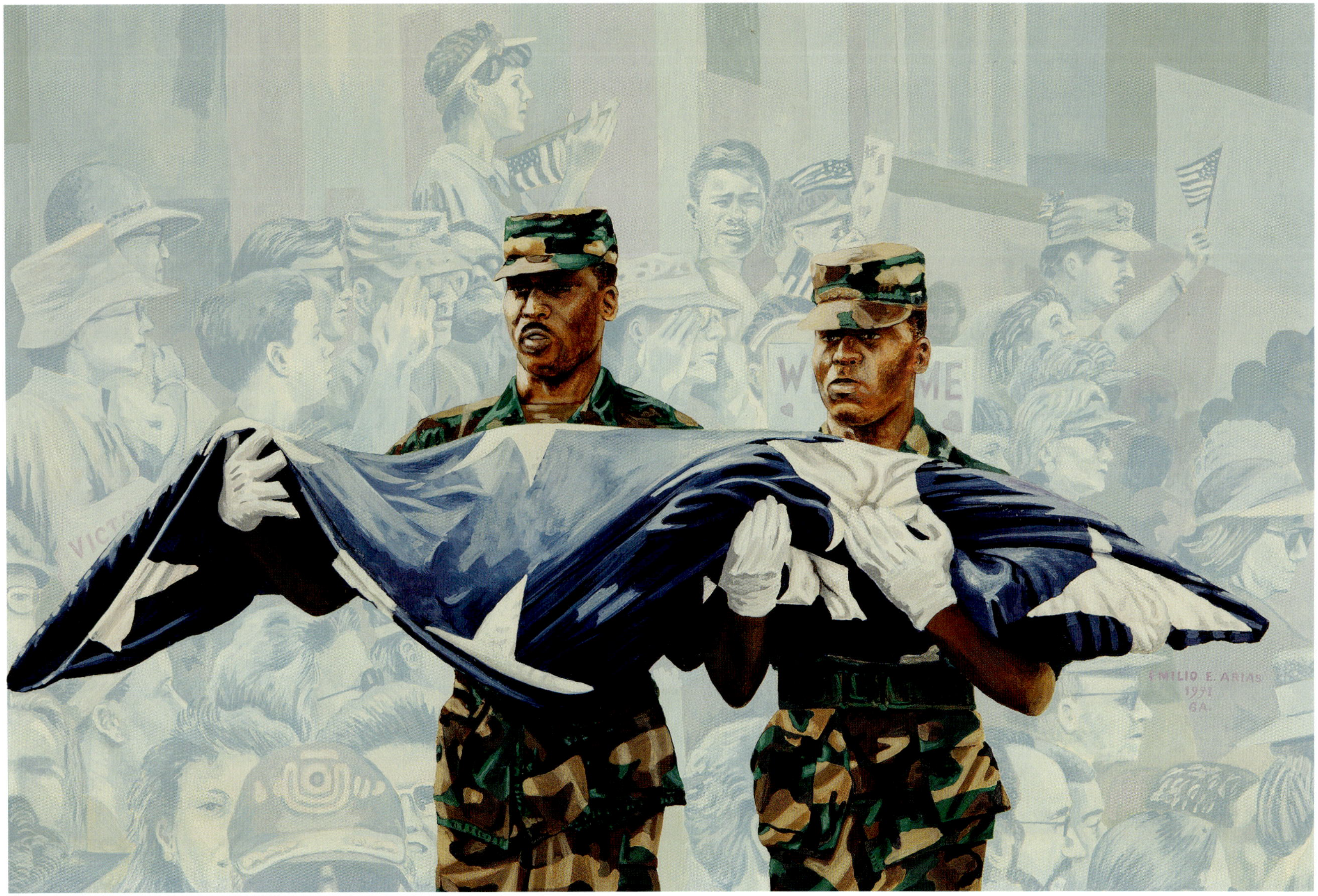

JULY 4TH, 1991
Atlanta, Georgia
Emilio E. Arias, 1991
Gouache on board, 12½" x 18¼"

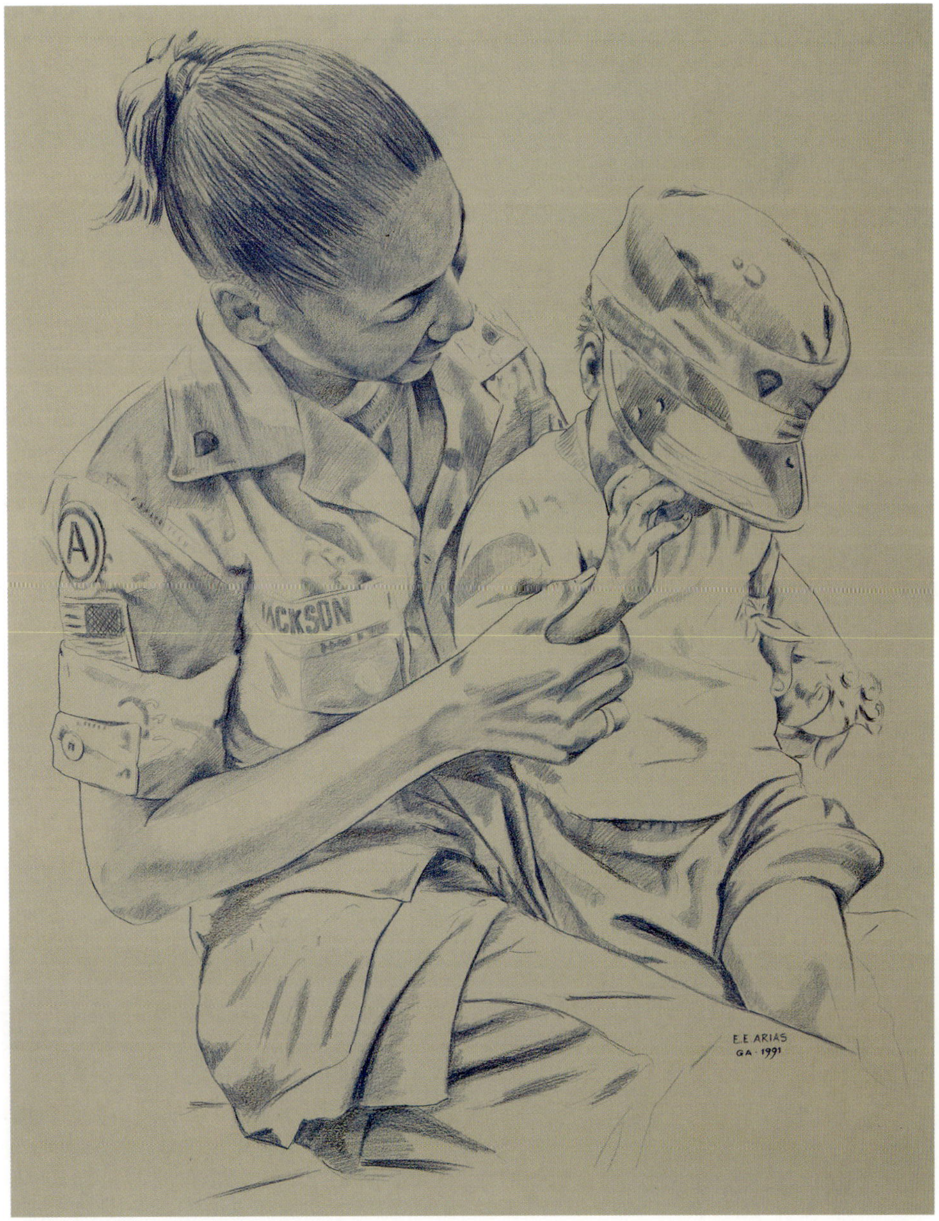

MOTHER WITH CHILD
Atlanta, Georgia
Emilio E. Arias, 1991
Colored pencil on board, 30" x 20¼"

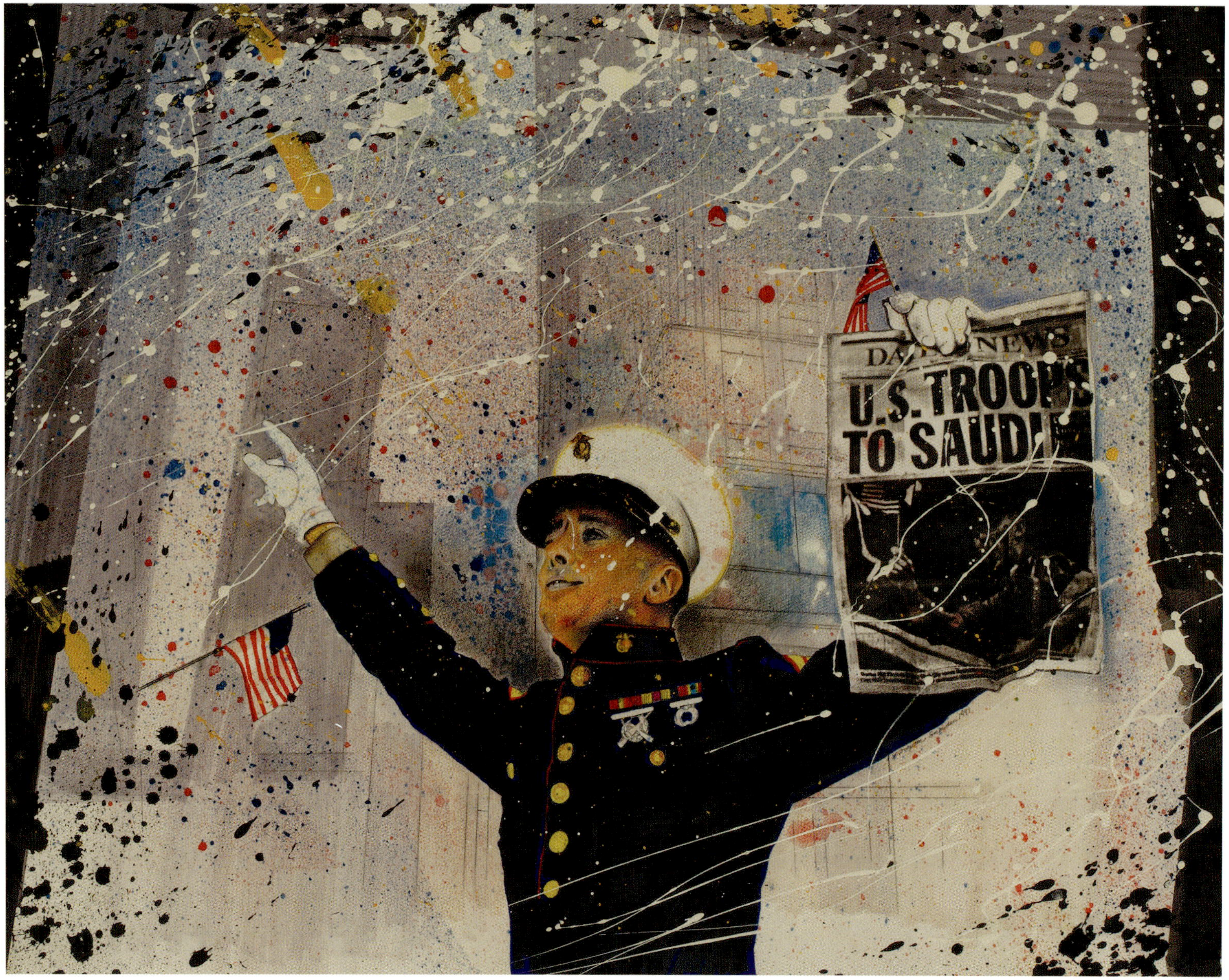

NEW YORK WELCOME
New York City
Henrietta M. Snowden, 1991
Enamel and watercolor on board, 20" x 22½"

Artists-in-Residence
The 1990s

With the establishment of the Artist-in-Residence Program in 1991, artists began grappling with how to approach the monumental task of chronicling a significant portion of history. Though their role was steeped in tradition dating back to World War I, this period represents the first time that an artist was responsible for documenting the climate of an era rather than a war or a specific assignment. Where does one begin? How does one determine moments of historical importance as they are happening? With only one artist, what is recorded and what is omitted?

More than at any other time, the Army represented a cross-section of the U.S. population. In the past, diversity was often highlighted for its own sake. In the Army of the 1990s, soldiers of various ethnicities and genders are depicted because of the jobs they are doing, rather than for being members of an ethnic group or a particular gender, as seen in Sfc. Christopher W. Thiel's *Loading Rail: Preboard Inspection* (page 144). With increased freedom to decide for themselves what should be recorded and remembered, artists often choose to emphasize the individual soldier as the core of the Army. As in the 1980s, the Army's recruiting slogan is still "Be All You Can Be," and this striving for individual accomplishment is reflected in the art. The subject matter is diverse and covers soldiers executing a wide variety of tasks, some of which had never been documented in Army art before. Some of the art from this period portrays soldiers performing critical, but less-than-glamorous, roles, with an increased focus on supply and support.

During this era, the Army expanded artistic coverage of humanitarian missions and relief efforts both at home and overseas. With the creation of the Artist-in-Residence position, it was logistically easier to send an artist out quickly to document disasters and relief missions. With a deployable and available artist, there was no need to recruit artists, review applications, and request temporary assignment of soldiers to artist teams, as had been done in the past. This allowed artists to promptly respond to disasters such as Hurricane Andrew in Florida in 1992, as seen in Sfc. Peter G. Varisano's *Loading Up* (page 131), and to document humanitarian missions such as Operation RESTORE HOPE in Somalia, as seen in Sgt. Jeffrey T. Manuszak's *Road March* (page 133).

Overall, the style of Army art in the 1990s is a departure from the commercial look of the DESERT STORM artwork and a return to traditional draftsmanship. As though they were trying to prove that soldiers can be fine artists, they strove to be all that they could be, focusing on the artistic basics of color, form, and perspective. Their palette is generally earthy, with bright colors used sparingly. The art of this period strongly reflects a soldier's view of what is important to the Army's history: the many personalities that make up the Army and define its character.

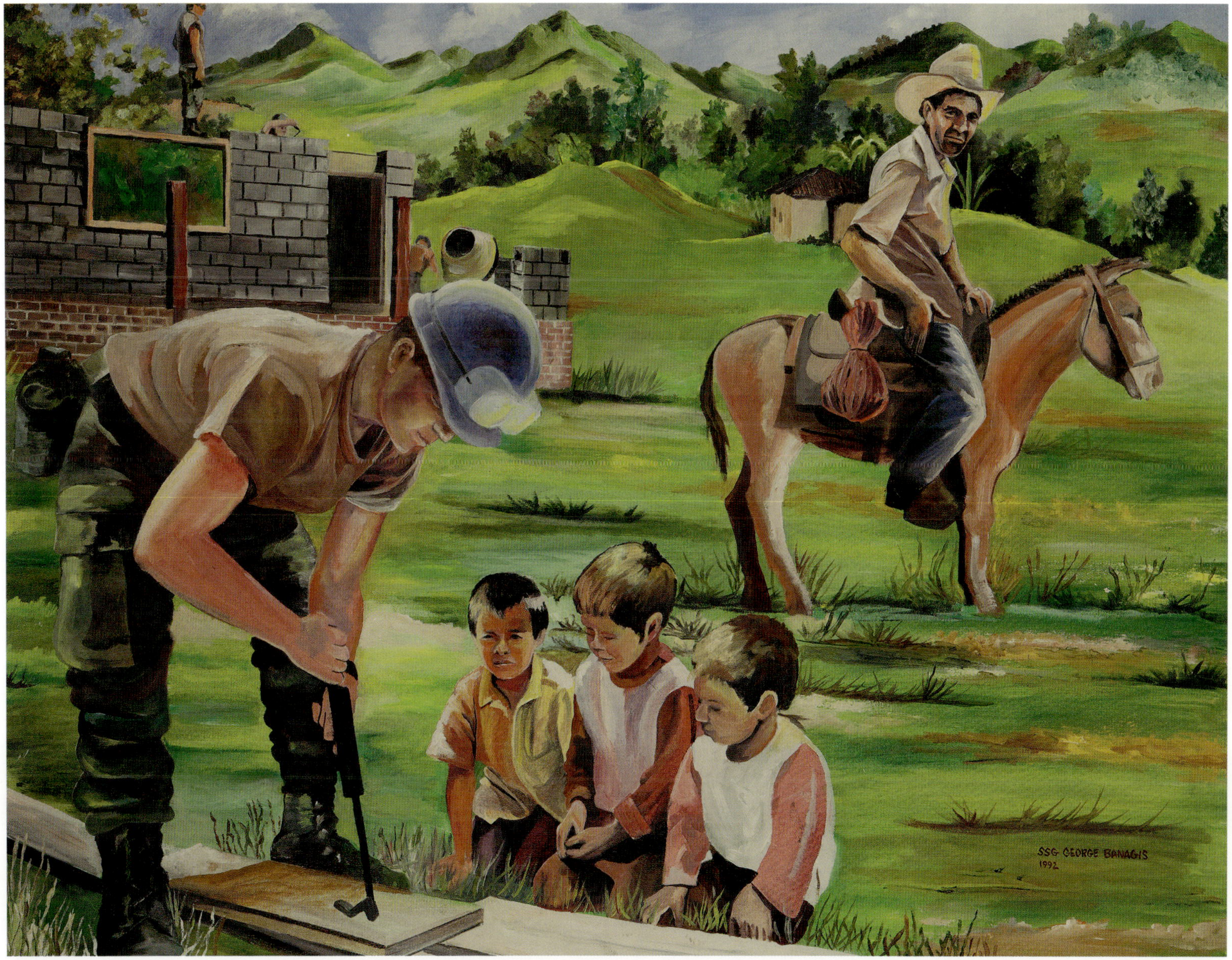

NEW SCHOOL IN EL SOCORRO, HONDURAS
El Socorro, Honduras
George Banagis, 1992
Oil on canvas, 38" x 48"

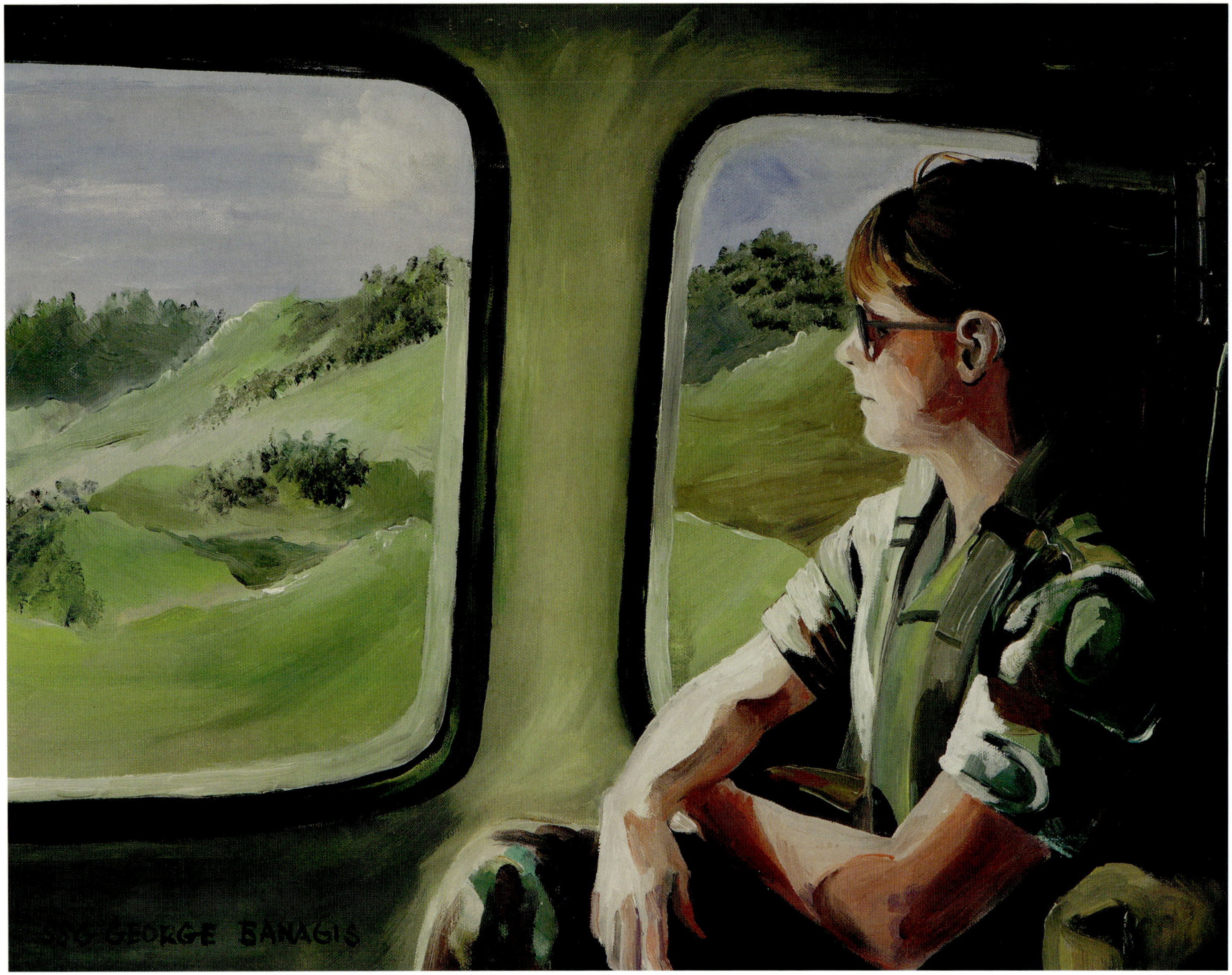

COMPANY COMMANDER'S RETURN FLIGHT
Honduras
George Banagis, 1992
Oil on canvas, 16" x 20"

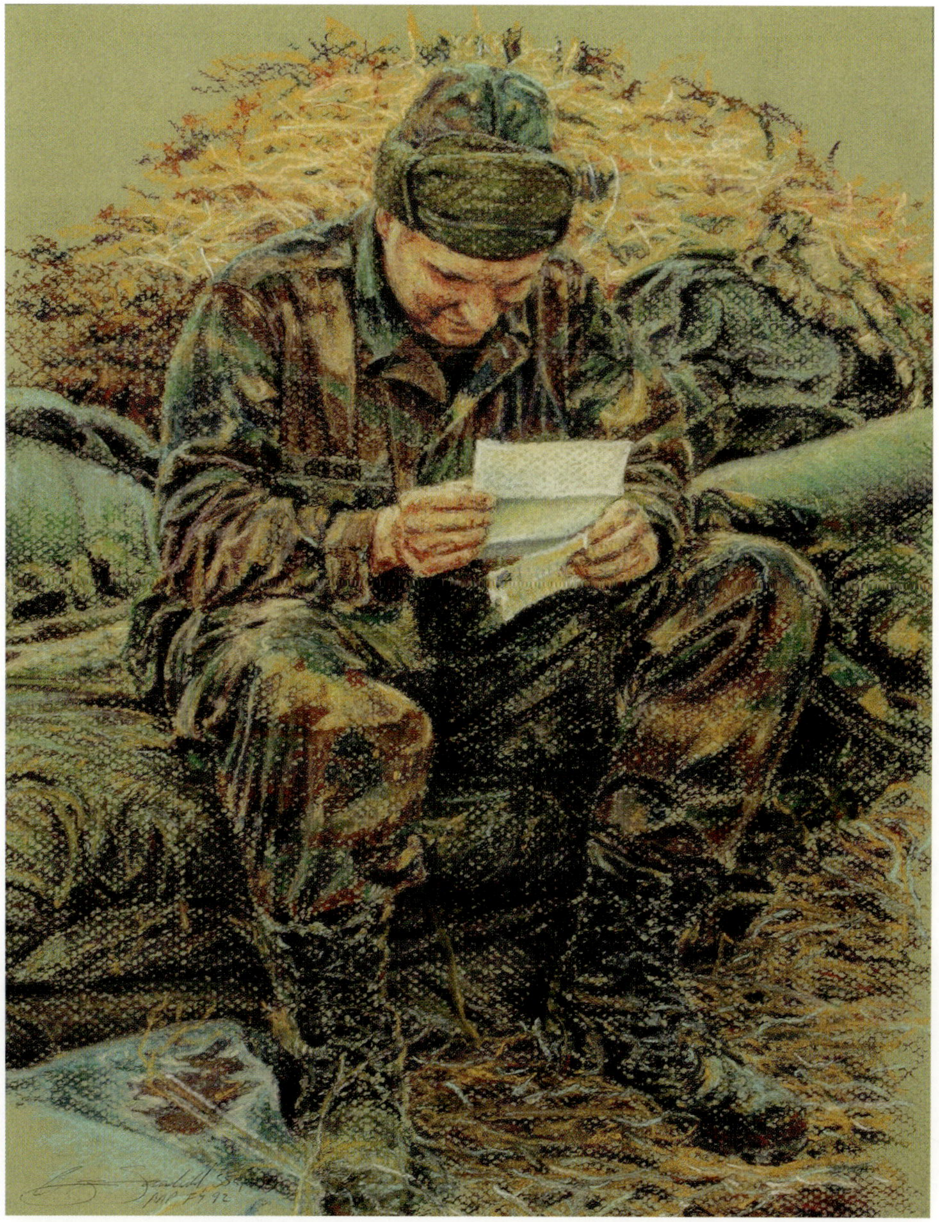

"His squad had just negotiated an obstacle course and finished a nine mile road march. A vehicle approached the squad; a soldier quickly got out and distributed a handful of mail. Those soldiers lucky enough to get a personal letter read them where they stood or plopped."
—*Brian D. Fairchild*

2 MINUTES AT HOME
Korea
Brian D. Fairchild, 1992
Pastel on paper, 17" x 12¾"

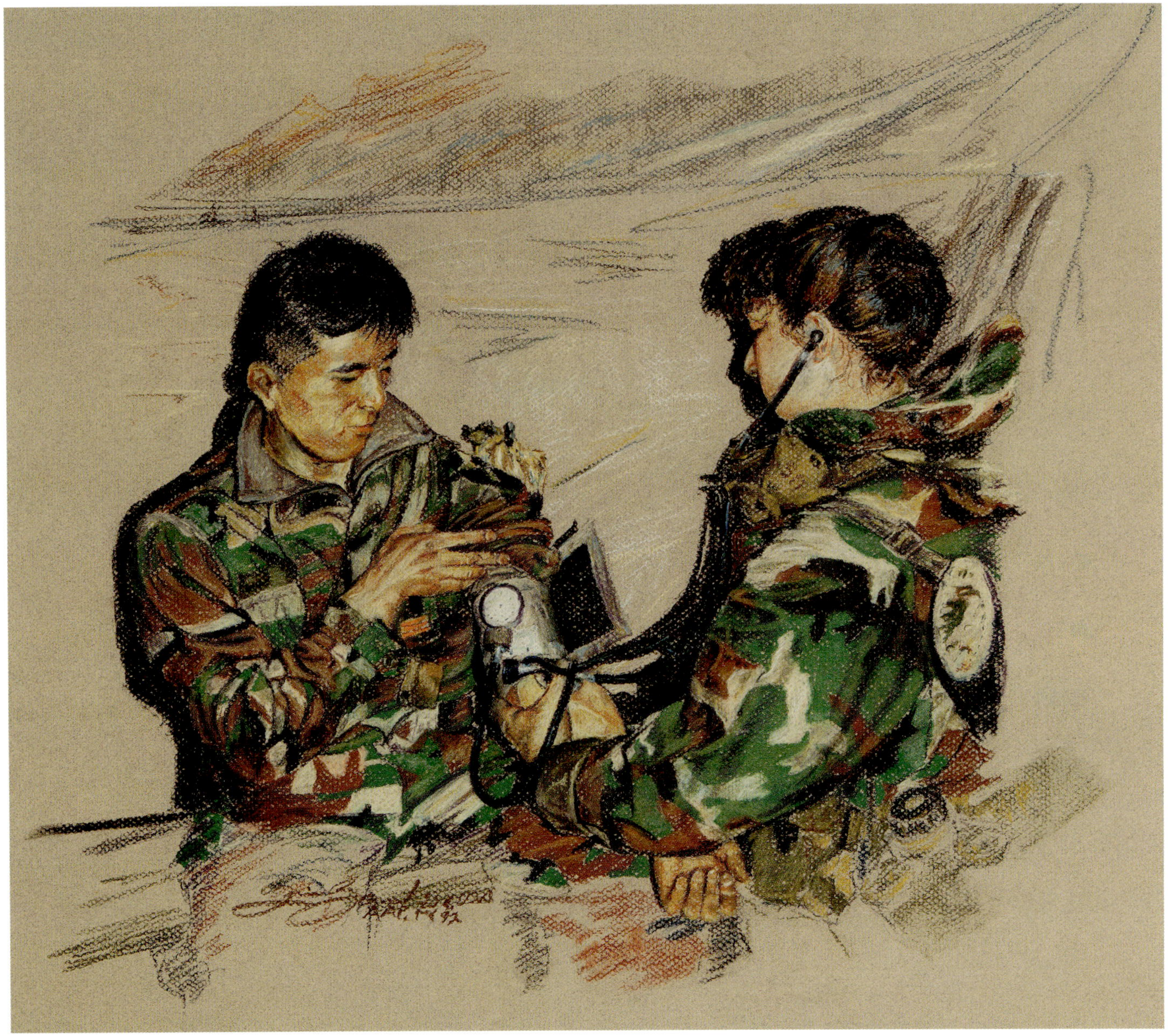

FORWARD SUPPORT
Korea
Brian D. Fairchild, 1992
Pastel on paper, 19¾" x 25½"

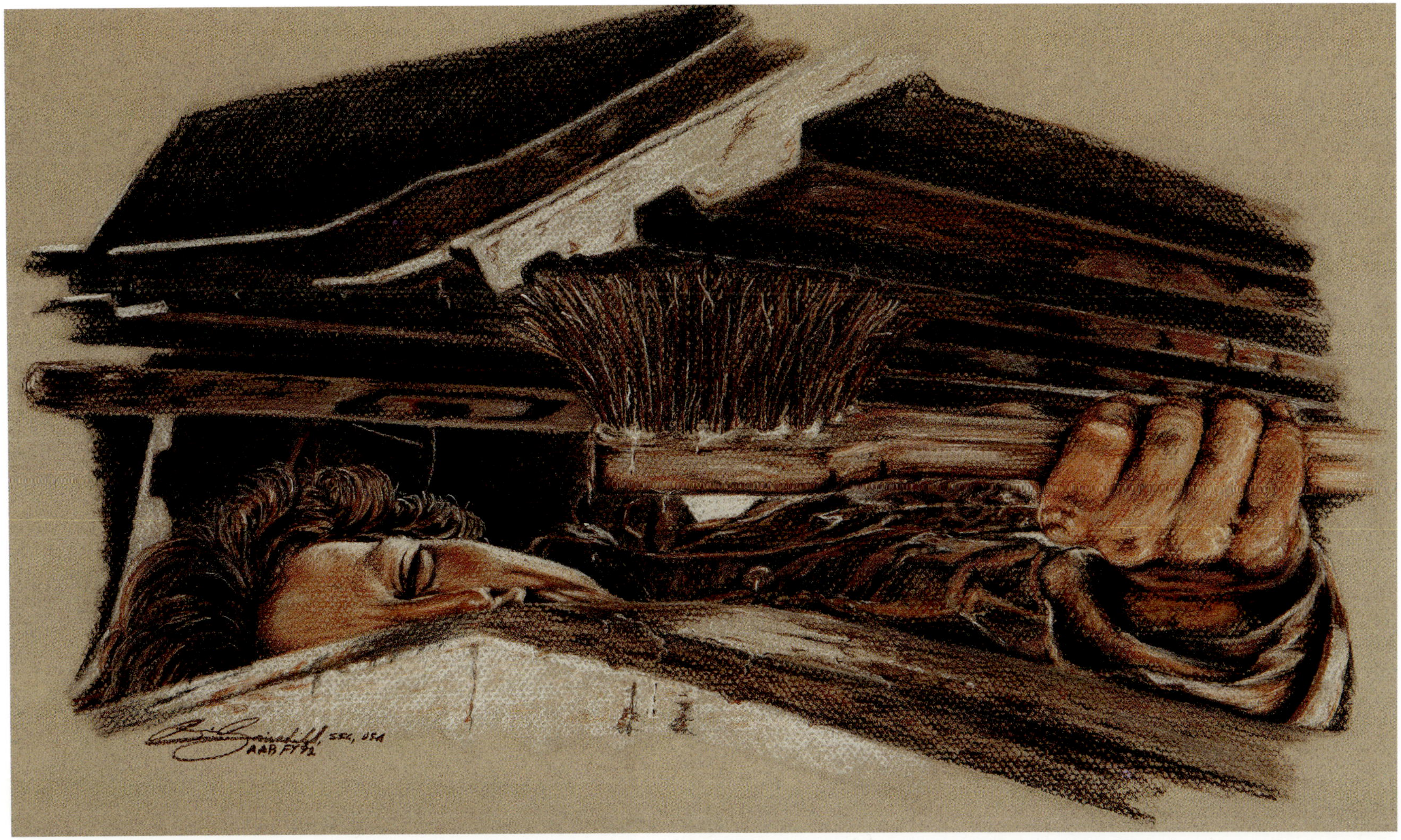

UNDER A 'HUMMER'
Taegu, Korea
Brian D. Fairchild, 1992
Conte lead on paper, 12¾" x 20"

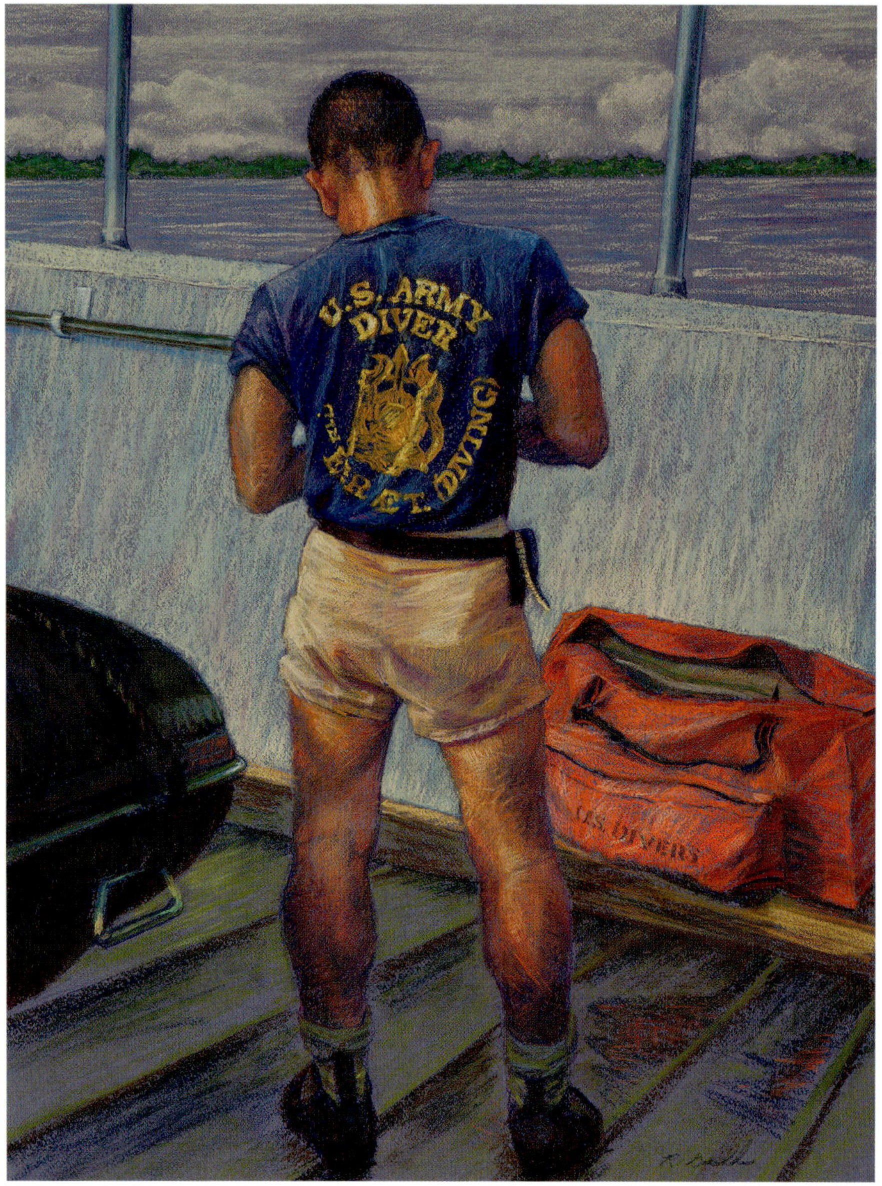

U.S. ARMY DIVER
Panama
Roberta Lynn Goschke, 1992
Pastel on paper, 25" x 19"

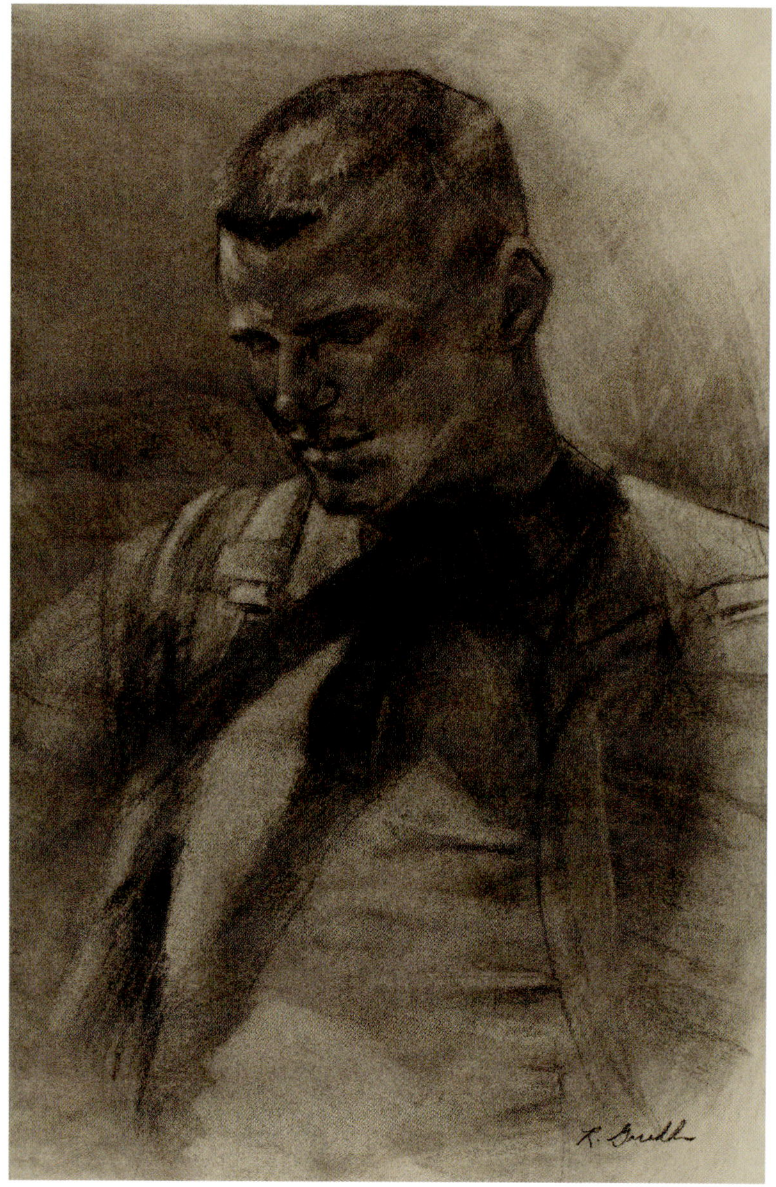

YOUNG SOLDIER
Panama
Roberta Lynn Goschke, 1992
Charcoal on paper, 22½" x 15¼"

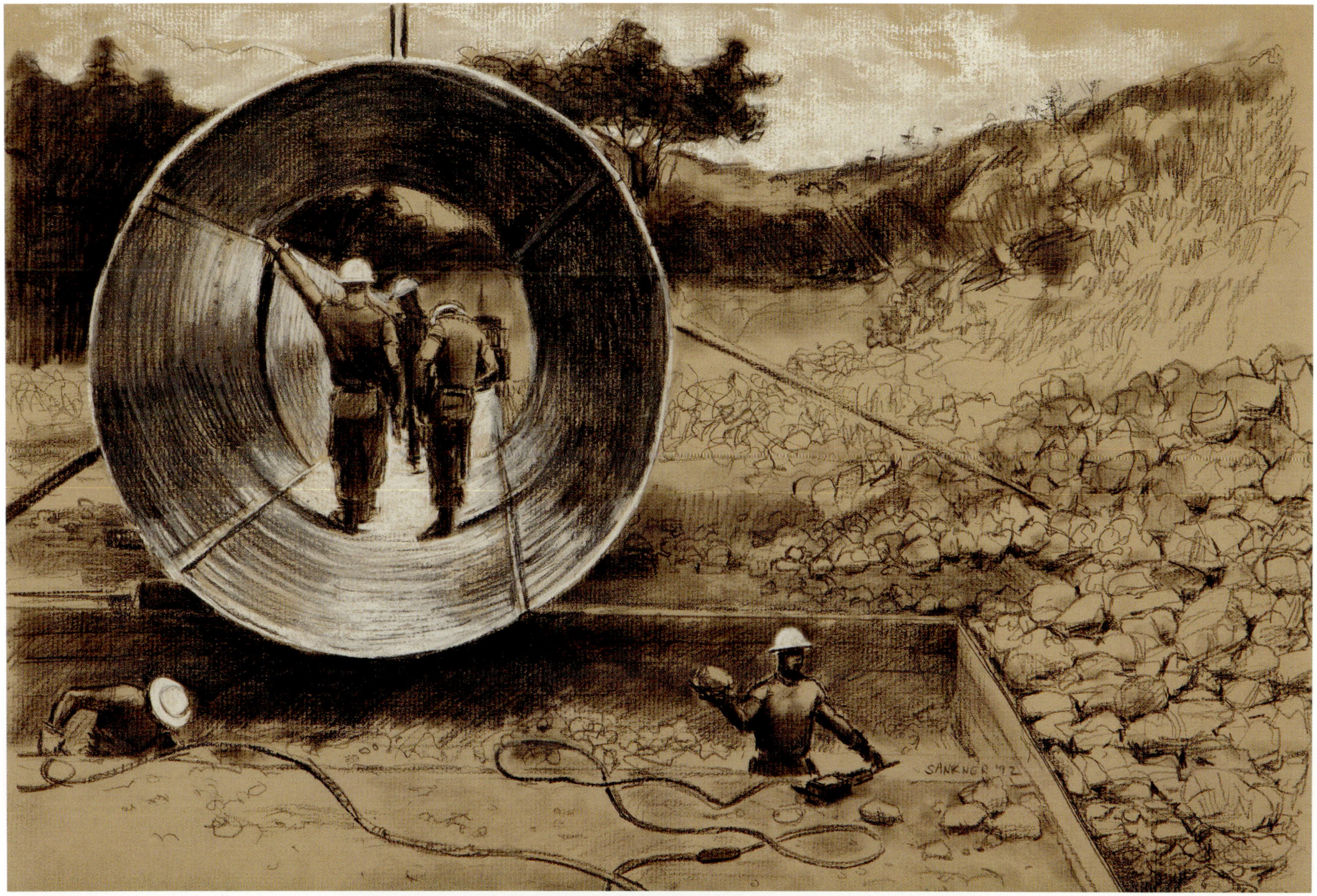

CONDUIT
Panama
Robert Sankner, 1992
Chalk and charcoal on paper, 12" x 17½"

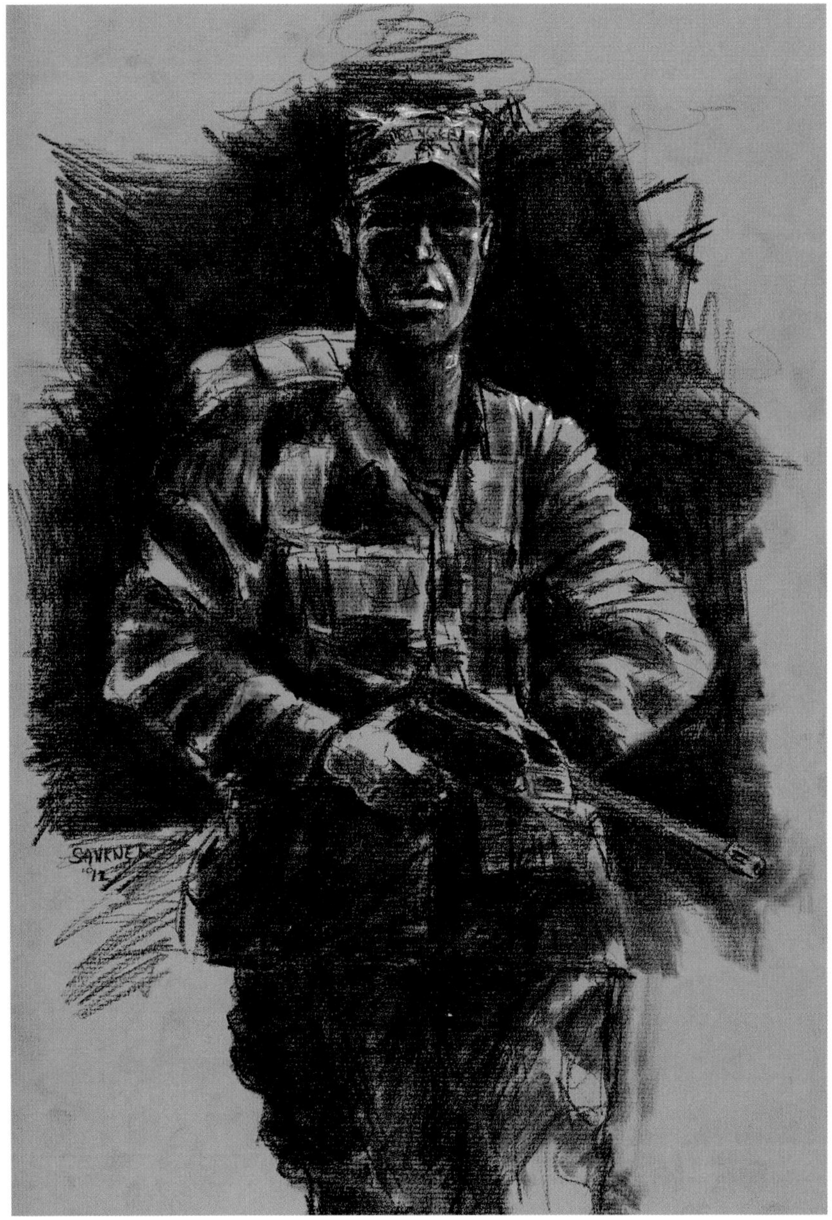

RANGER
Panama
Robert Sankner, 1992
Charcoal on paper, 17¾" x 12"

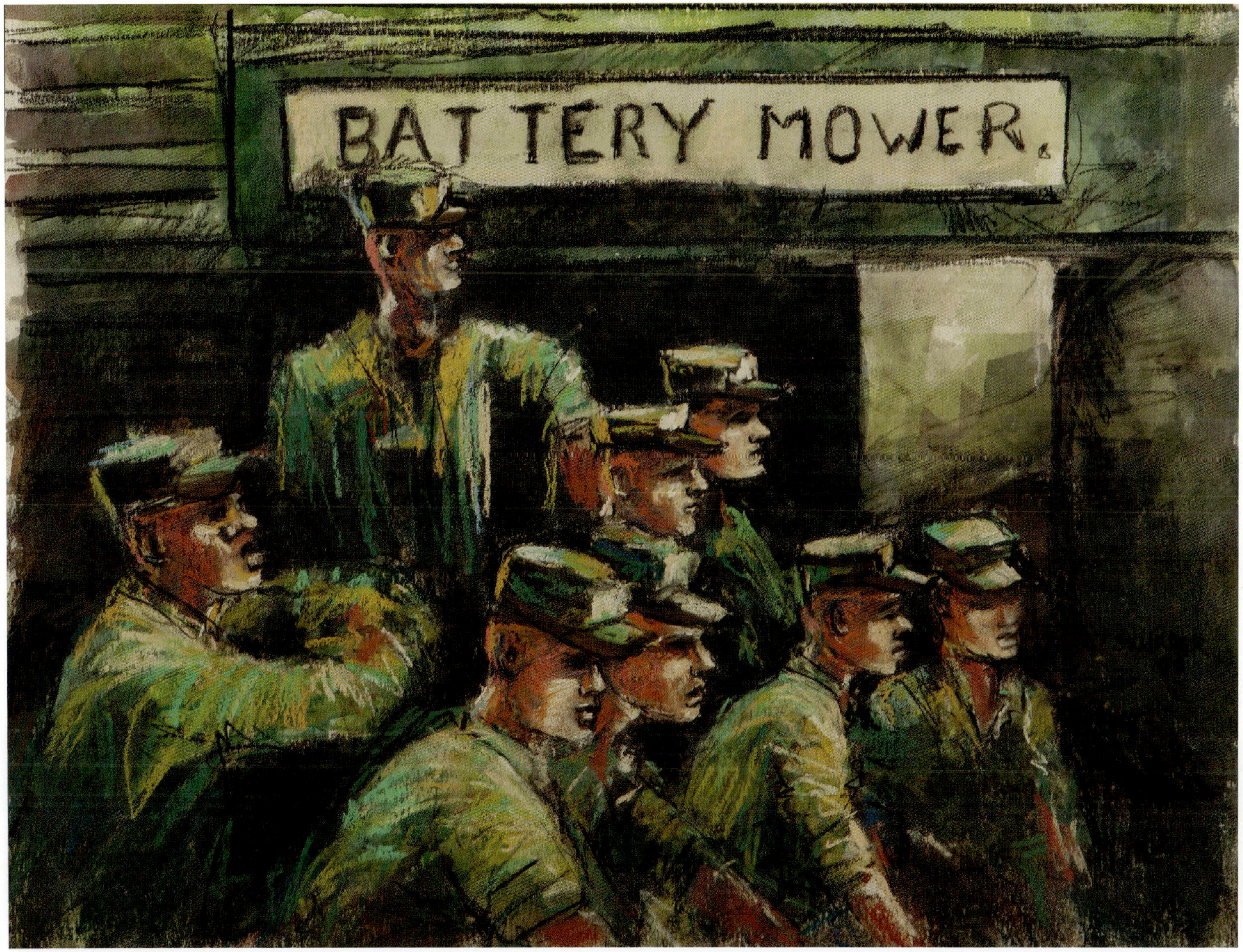

BATTERY MOWER
Panama
Robert Sankner, 1992
Pastel and watercolor on paper, 13¼" x 17"

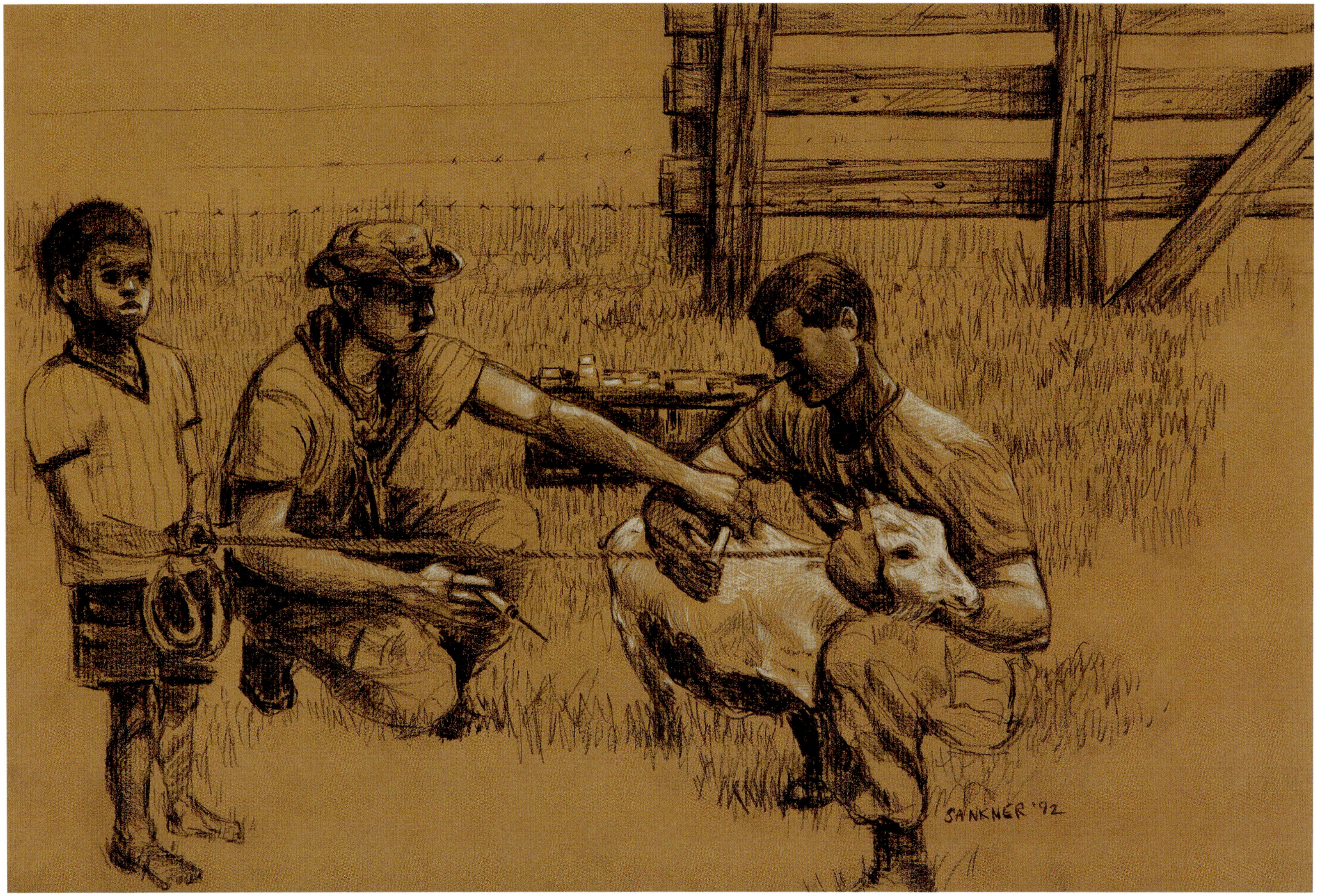

INOCULATING A GOAT
Panama
Robert Sankner, 1992
Charcoal and gouache on paper, 12" x 17¾"

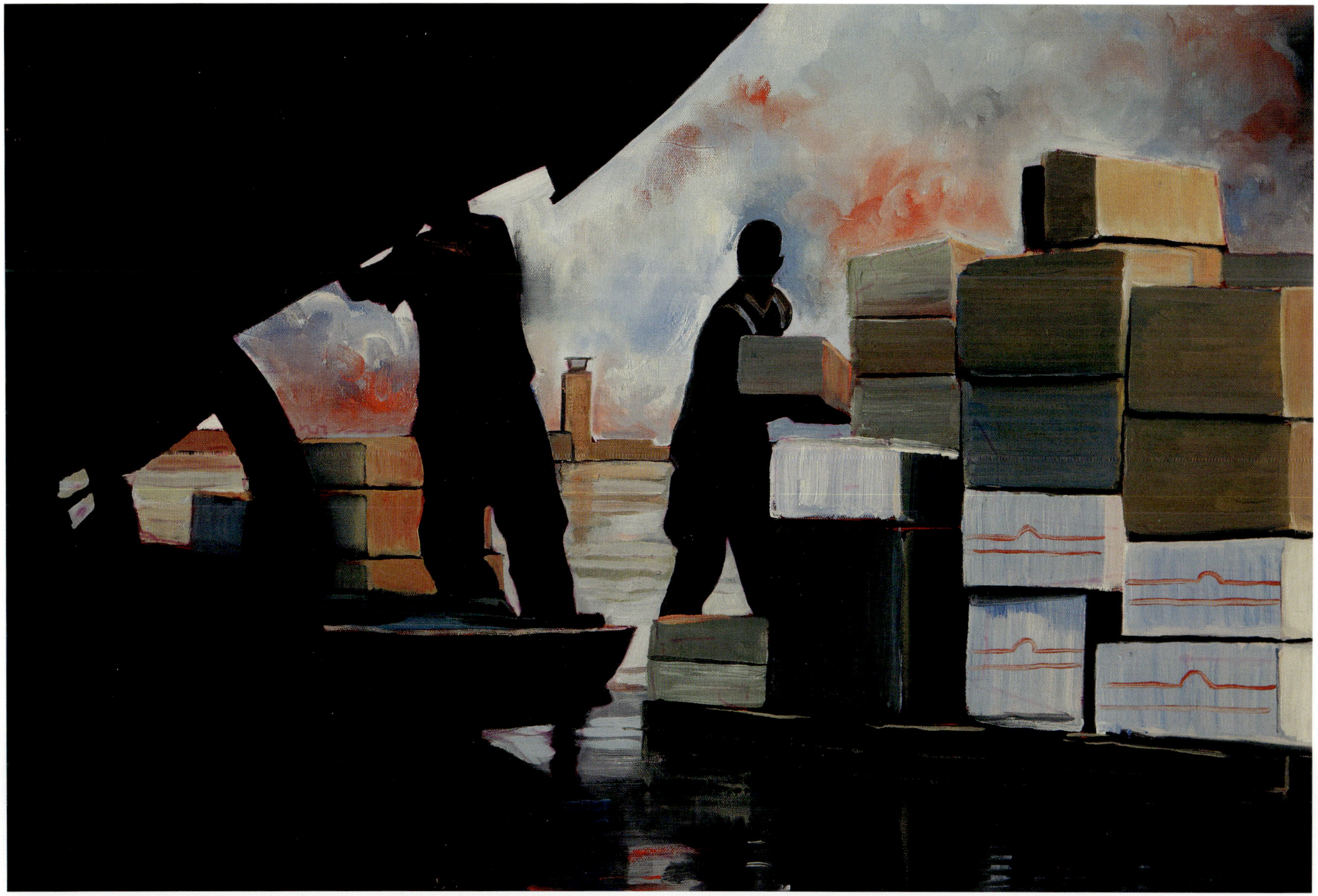

LOADING UP
Florida, Hurricane Andrew Relief
Peter G. Varisano, 1992
Oil on canvas, 20" x 30"

"The color contrast here is remarkable. The orange and white sand, the rich blue Indian Ocean, the sky is always presenting a tempting view. The people are the best thing. The Somalis are intelligent, beautiful and most can speak 2 or 3 languages."
—Peter G. Varisano

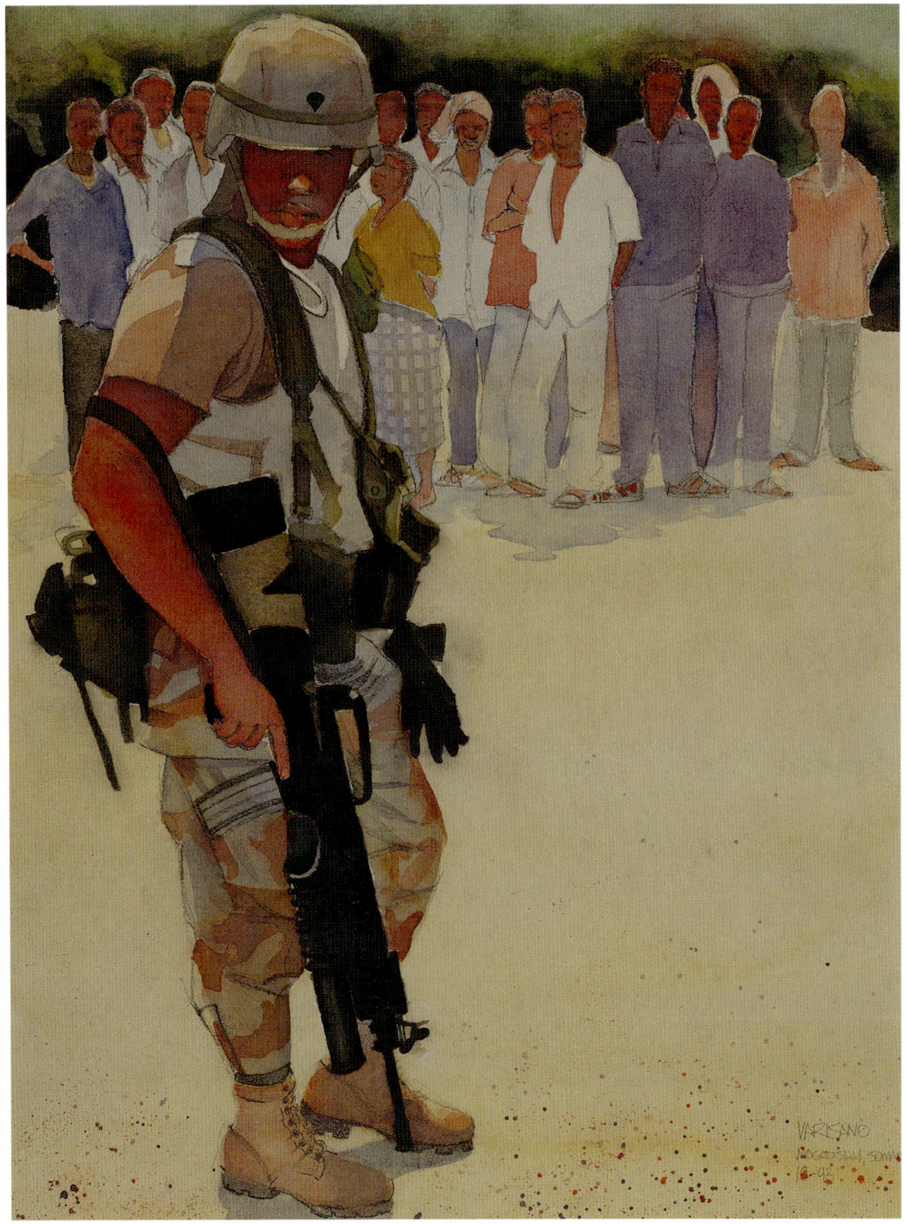

MOGADISHU, SOMALIA
Somalia, Operation Restore Hope
Peter G. Varisano, 1993
Pencil and watercolor on paper, 23" x 17¼"

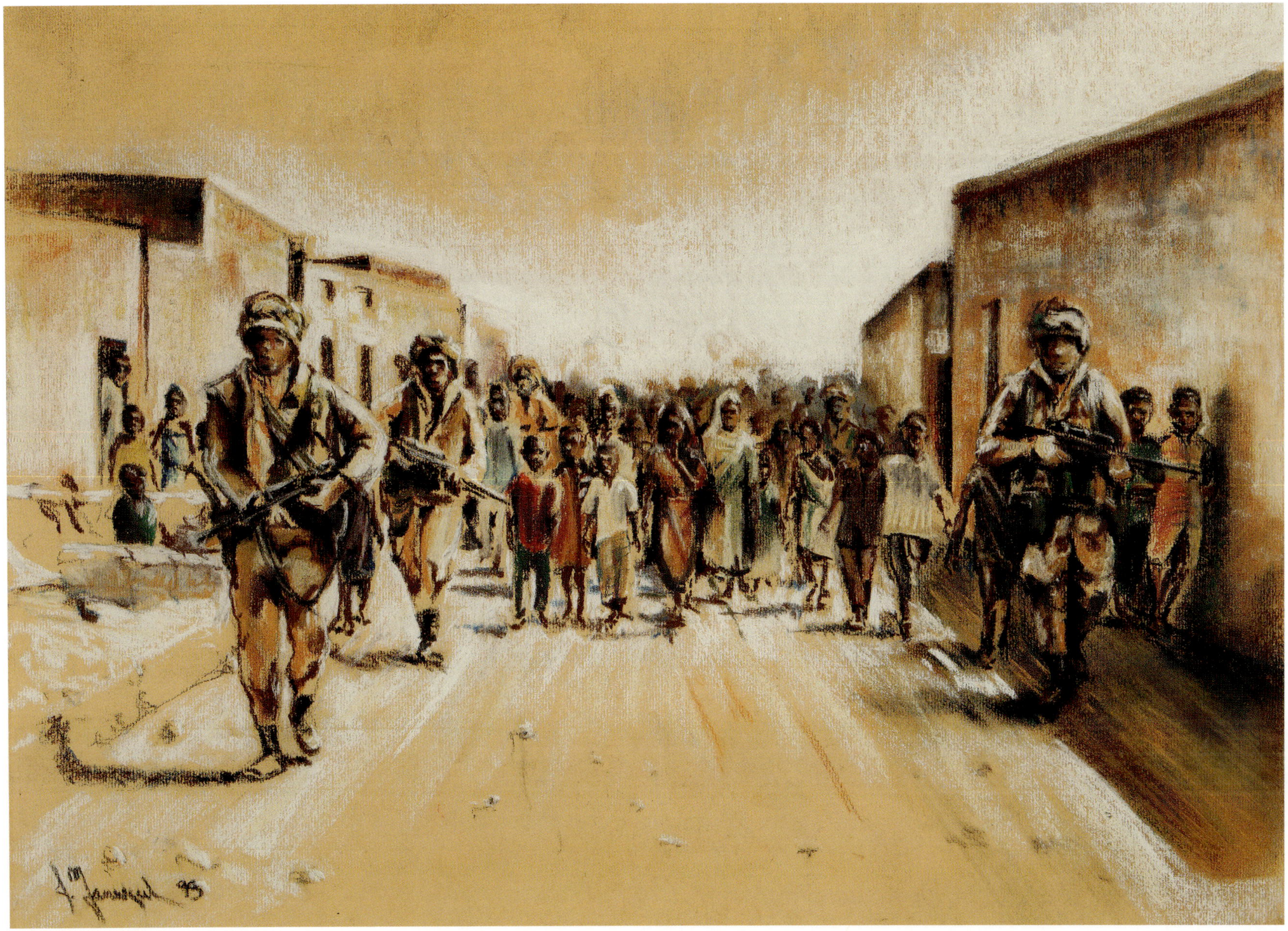

"The way a historian writes about history—we paint. We can make places look hotter, rucksacks bigger, men more tired. We still have to remain historically correct, but we can exaggerate using the elements of art. That's what makes it so different from a photograph. We can tell a story through painting."
—*Jeffrey T. Manuszak*

ROAD MARCH
Somalia, Operation Restore Hope
Jeffrey T. Manuszak, 1993
Pastel on paper, 19¾" x 25½"

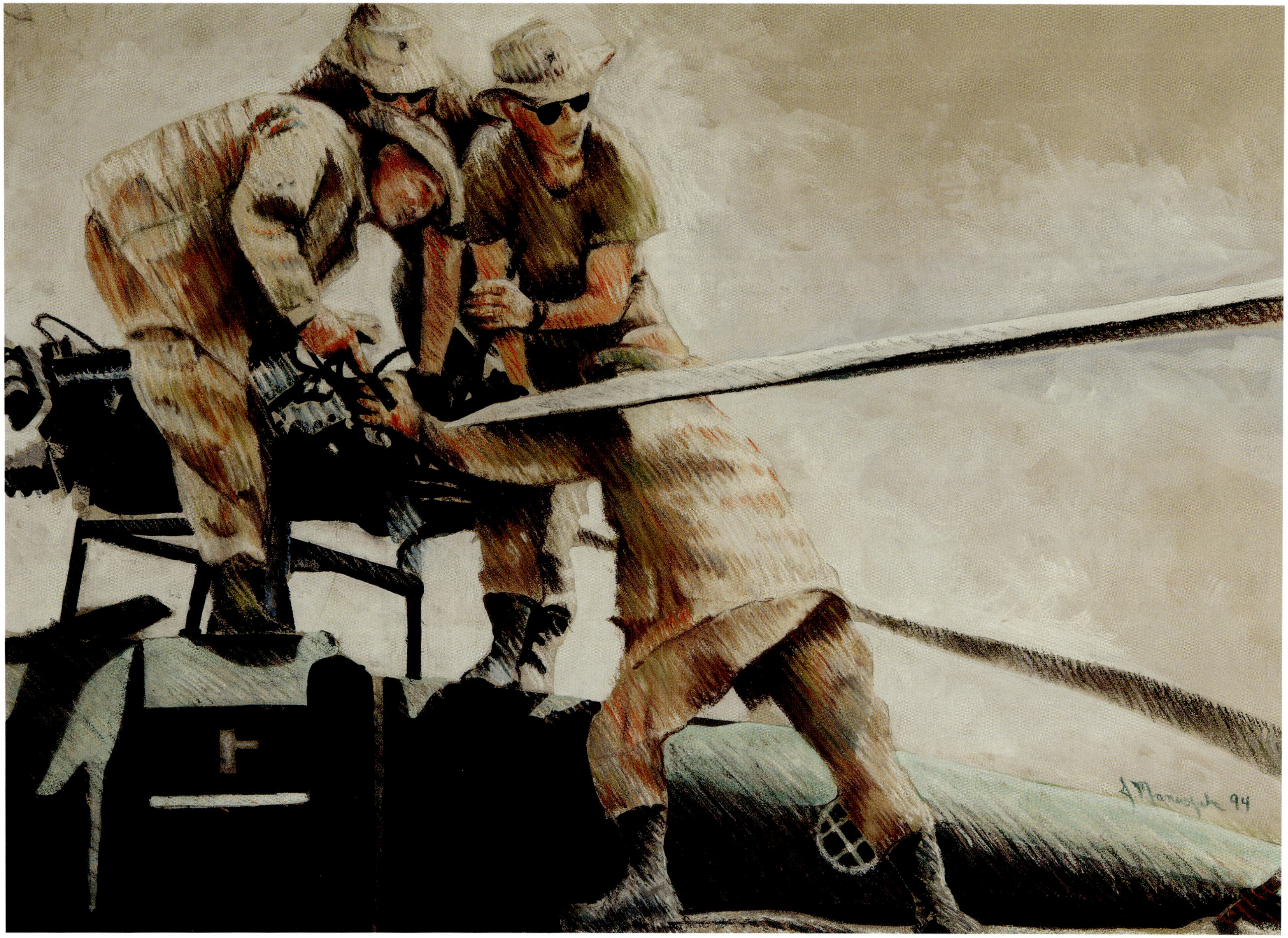

BLADE ASSEMBLY
Somalia, Operation RESTORE HOPE
Jeffrey T. Manuszak, 1994
Pastel on paper, 20" x 25½"

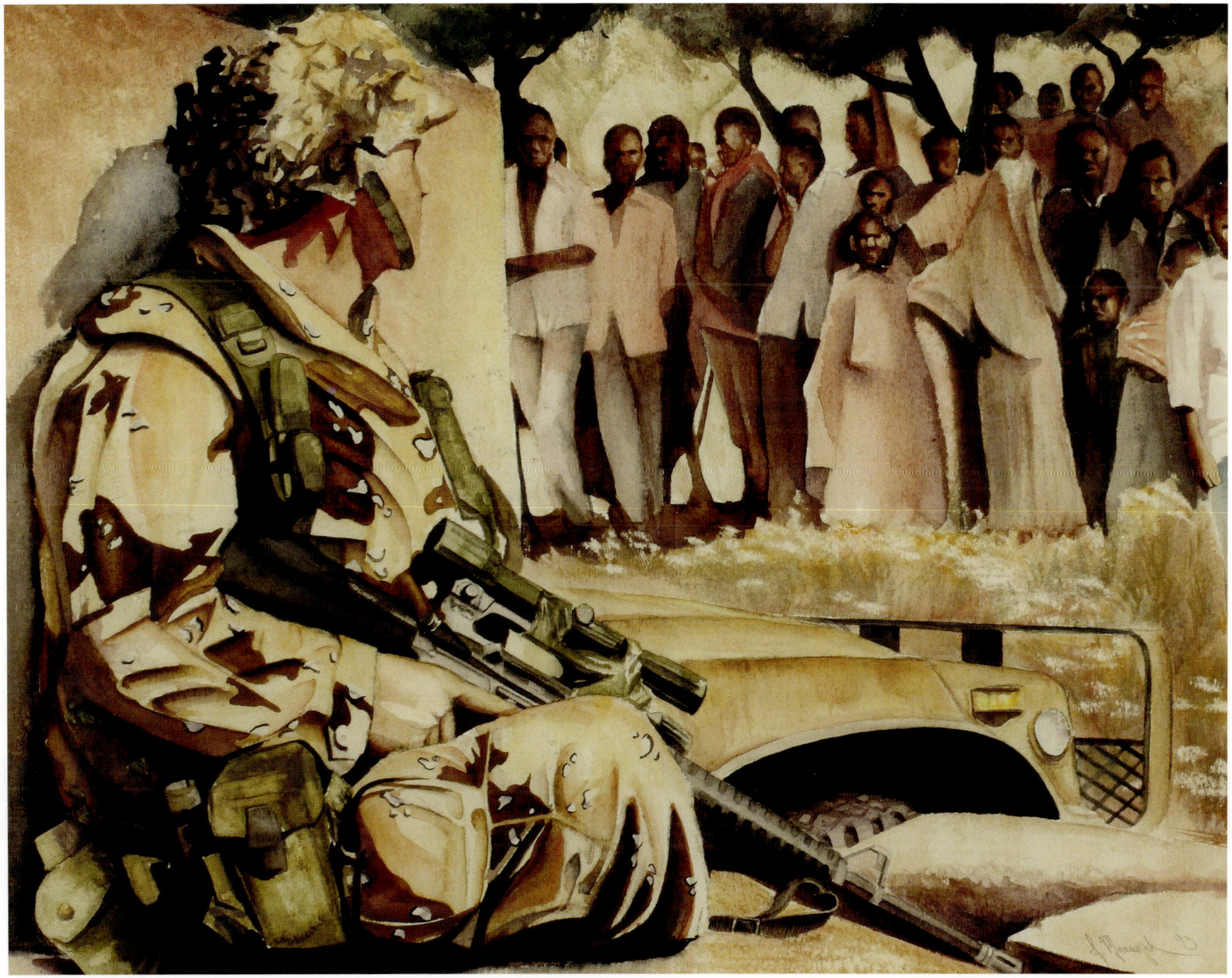

ON WATCH
Somalia, Operation Restore Hope
Jeffrey T. Manuszak, 1993
Watercolor on paper, 14¾" x 19"

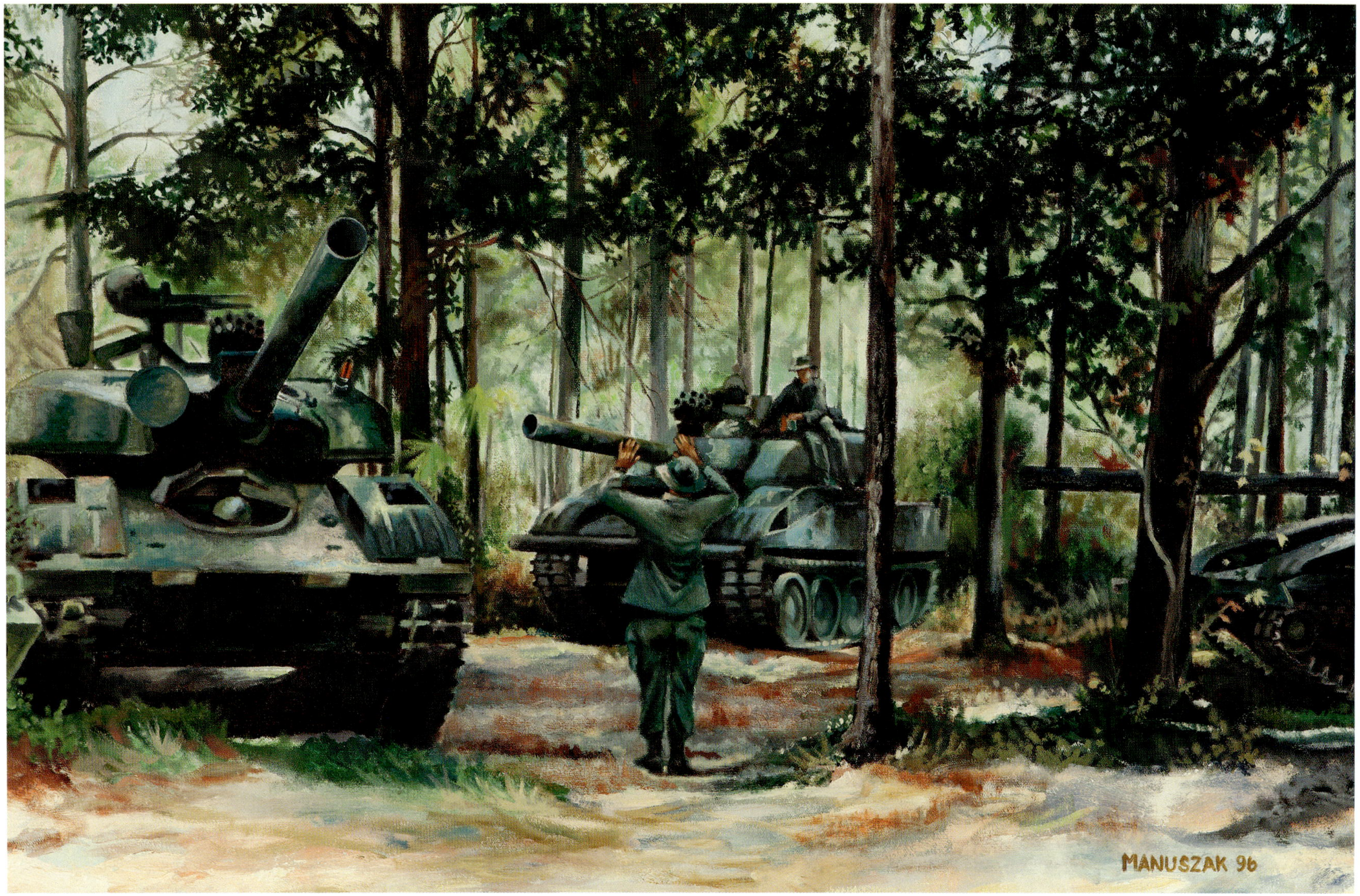

SYMPHONY OF STEEL
Fort Polk, Louisiana
Jeffrey T. Manuszak, 1996
Oil on canvas, 20" x 30"

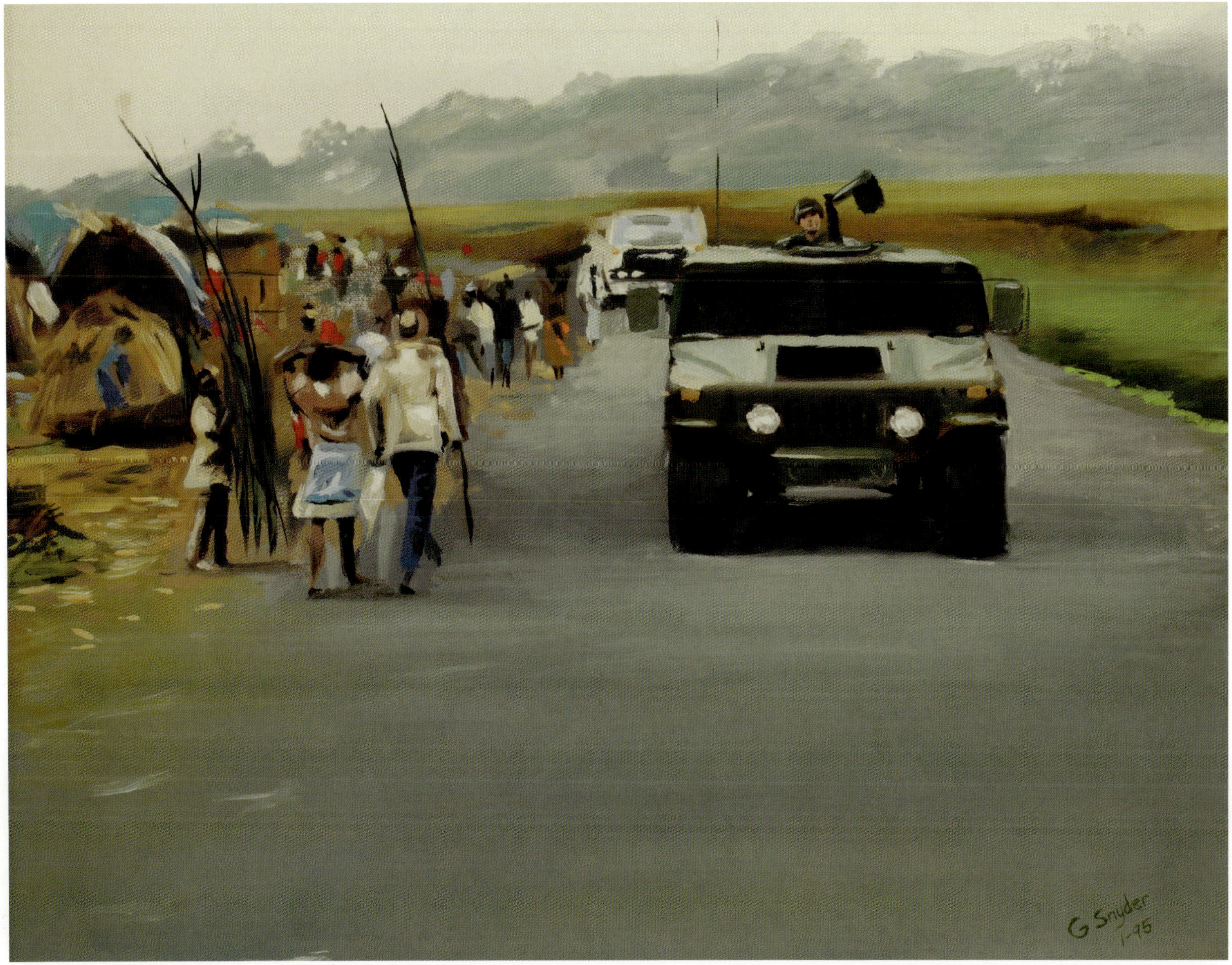

"This is for me. To be able to document the unsung heroes in the mud, grease and sweat—it's poetry. A hundred years from now our work will reflect the Army at the beginning of the 21st century."
—Carl E. "Gene" Snyder

RWANDA PATROL
Rwanda
Carl E. "Gene" Snyder, 1995
Oil on canvas, 16" x 20"

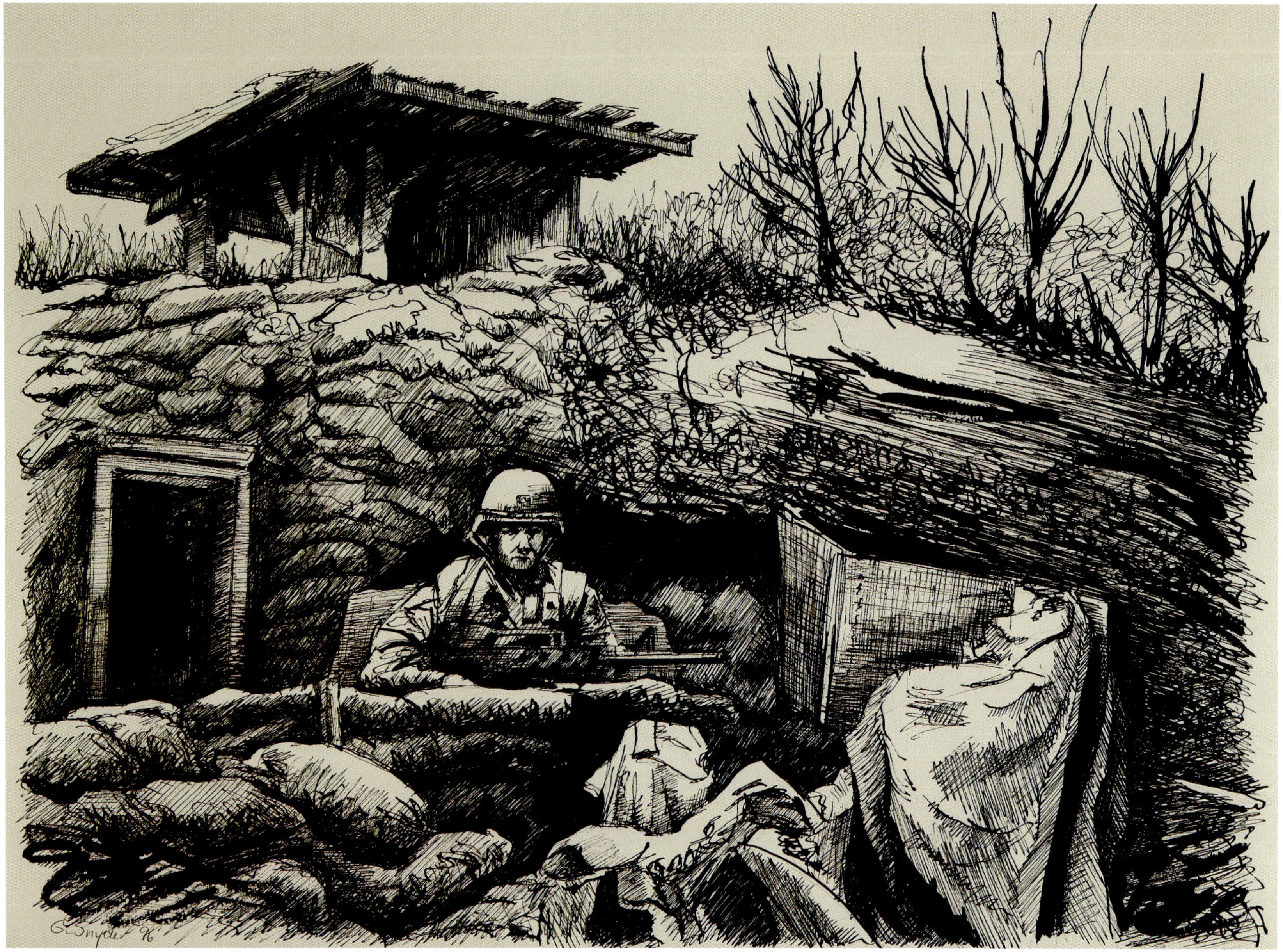

ON WATCH IN BOSNIA
Bosnia-Herzegovina, Operation Joint Endeavor
Carl E. "Gene" Snyder, 1996
Ink on paper, 18" x 24"

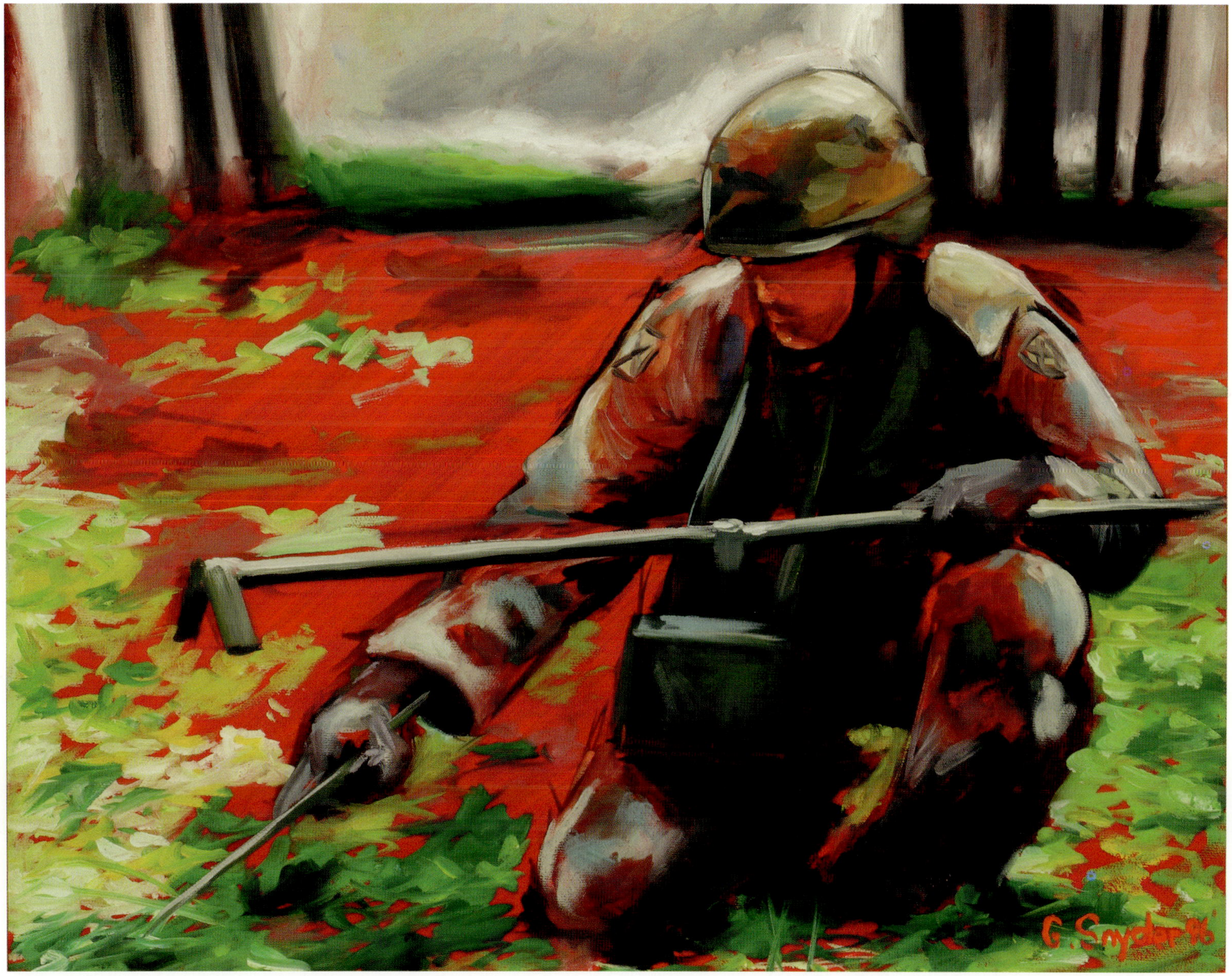

PROBING FOR LAND MINES IN BOSNIA
Bosnia-Herzegovina, Operation Joint Endeavor
Carl E. "Gene" Snyder, 1996
Oil on canvas, 16" x 20"

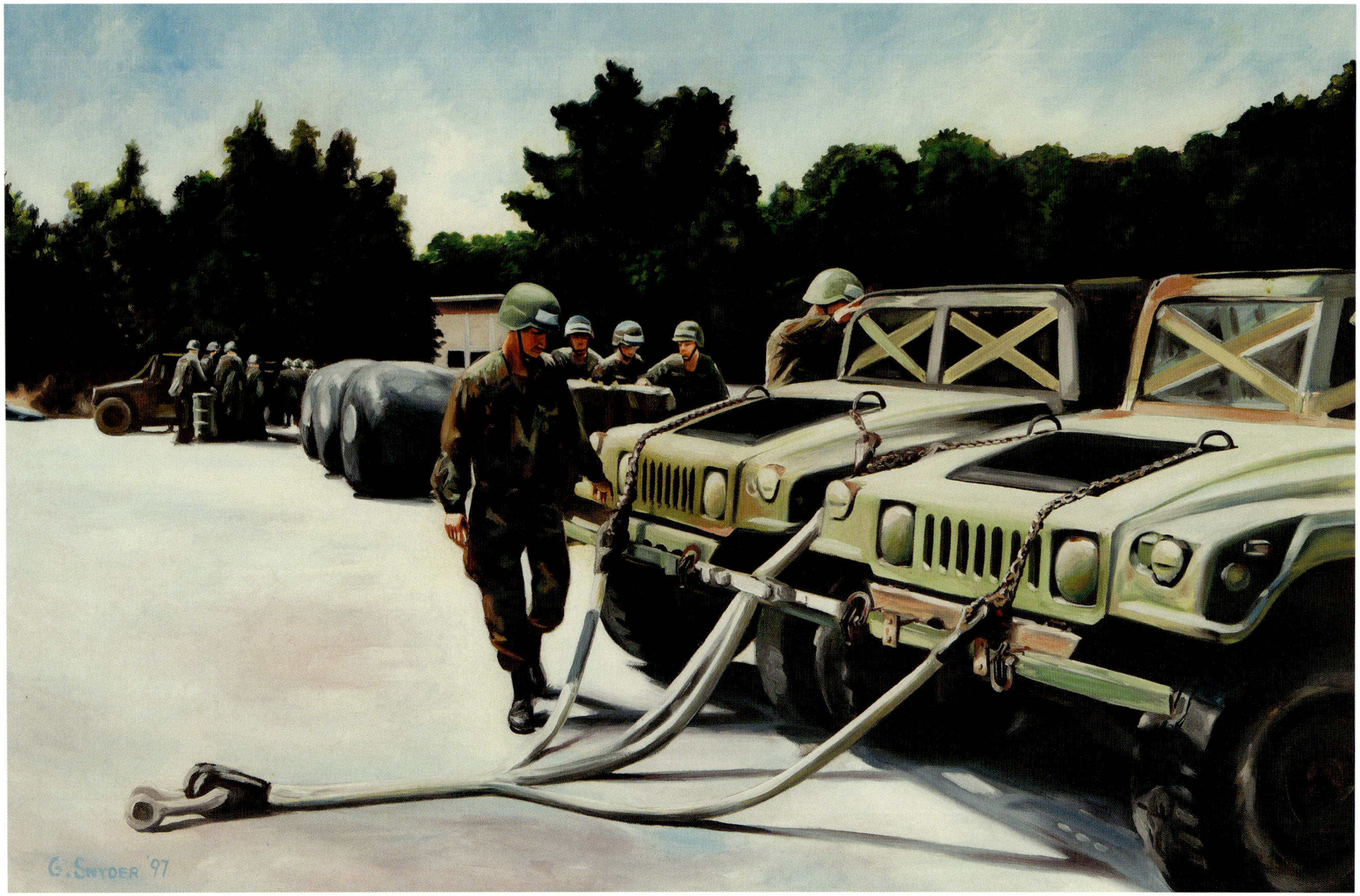

TEST DAY, PHASE II
Fort Campbell, Kentucky
Carl E. "Gene" Snyder, 1997
Oil on canvas, 24" x 36"

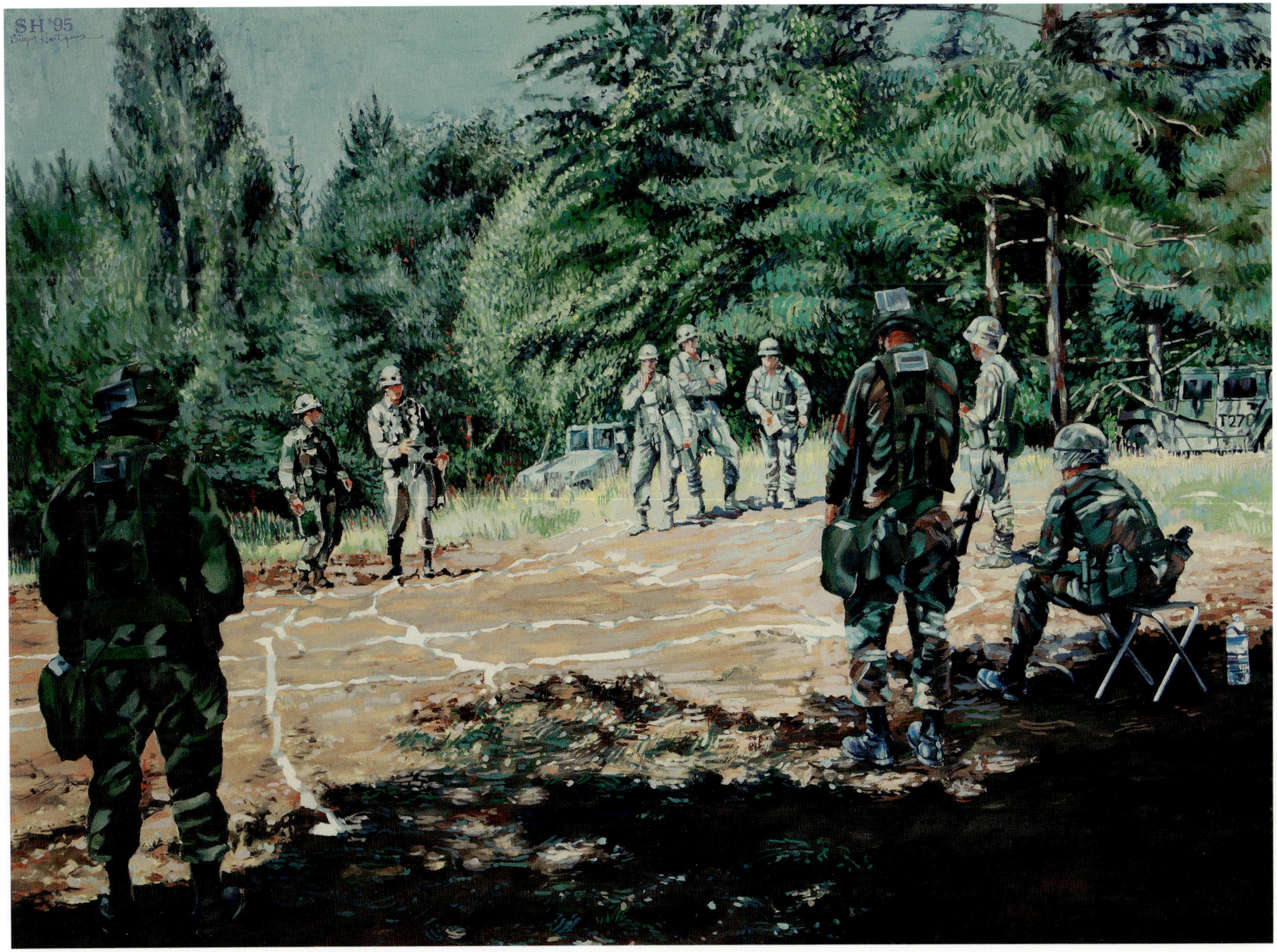

"I like painting and I like the Army. This is the best job for me and the Army. I consider myself extremely lucky to do this for a living. . . . To be an Army Artist, you have to be a Soldier first. That is always the bottom line."
—*Sieger Hartgers*

ROCK DRILL
Hohenfels, Germany
Sieger Hartgers, 1995
Oil on canvas, 30" x 40"

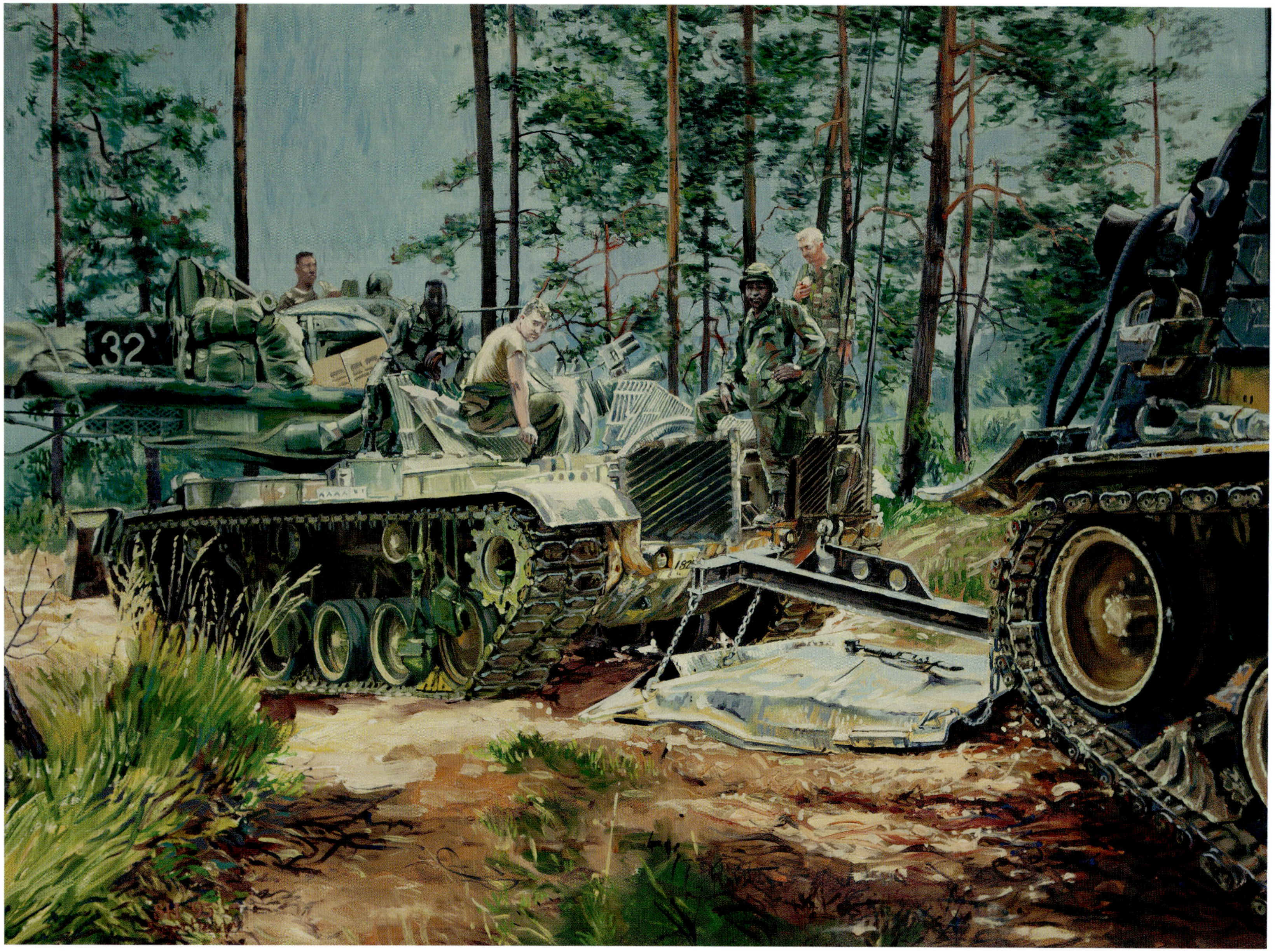

PULLING THE ENGINE
Hohenfels, Germany
Sieger Hartgers, 1995
Oil on canvas, 30" x 40"

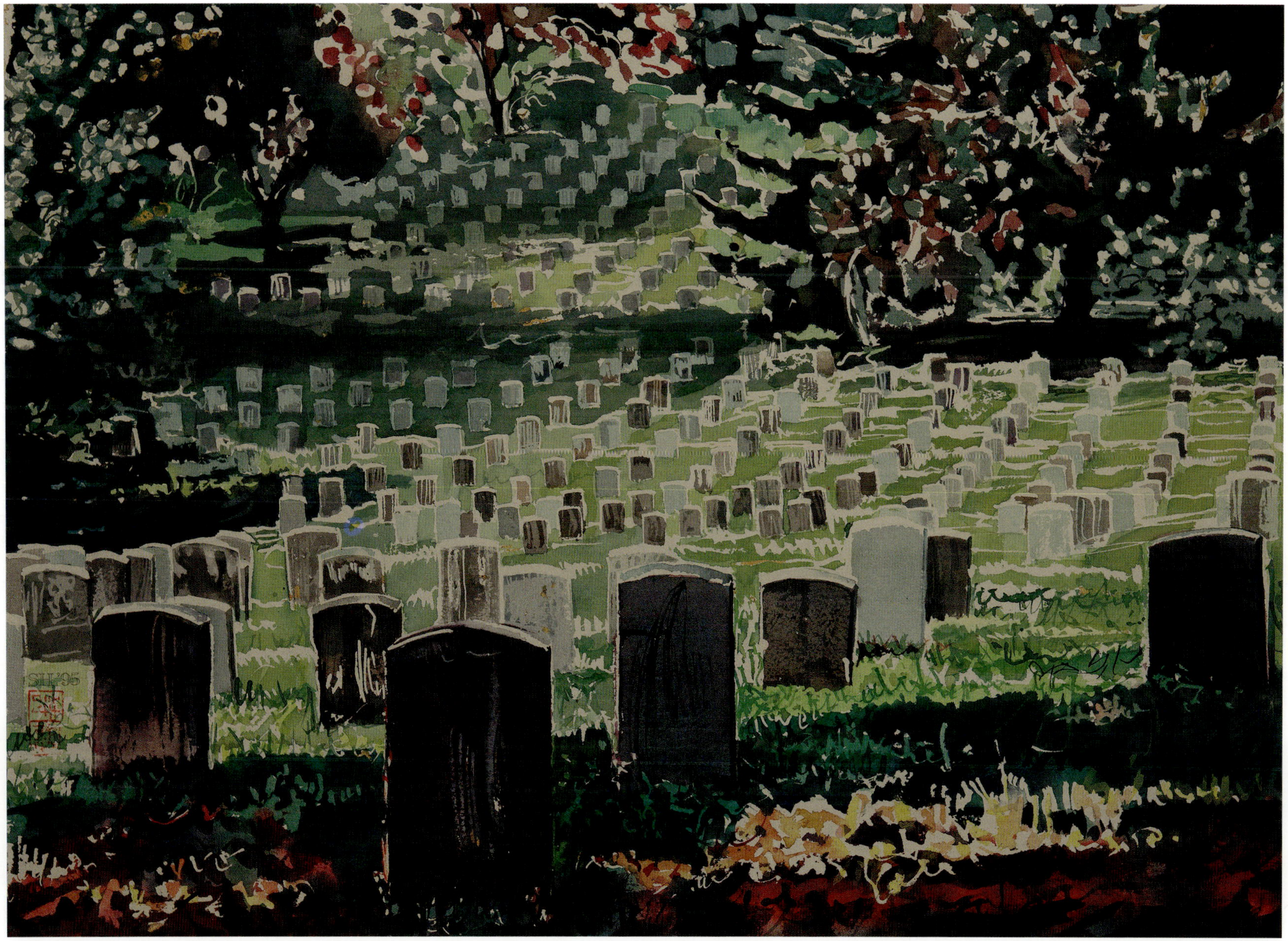

SO MANY GRAVES
Arlington National Cemetery
Sieger Hartgers, 1995
Watercolor on paper, 23" x 30½"

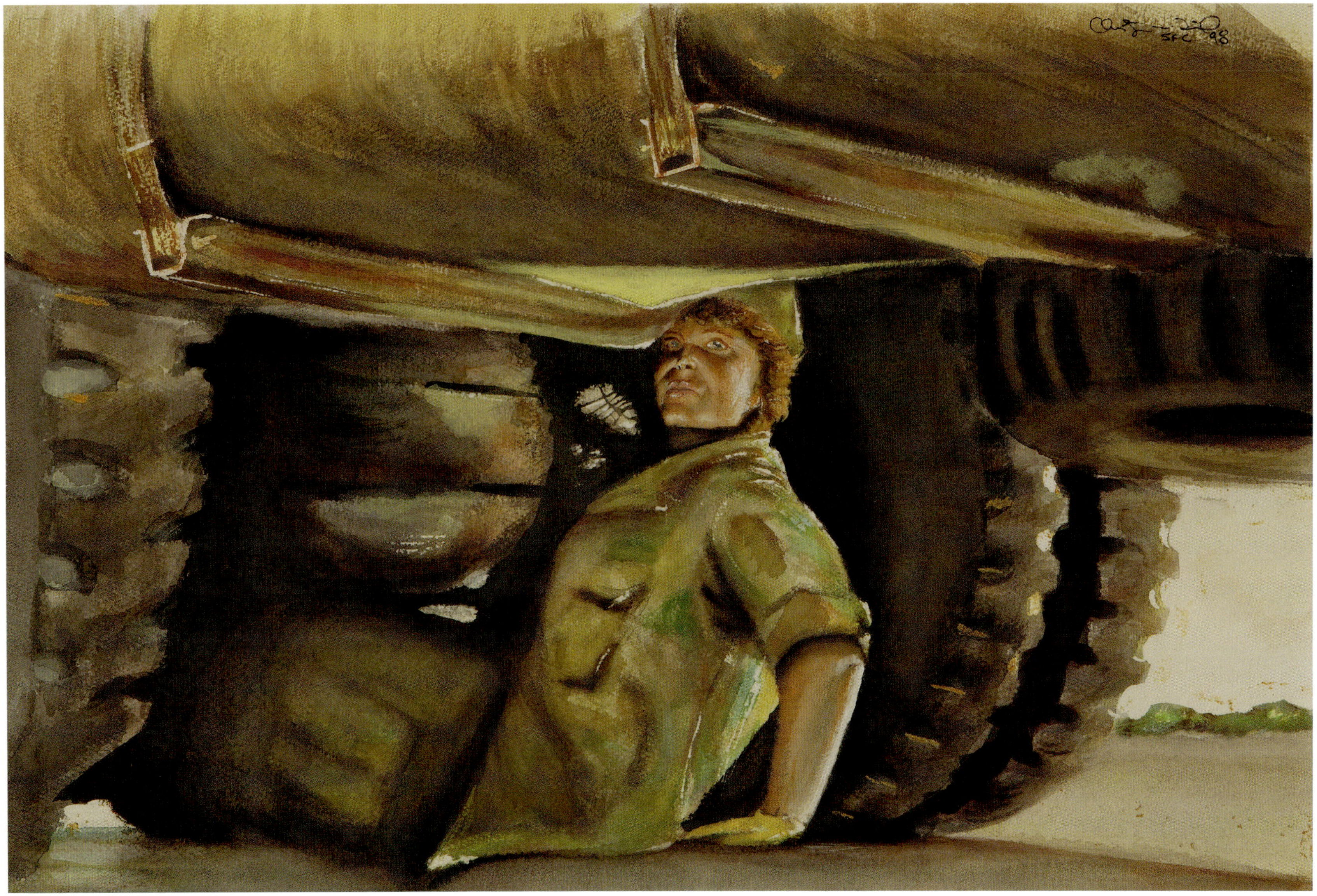

LOADING RAIL: PREBOARD INSPECTION
Oakland, California
Christopher W. Thiel, 1998
Watercolor on paper, 15" x 22¼"

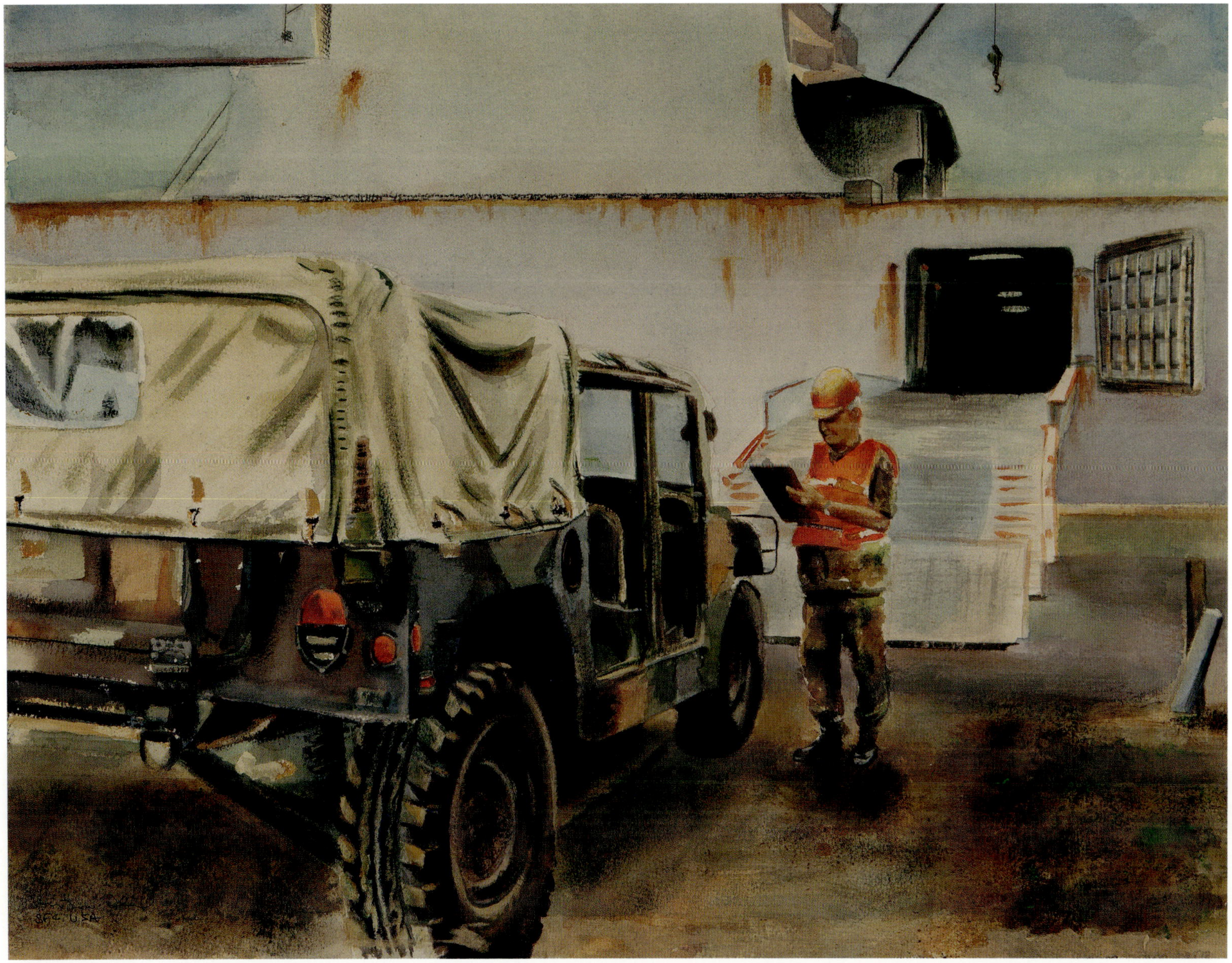

JOURNEY BOUND MOTBY
Bayonne, New Jersey
Christopher W. Thiel, 1998
Watercolor on paper, 18" x 24"

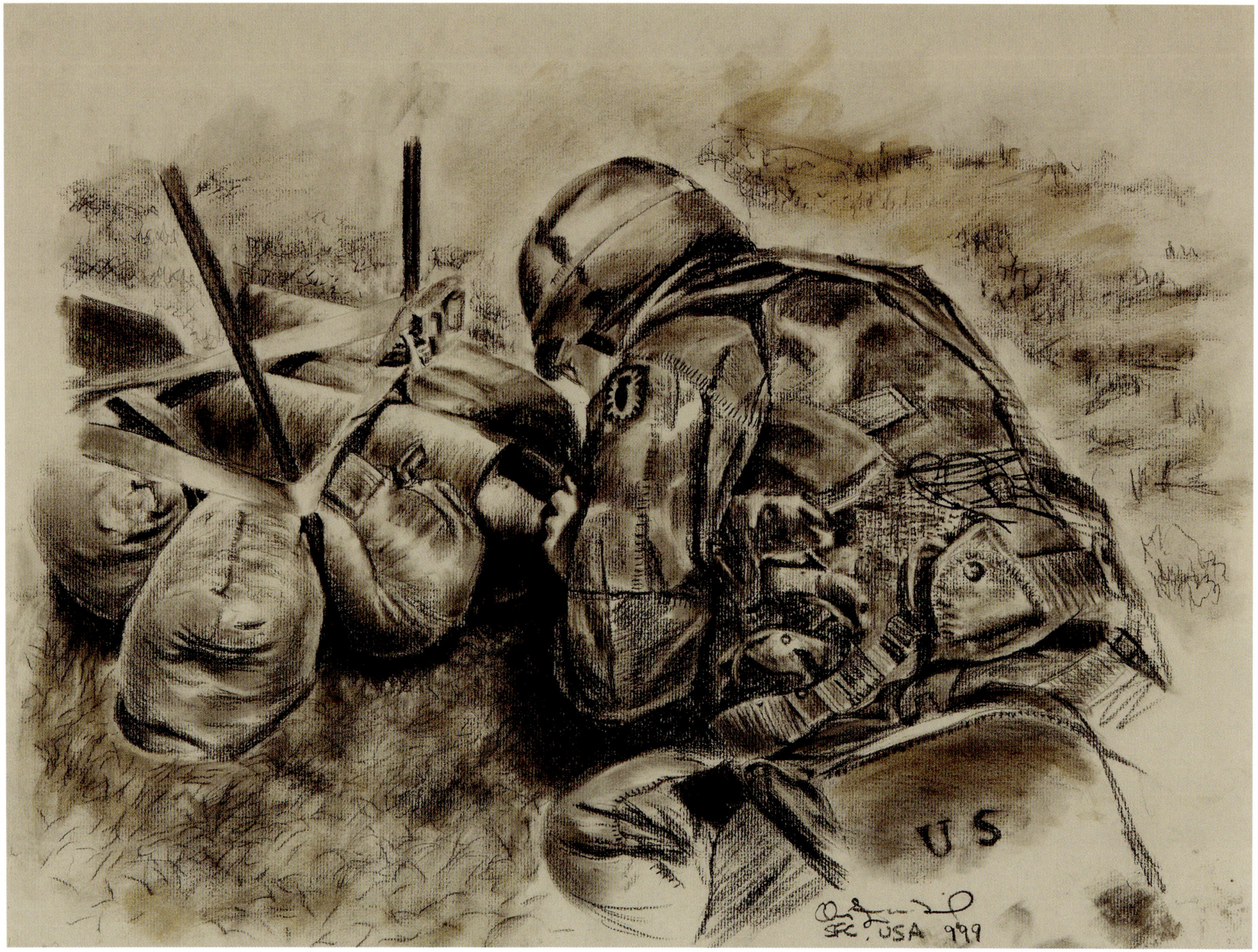

DISARMING THE SWORD—744TH EOD
1999
Christopher W. Thiel
Charcoal on paper, 24" x 18"

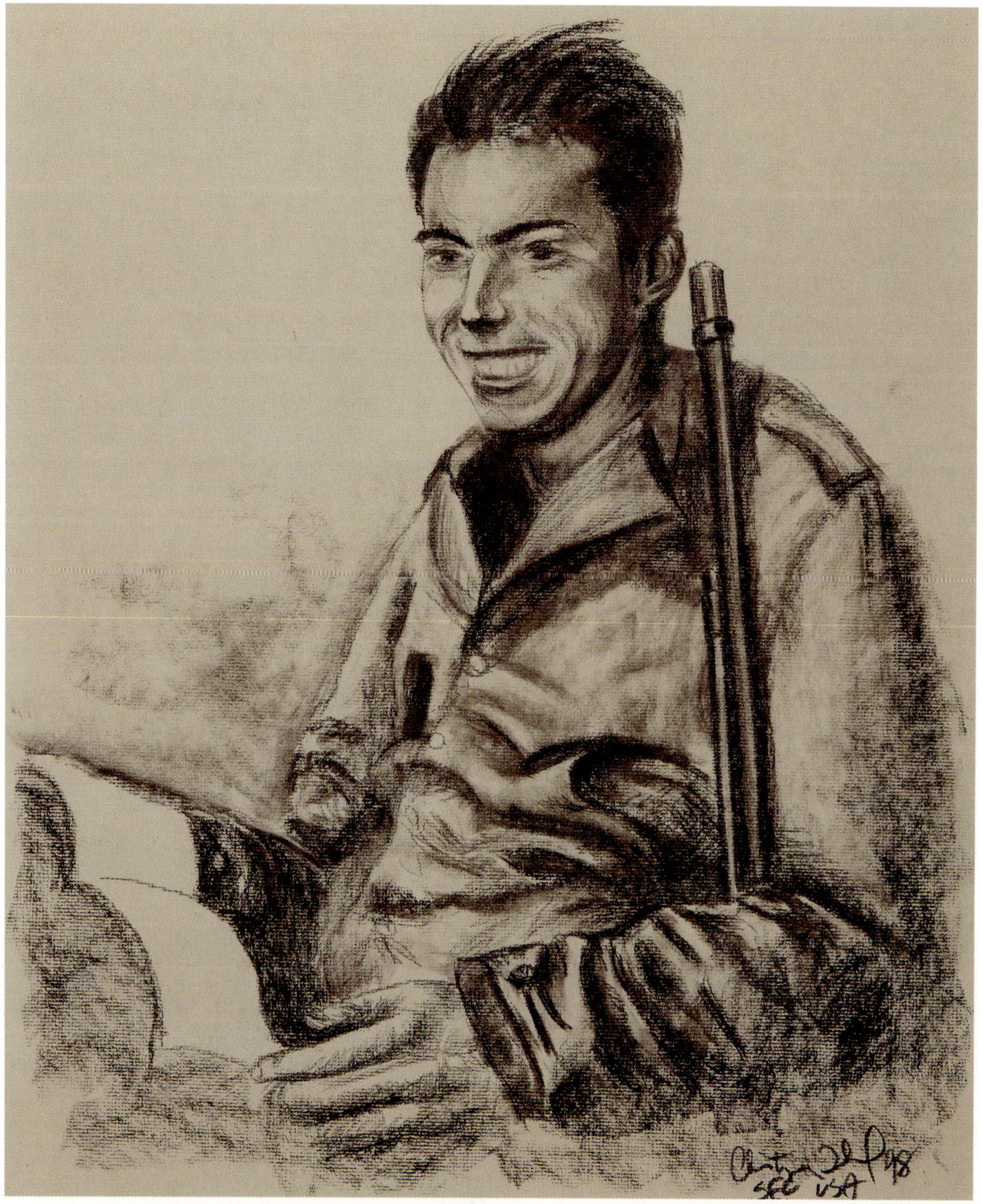

SGT F. CURREY, MEDAL OF HONOR
Awarded in 1945
Christopher W. Thiel, 1999
Charcoal on paper, 30" x 20"

"It's the kind of job that illustrators look forward to . . . you cannot replace the experience of being on the ground alongside the Soldier who's working 24/7."
—Henrietta M. Snowden

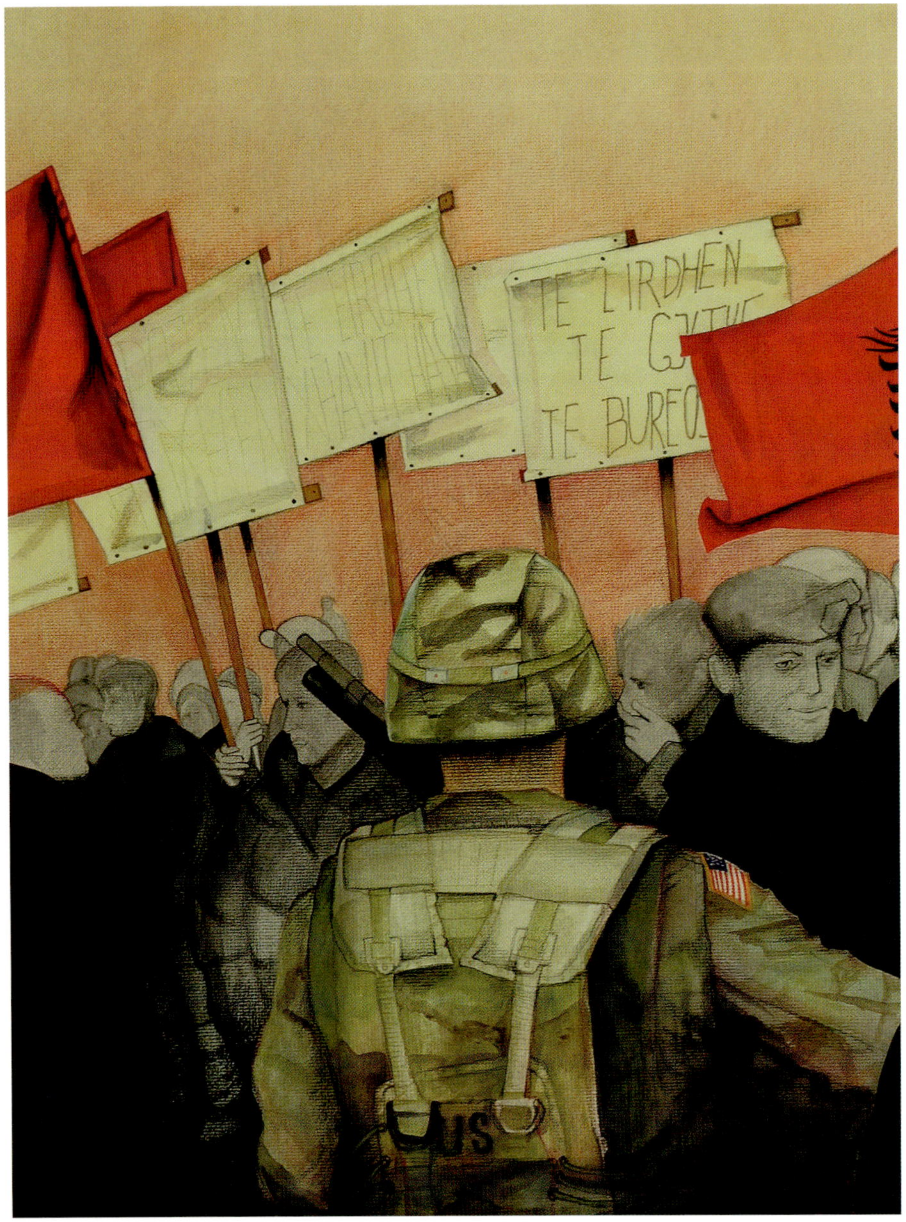

PEACE KEEPER
Kosovo, Operation JOINT GUARDIAN
Henrietta M. Snowden, 2001
Colored pencil and watercolor on paper, 24" x 18"

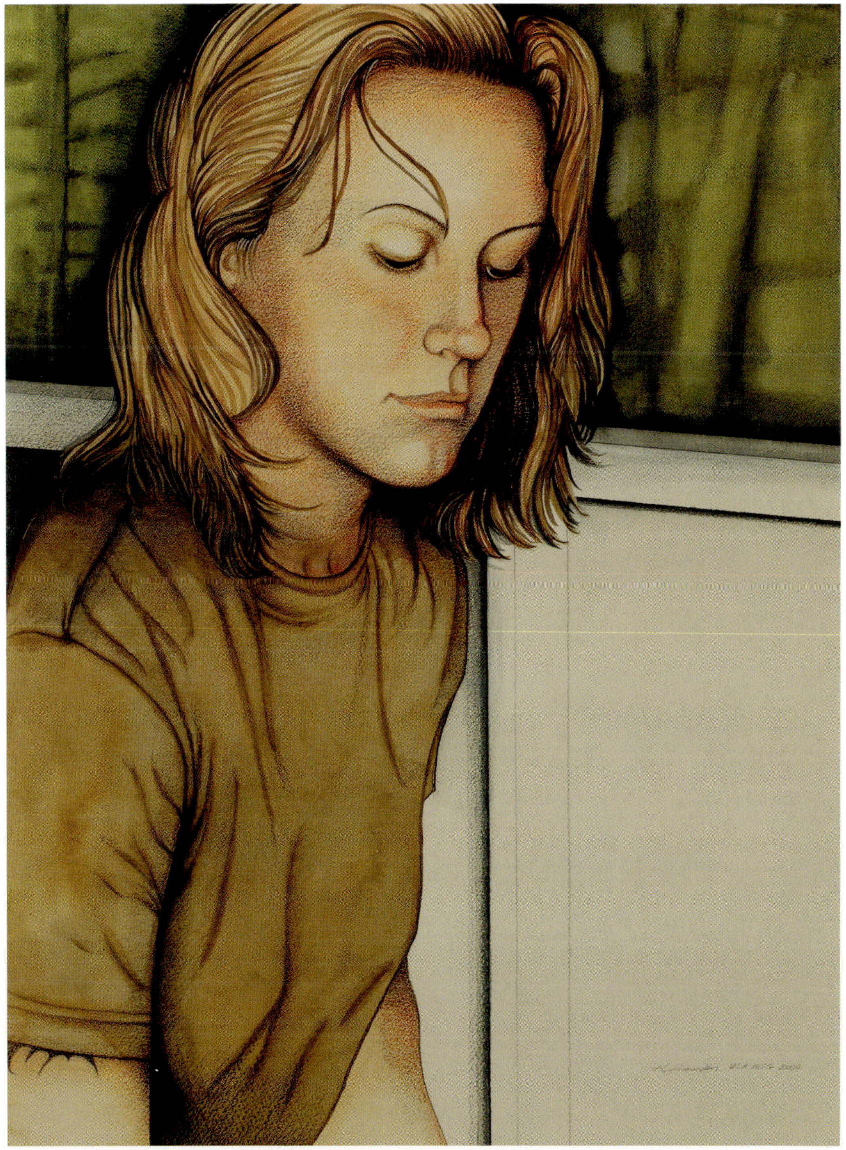

0530 WAKE-UP CALL
Camp Bondsteel, Kosovo, Operation Joint Guardian
Henrietta M. Snowden, 2000
Colored pencil and watercolor on paper, 24" x 18⅛"

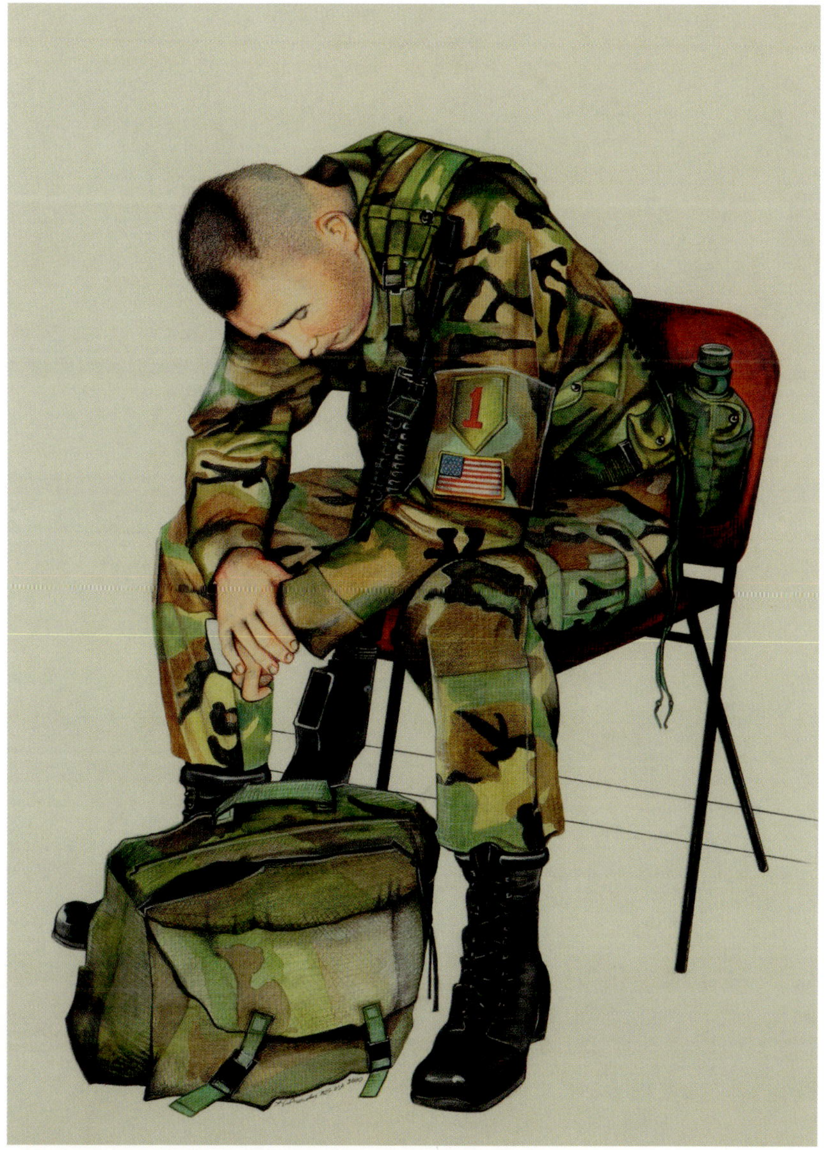

WAITING TO PHONE HOME
Camp Bondsteel, Kosovo, Operation Joint Guardian
Henrietta M. Snowden, 2000
Colored pencil and watercolor on paper, 24" x 18¼"

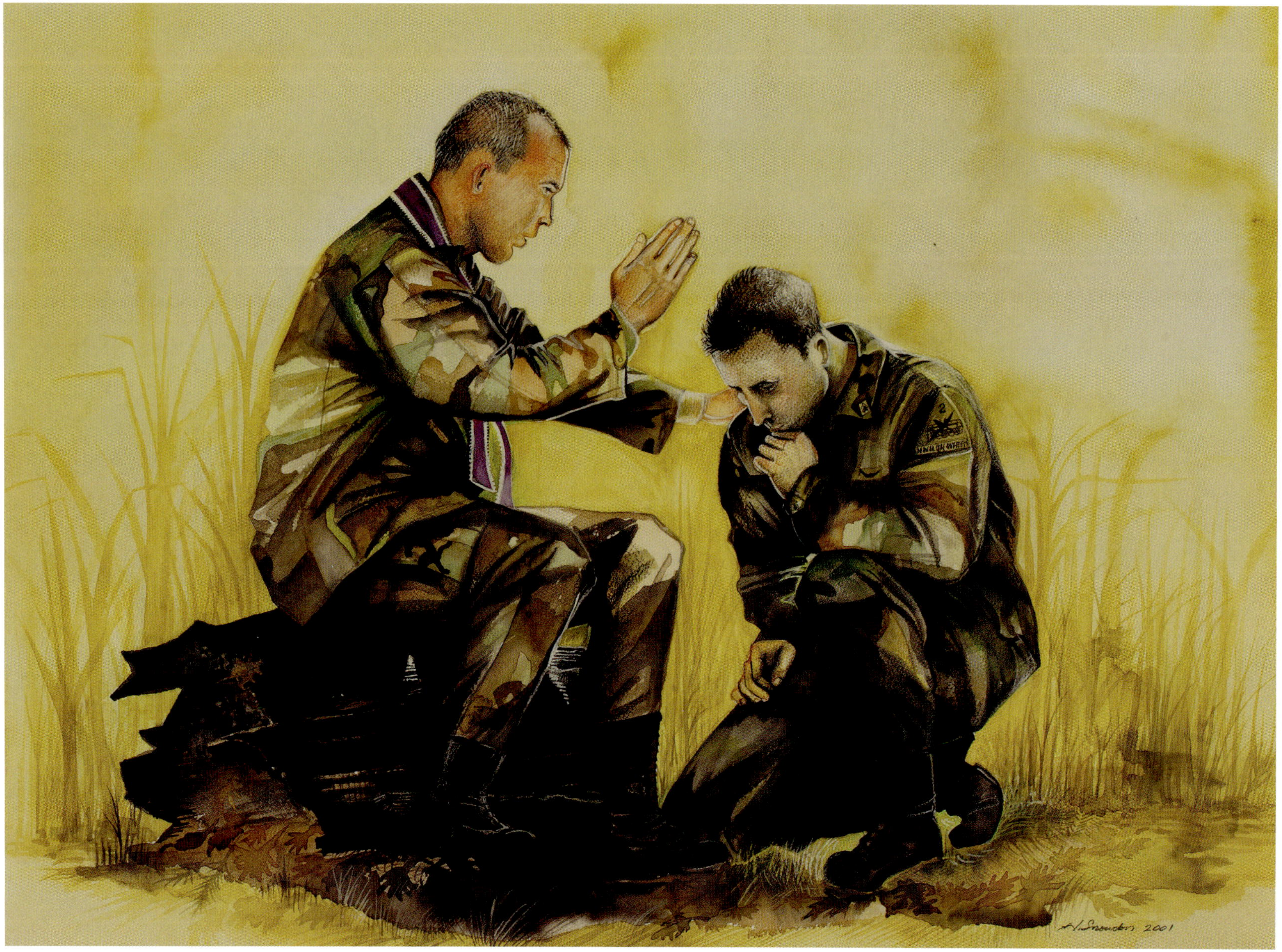

MINISTRY IN THE FIELD
2001
Henrietta M. Snowden
Colored pencil and watercolor on paper, 18" x 24"

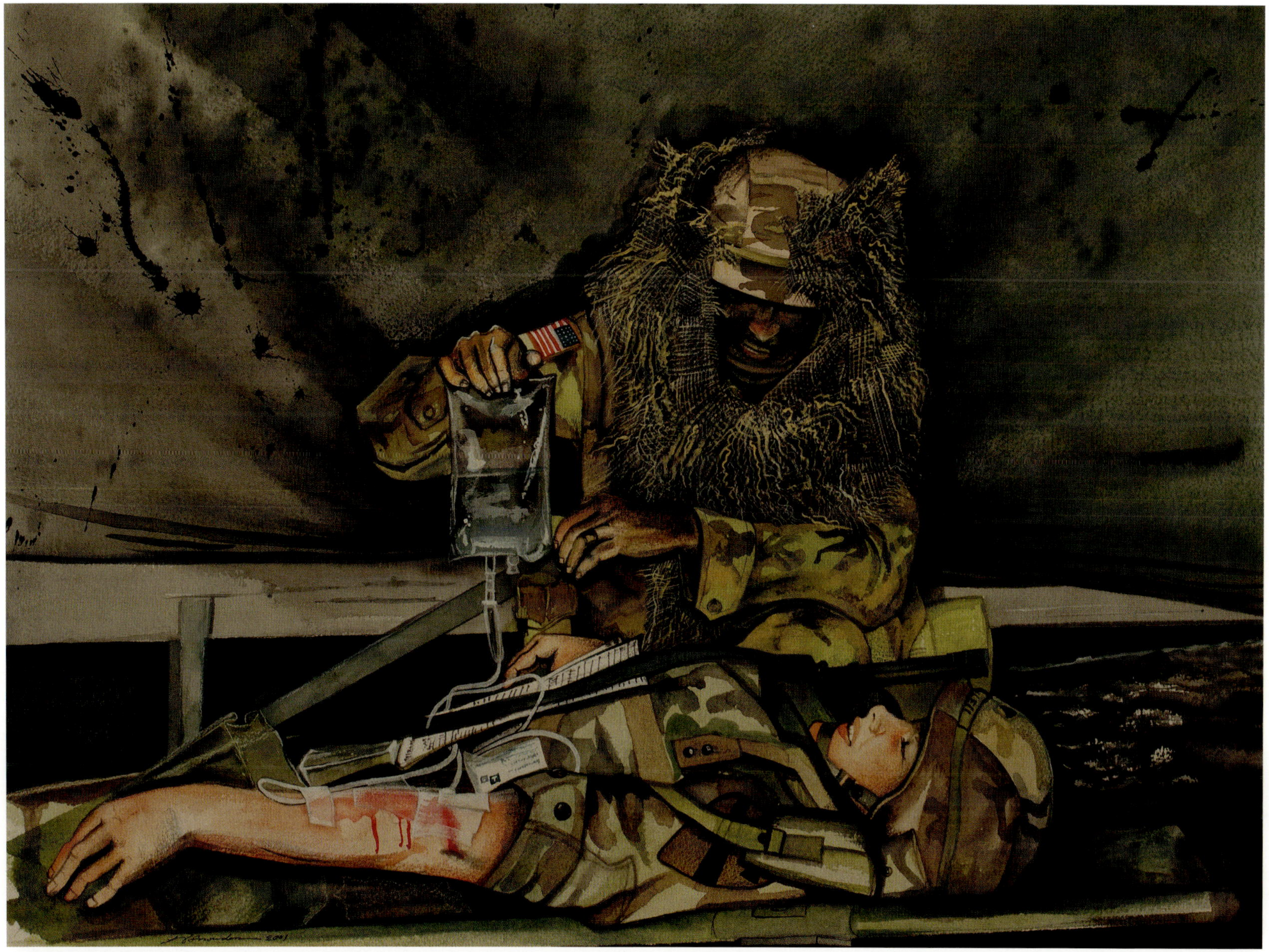

MEDICINE MAN
Camp Dobol, Bosnia-Herzegovina, Operation Joint Endeavor
Henrietta M. Snowden, 2001
Colored pencil and watercolor on paper, 18" x 24"

Global War on Terrorism

The Global War on Terrorism (GWOT) marks a maturity of the Artist-in-Residence Program. The artists of this period inherited a well-established mission from their predecessors in the 1990s, who defined the role of artists as documenters of the Army's contemporary history. Members of the Artist-in-Residence Program serve in a highly prized and competitive assignment and are the elite of their military occupational specialty, often chosen for the position by their reputation within the career field and vetted by the U.S. Army Center of Military History.

In GWOT art, the atmosphere is charged, such as in Sfc. Elzie Golden's *The Hizara Province* (page 157). The focus is on the soldiers and their mission. Soldiers wait in anticipation, ready to spring into action at a moment's notice, such as in M. Sgt. Martin J. Cervantez's *Tailgating Over the Valley* (page 178). Isolated instances of frenzied action occur, as in Golden's *Street Fight* (page 159). There are very few direct portrayals of an enemy, though the enemy is referenced indirectly in Golden's *Martyrdom Denied* (page 161). The surroundings are austere and often barren, as in Cervantez's *Dust Devils* (page 181).

Compared to World War II, this global conflict has been recorded by far fewer artists. As a result, this period so far has been defined by the singular vision of a small number of dedicated painters and illustrators. The awareness that he or she is solely responsible for such a vast mission creates great pressure on the serving resident artist to cover the most important aspects of the time, often to the exclusion of all else. Creatively, these artists find themselves working in isolation, as opposed to the team atmosphere of the art program during the world wars and Vietnam.

With the resident artist's focus in one direction and two wars taking place, the visual history of the Global War on Terrorism is told simultaneously through art and photography, which does not capture the ethos of the time. As a result, the resident artist is forced to concentrate the themes and message into high-quality compositions that epitomize the era. Of his approach to his work, M. Sgt. Martin J. Cervantez, Artist-in-Residence from 2008 to 2012, concisely stated that he strove for soldiers to be able to point to his art and say, "That's what I did, that's where I was, and that's what it looked like." Distilling the feel of a period of history into a small body of artwork requires the artist to bring a great confidence to the creation of each piece.

The overall style of this art demonstrates the artists' self-assurance. Each artist's individual approach is clear and well developed, instantly recognizable from work to work. Despite the monochromatic landscapes of Iraq and Afghanistan, bright colors often make an appearance in the art, as seen in Cervantez's *Treacherous Corner* (page 180). Numerous paintings and drawings of the period show that the artists shared a common interest in light. In 1st Lt. Heather C. Englehart's watercolor *Big Country* (page 166), backlighting is used to turn a personal moment into a universal one with which any soldier can identify. Dramatic lighting, reminiscent of chiaroscuro of Italian Baroque painting, is used to highlight an important collaboration in Cervantez's *A Huge Responsibility* (page 179). The green light of night-vision goggles sets a spooky atmosphere in M. Sgt. Christopher W. Thiel's *Bailey's Pre-Combat Checks* (page 177).

With the artists' increased confidence in their role comes a return to artistic convention as well. Golden's *Fallujah* (page 162) is a glimpse of barracks life, commonly seen in World War II Army art. The hustle and bustle featured in S. Sgt. Timothy B. Lawn's *1/206th Tactical Operations Center* (page 172) recalls sights documented by Army artists during the world wars. Englehart's *Laundry Day* (page 164) watercolor is a traditional still-life and reminds the viewer of scenes in World War II and Vietnam Army art. Cervantez's *Tailgating Over the Valley*, which portrays soldiers riding in the back of a CH–47 Chinook helicopter above northeastern Afghanistan, is an eerie echo of Al Sprague's *Cruising the Panama Canal (Chinook)* (page 87), a similar depiction of soldiers engaged in the same activity twenty years earlier. The fact that both artists experienced the identical incident and that both were inspired to commit it to canvas is indicative of its importance to the history of soldiers. Technology and tactics shared across time frame the event, which unfolds visually through repeated themes like this one.

The talented artists assigned to the Artist-in-Residence Program continue to support the Army by documenting the military withdrawal from Afghanistan. They are prepared to deploy to future operations and meet the challenges of the next era of American conflict and peacekeeping.

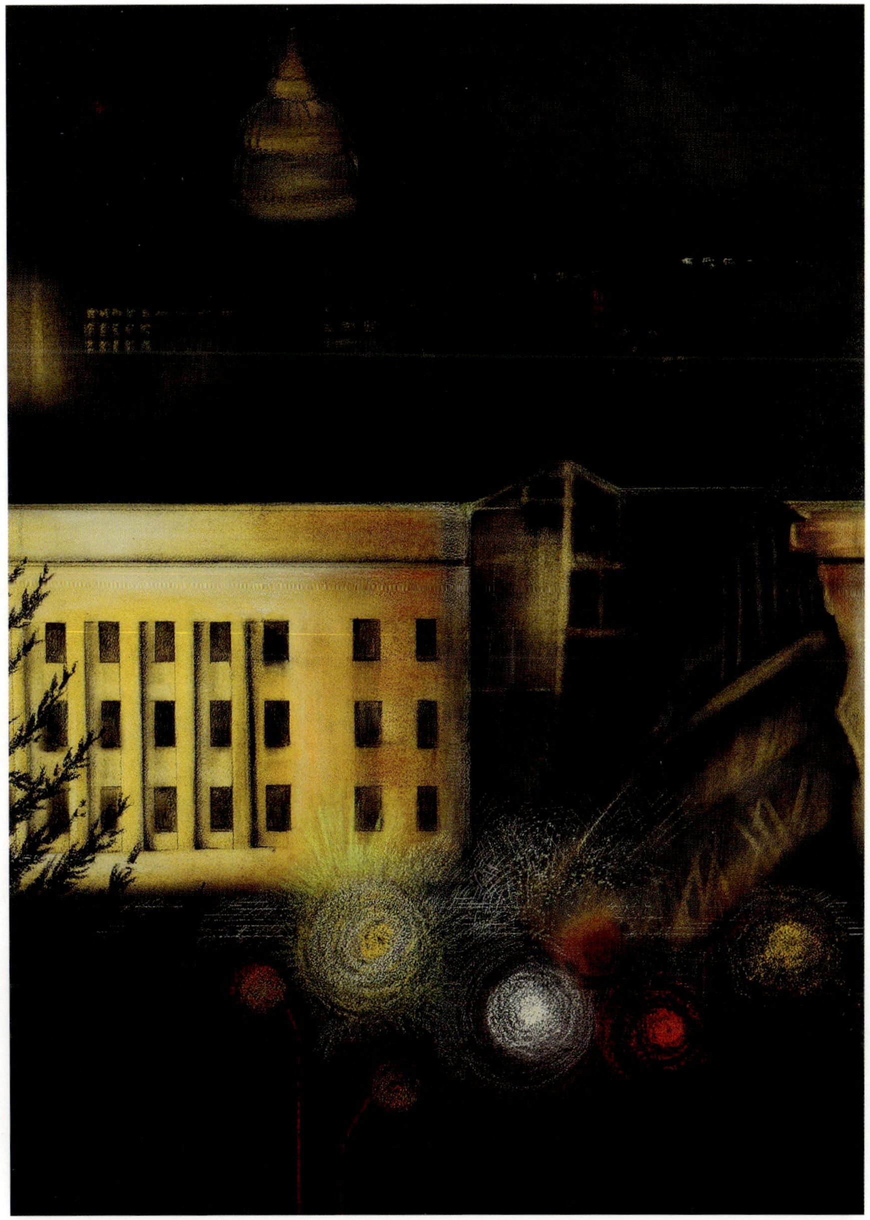

"No one tells me that I have to do this or I have to do that, or I have to use a certain medium or subject matter, so I do have a great deal of freedom."
—Henrietta M. Snowden

SEPTEMBER 11
Pentagon, Arlington, Virginia
Henrietta M. Snowden, 2001
Pastel on paper, 30" x 22½"

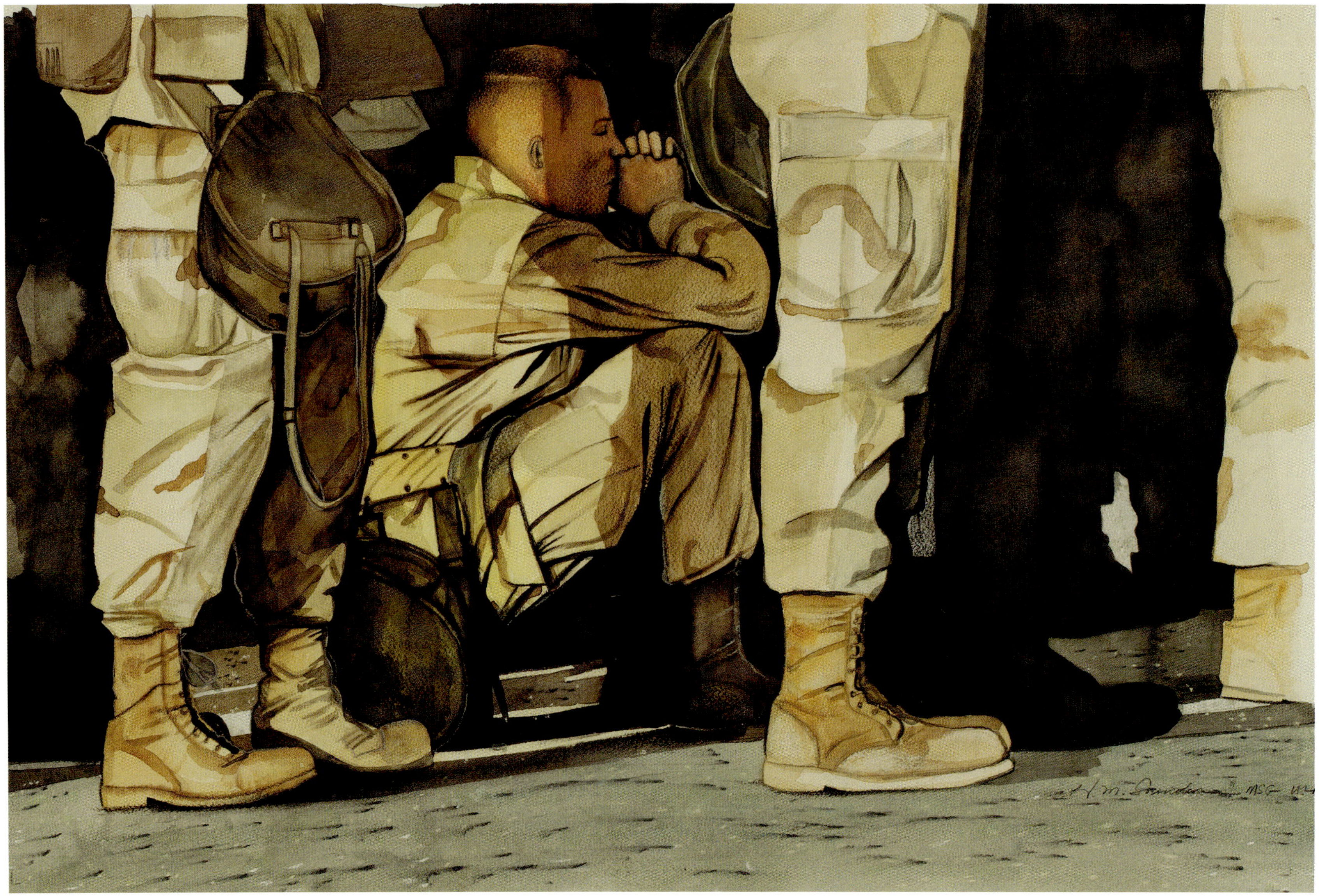

WAITING TO GO TO WAR, 2002
Operation Enduring Freedom
Henrietta M. Snowden, 2002
Colored pencil and watercolor on paper, 18" x 24"

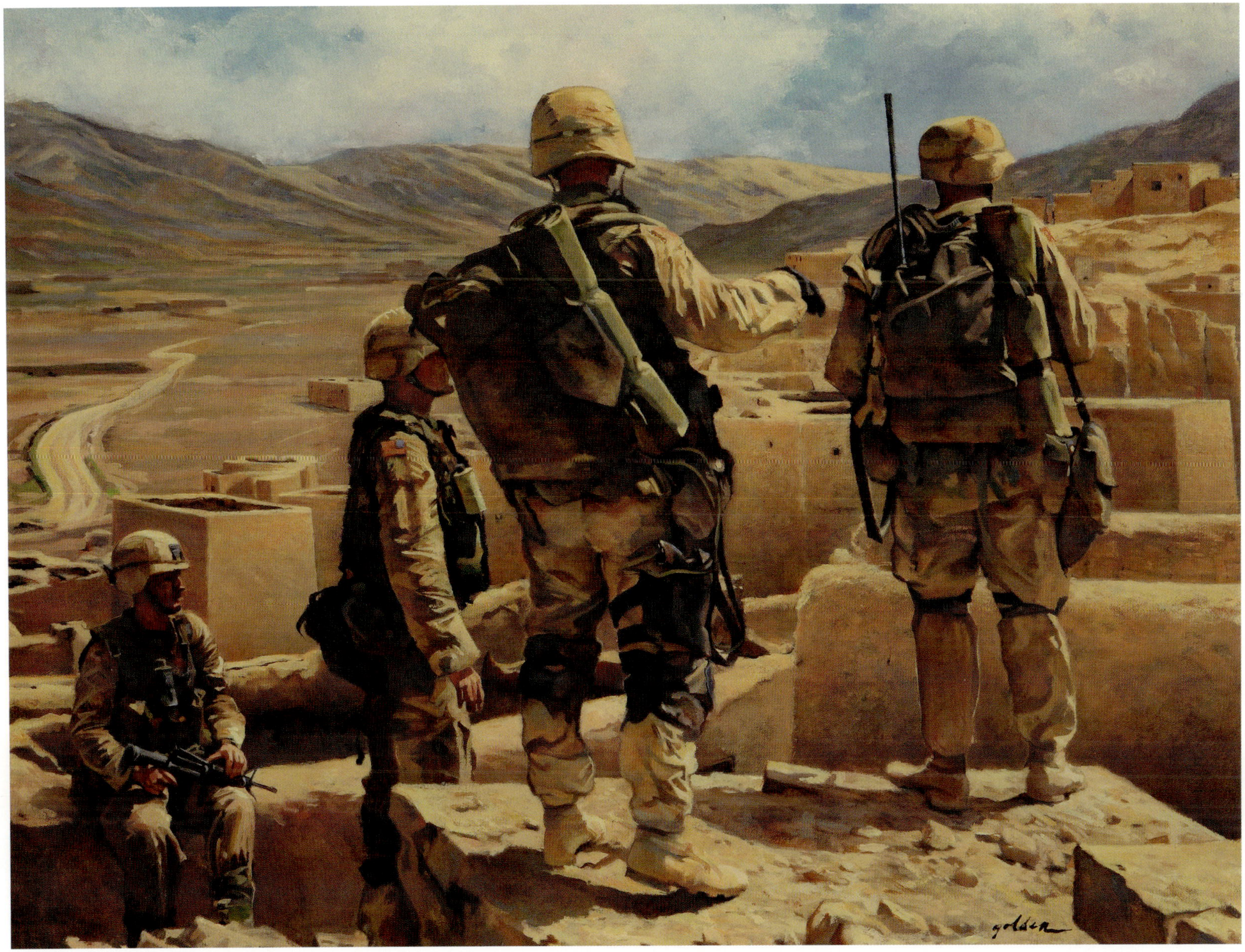

THE HIZARA PROVINCE
Afghanistan, Operation Enduring Freedom
Elzie Golden, 2003
Oil on canvas, 24" x 30"

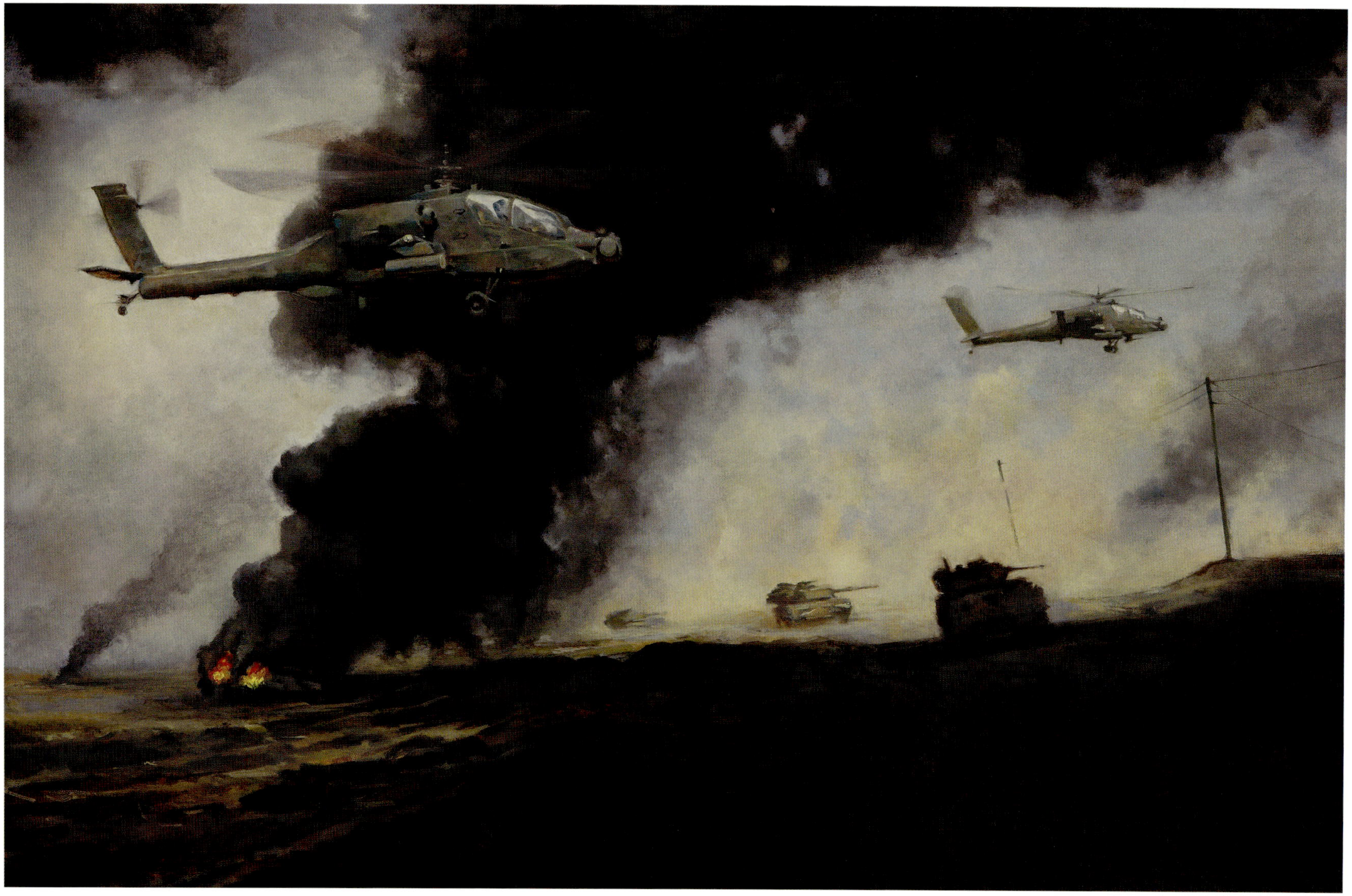

SATAN'S SANDBOX
Karbala, Iraq, Operation Iraqi Freedom
Elzie Golden, 2003
Oil on canvas, 24" x 36"

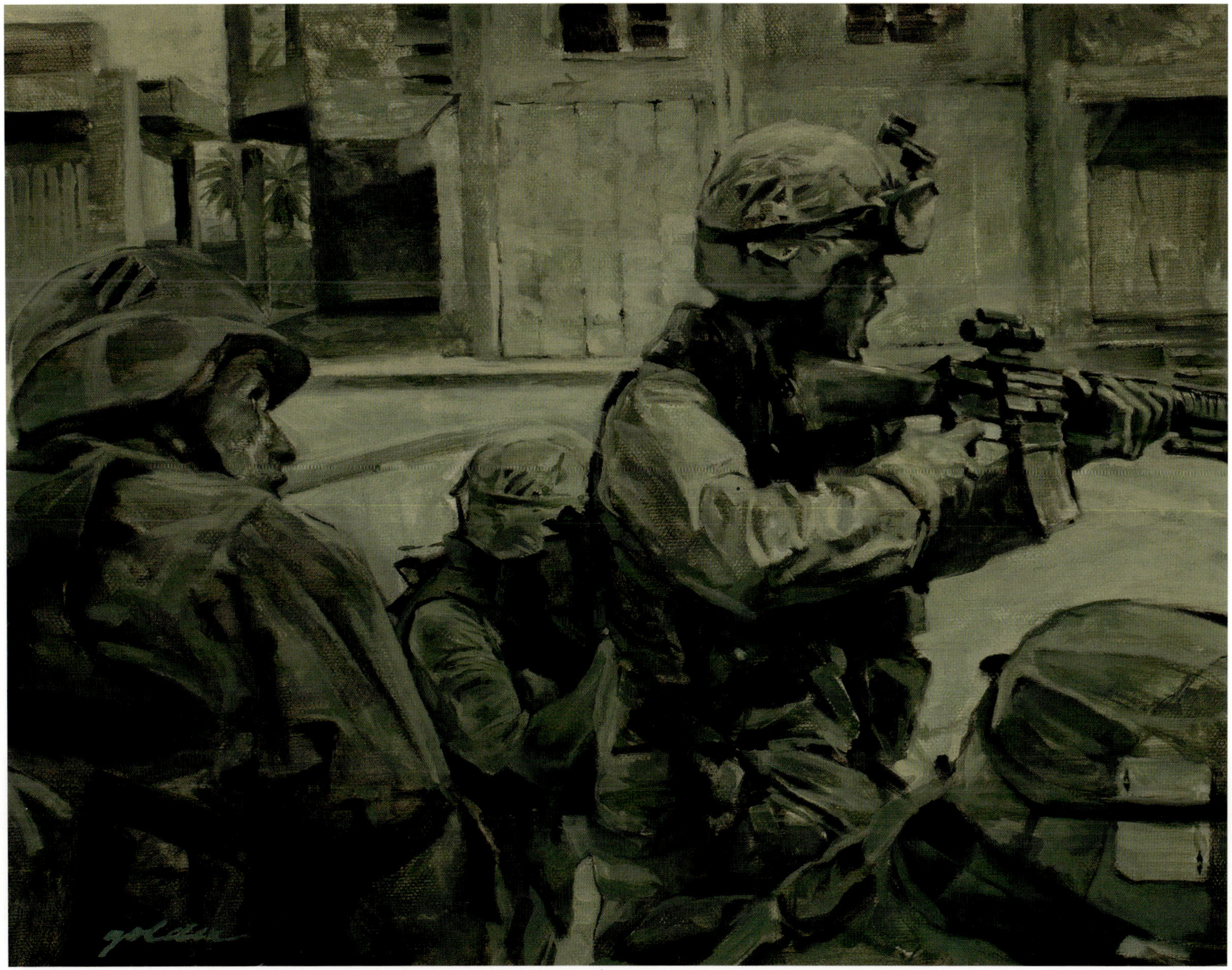

STREET FIGHT
An Najaf, Iraq, Operation Iraqi Freedom
Elzie Golden, 2003
Oil on canvas, 11" x 14"

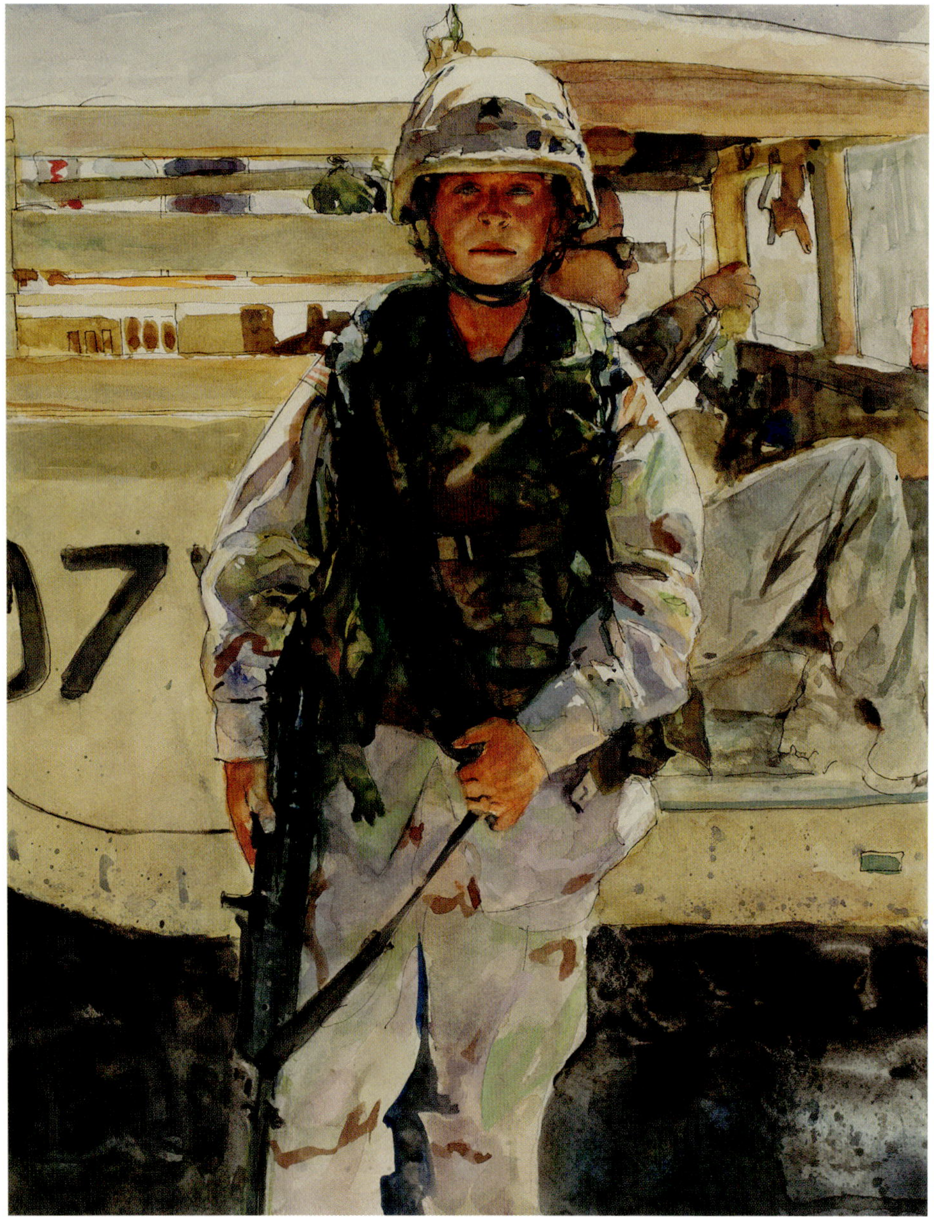

SGT VICKIE HODGE, 3D ID BAGHDAD AIRPORT
Baghdad, Iraq, Operation Iraqi Freedom
Elzie Golden, 2005
Ink and watercolor on paper, 17" x 13"

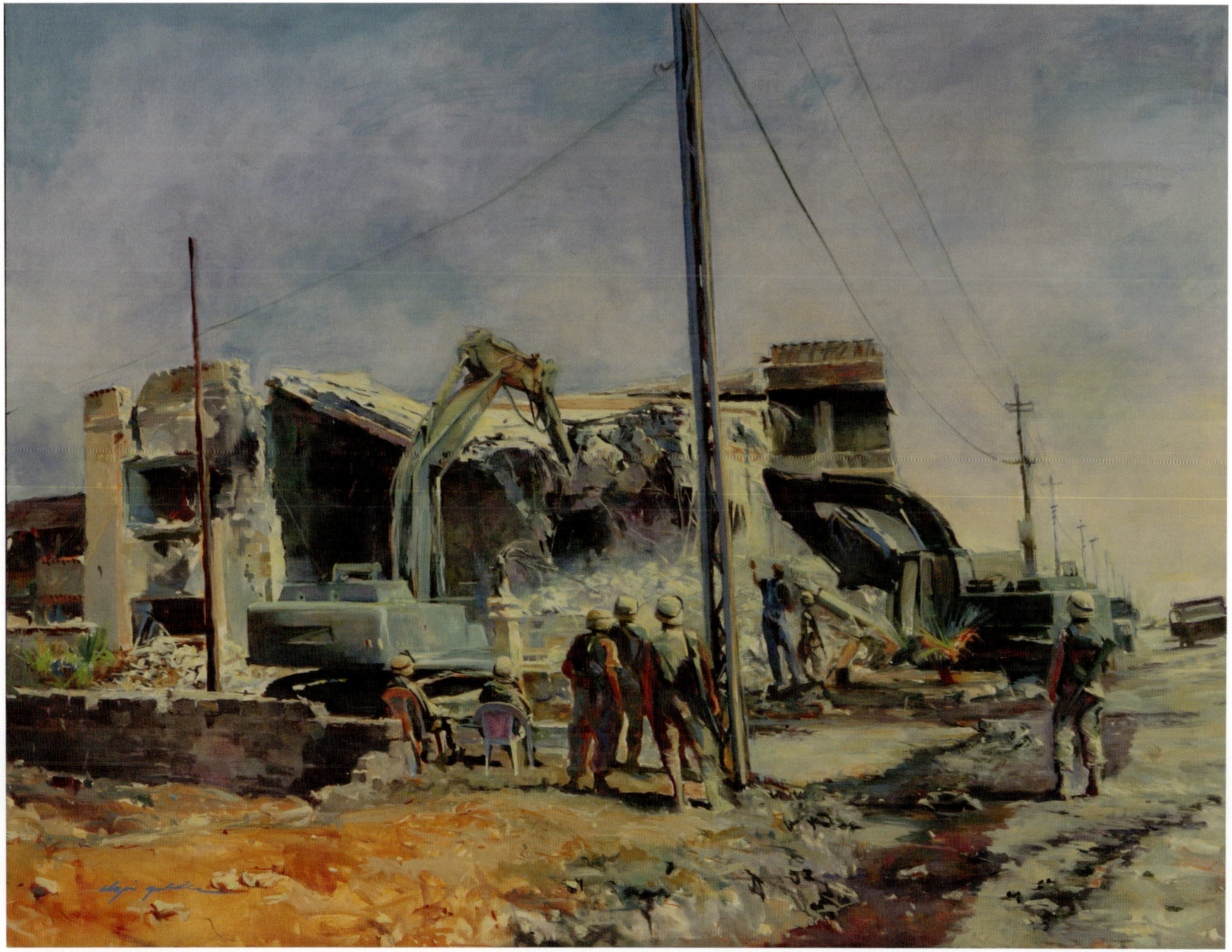

MARTYRDOM DENIED
Mosul, Iraq, Operation Iraqi Freedom
Elzie Golden, 2005
Oil on canvas, 30" x 40"

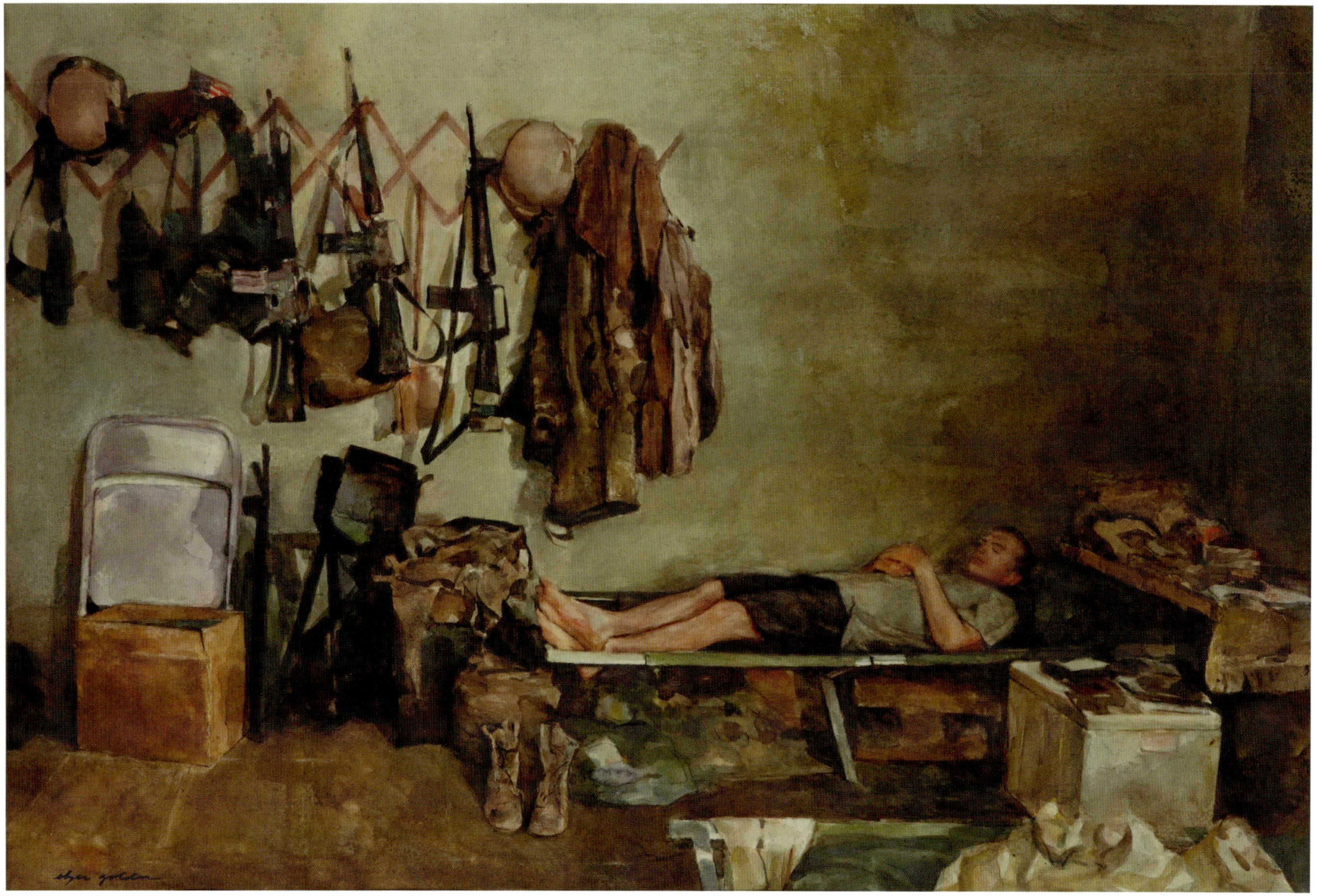

FALLUJAH
Al Fallujah, Iraq, Operation Iraqi Freedom
Elzie Golden, 2005
Watercolor on paper, 22¼" x 30"

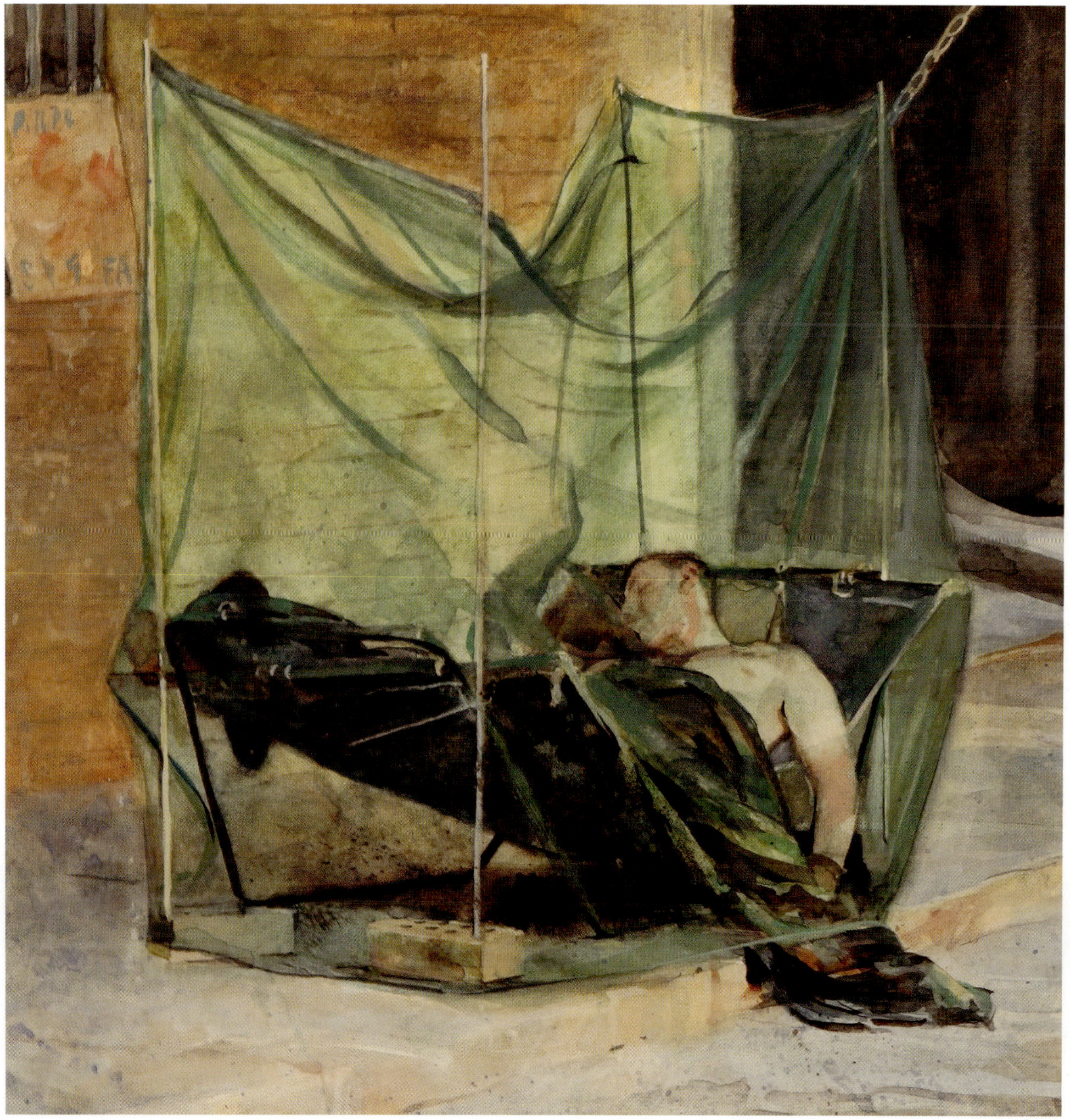

SLEEPING MECHANIC
Iraq, Operation Iraqi Freedom
Elzie Golden, 2005
Watercolor on paper, 12" x 11¼"

"550-cord, the military's duct tape, is utilized here as make-shift clothesline. The cord is tied off to two concrete barriers serving as walls of a bunker. This typical simple scene speaks volumes."
—Heather C. Englehart

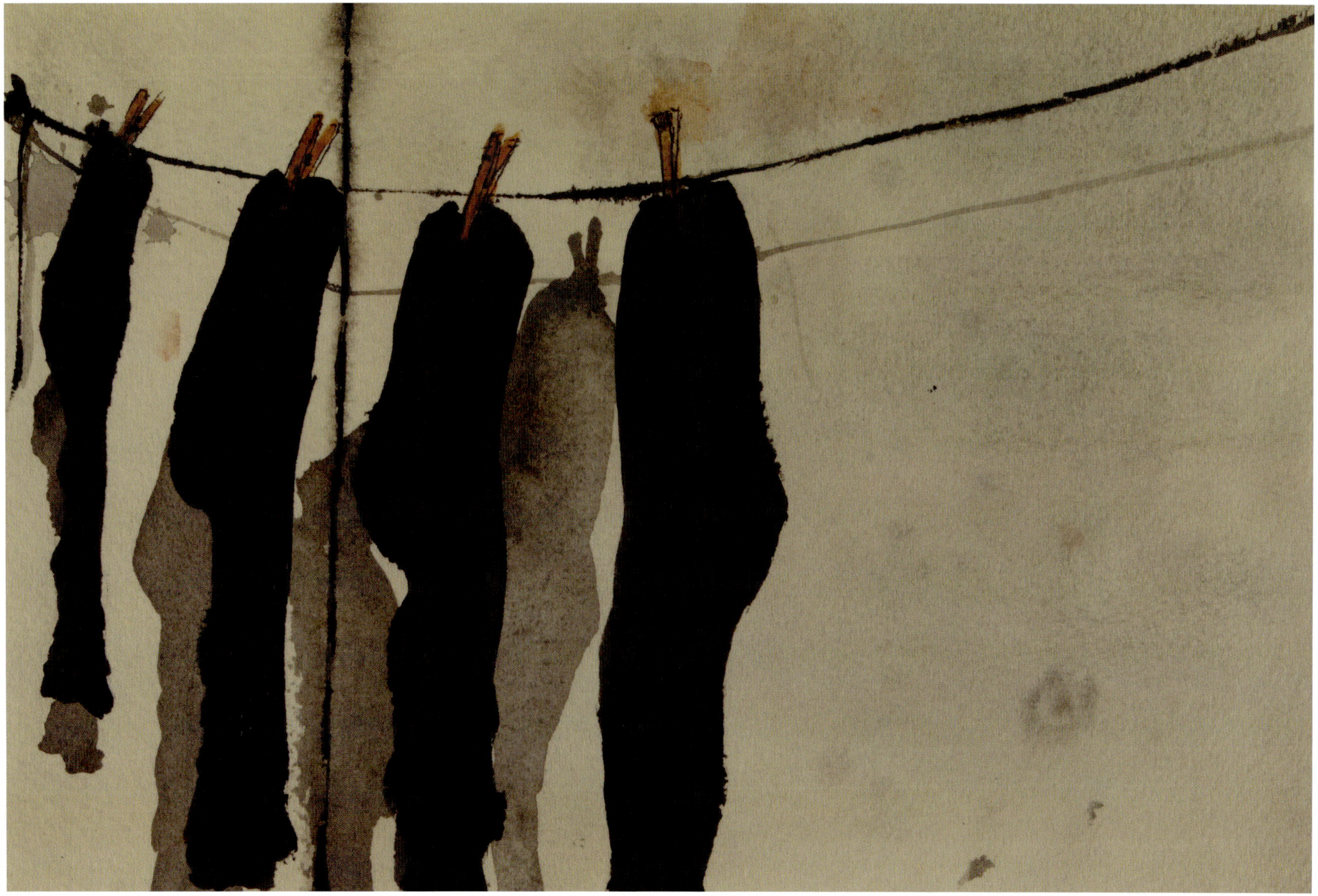

LAUNDRY DAY
Iraq, Operation Iraqi Freedom
Heather C. Englehart, 2004
Acrylic and watercolor on paper, 6" x 8¼"

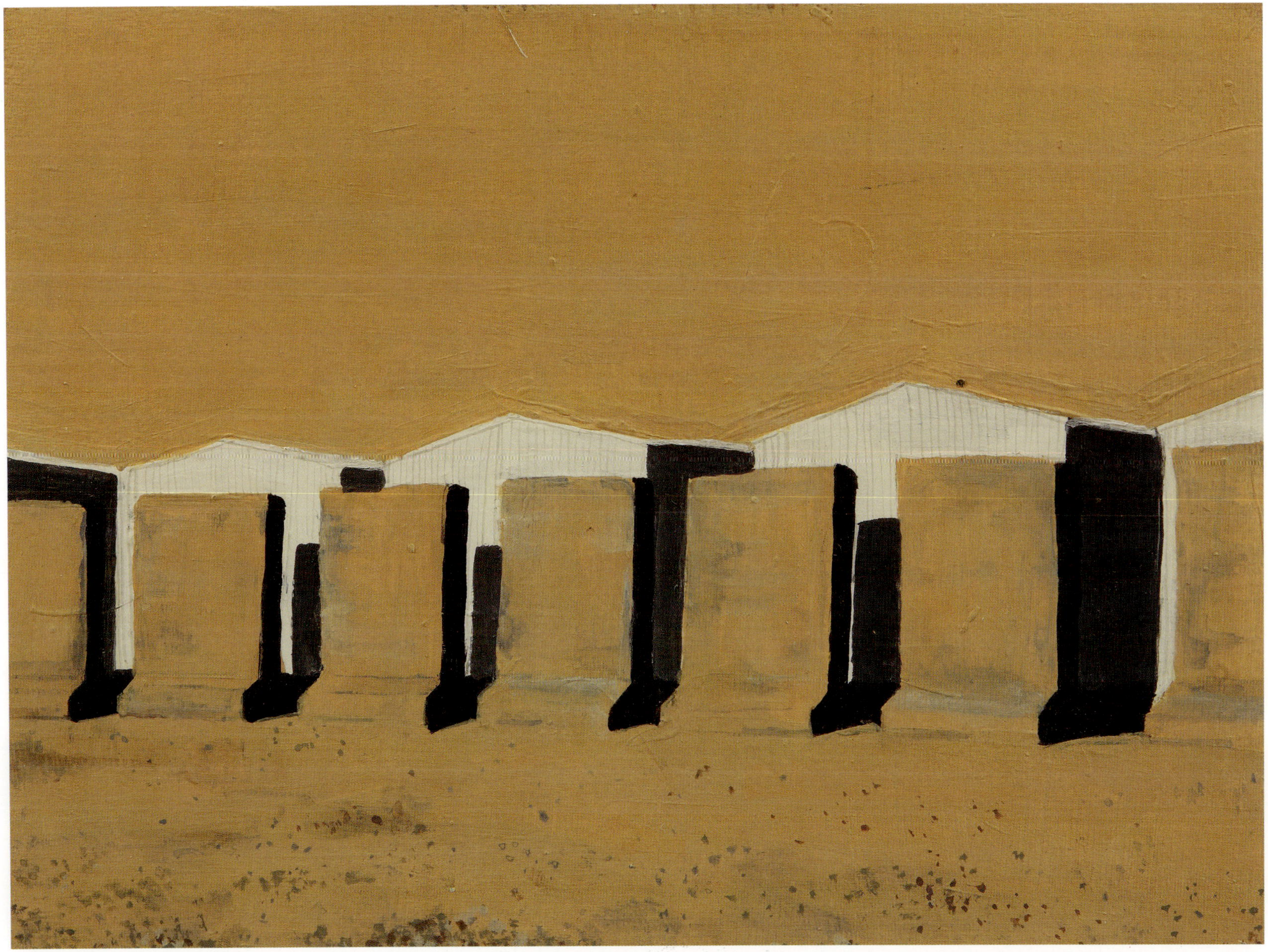

HOUSING AREA
Iraq, Operation Iraqi Freedom
Heather C. Englehart, 2004
Acrylic on paper, 9" x 11½"

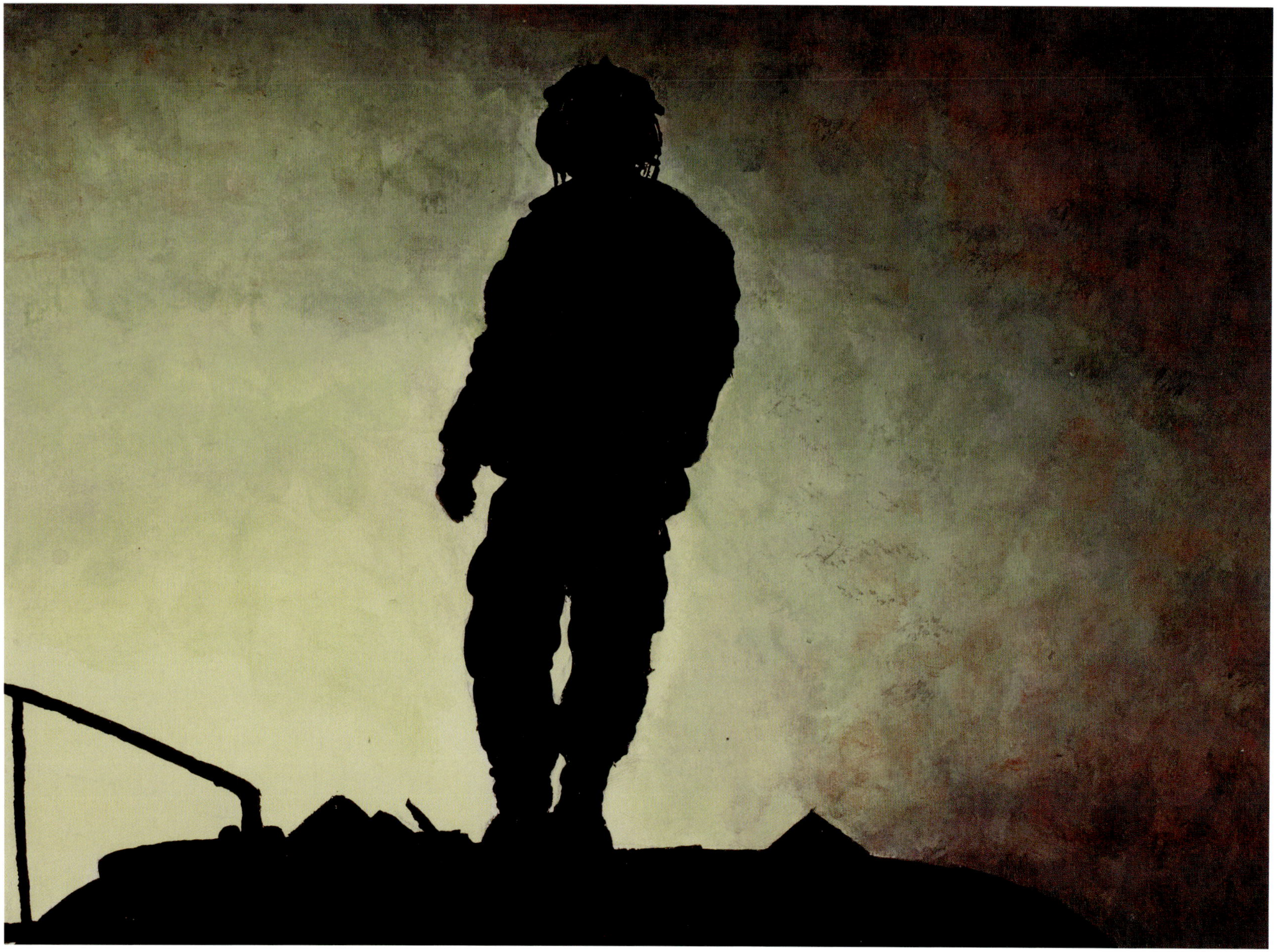

BIG COUNTRY
Camp Anaconda, Iraq, Operation Iraqi Freedom
Heather C. Englehart, 2004
Acrylic and watercolor on paper, 9" x 11½"

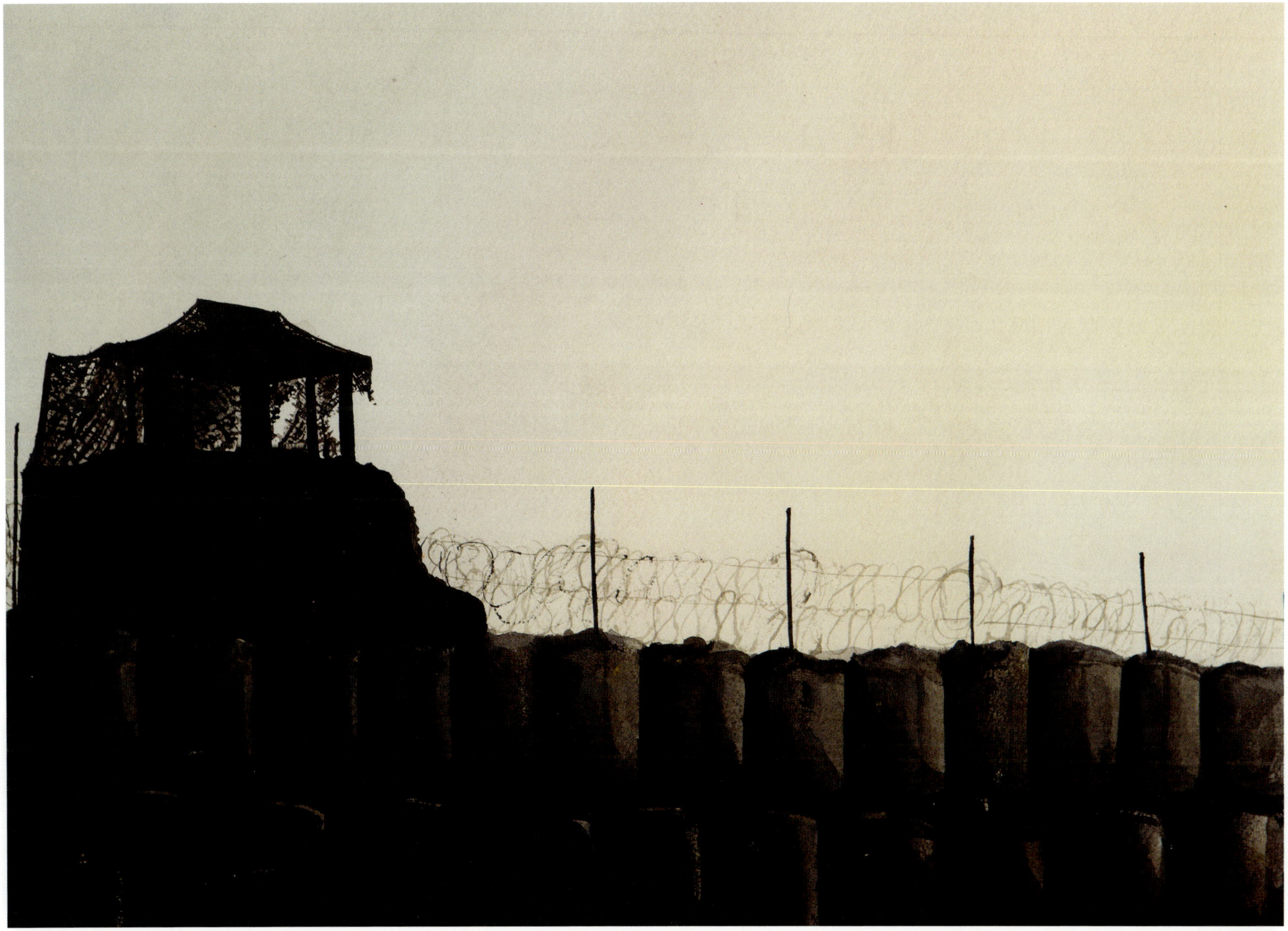

PERIMETER GUARD
Balad, Iraq, Operation Iraqi Freedom
Heather C. Englehart, 2004
Acrylic and watercolor on paper, 12¼" x 16¼"

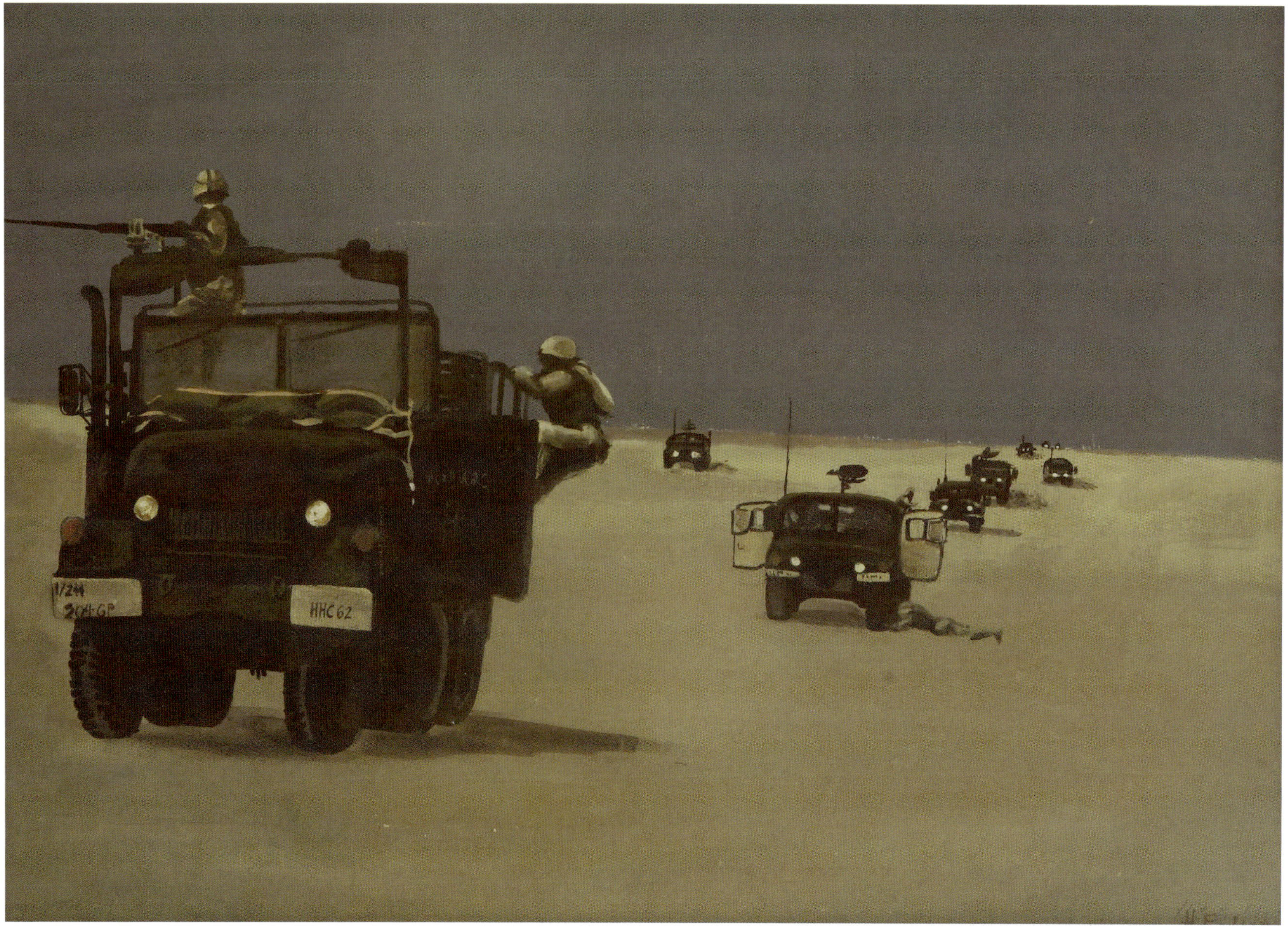

CONVOY LIVE FIRE
Udari, Kuwait, Operation Iraqi Freedom
Heather C. Englehart, 2004
Acrylic on tent canvas, 16" x 22"

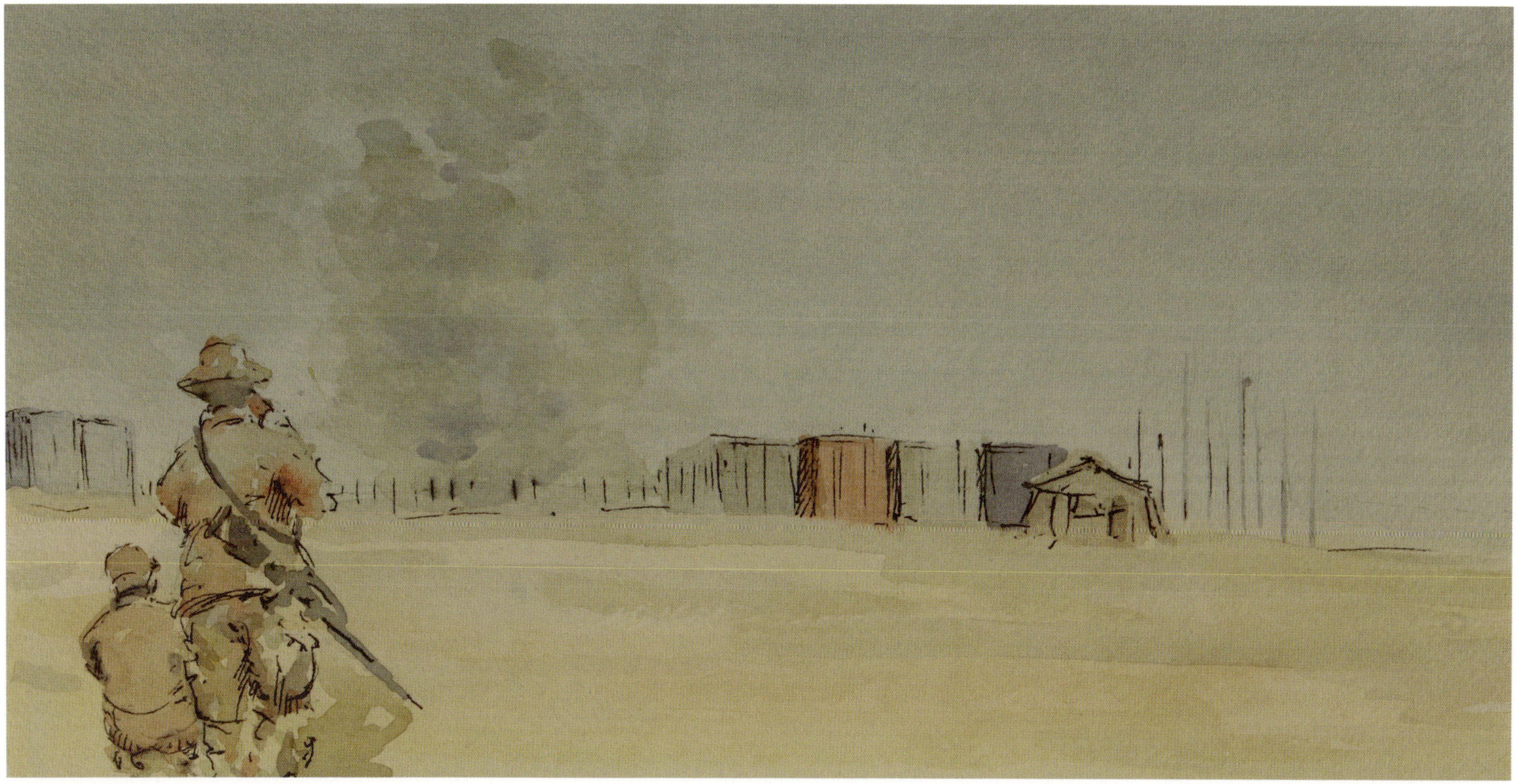

INCOMING
Camp Anaconda, Iraq, Operation Iraqi Freedom
Heather C. Englehart, 2005
Ink and watercolor on paper, 6¾" x 12"

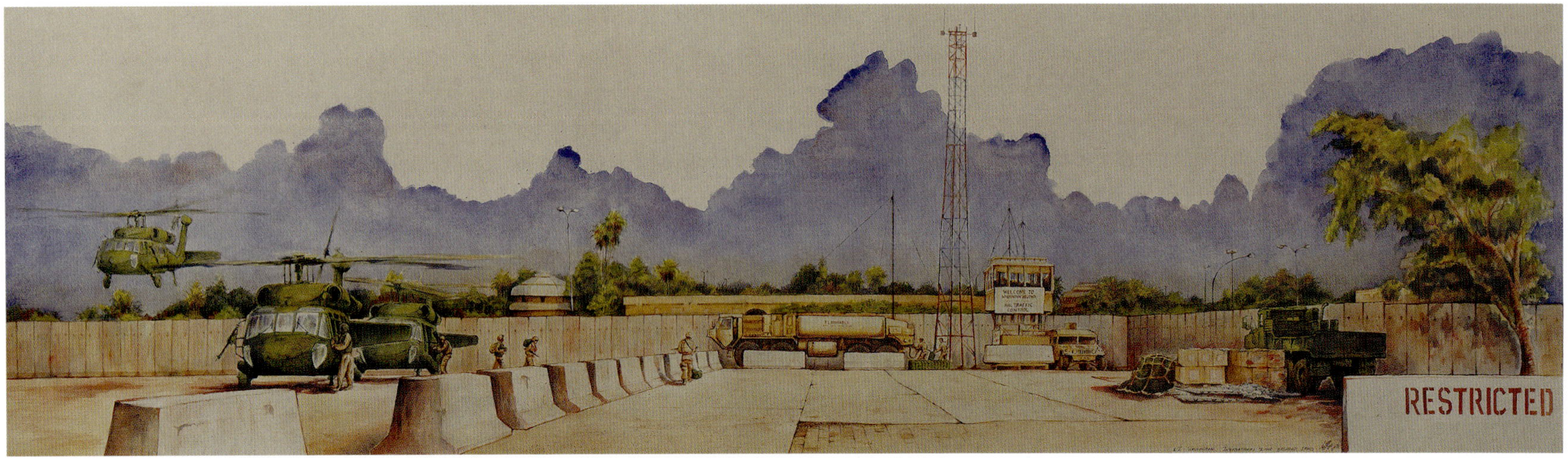

LZ WASHINGTON, U.S. EMBASSY, BAGHDAD, IRAQ
Baghdad, Iraq, Operation Iraqi Freedom
Timothy B. Lawn, 2005
Watercolor on paper, 22½" x 70"

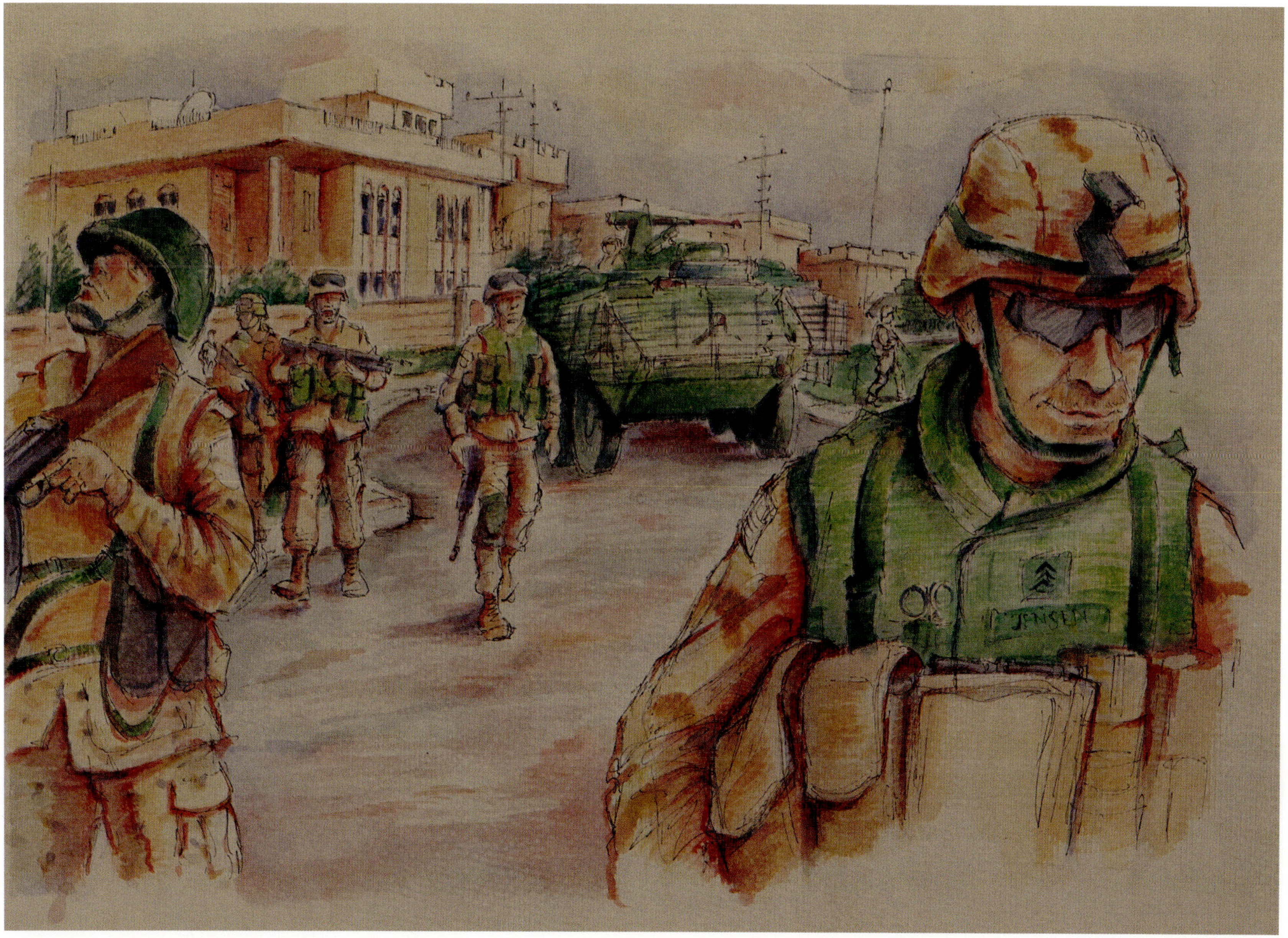

MARKET STREET PATROL, MOSUL, IRAQ
Mosul, Iraq, Operation Iraqi Freedom
Timothy B. Lawn, 2005
Ink and watercolor on paper, 10¾" x 15"

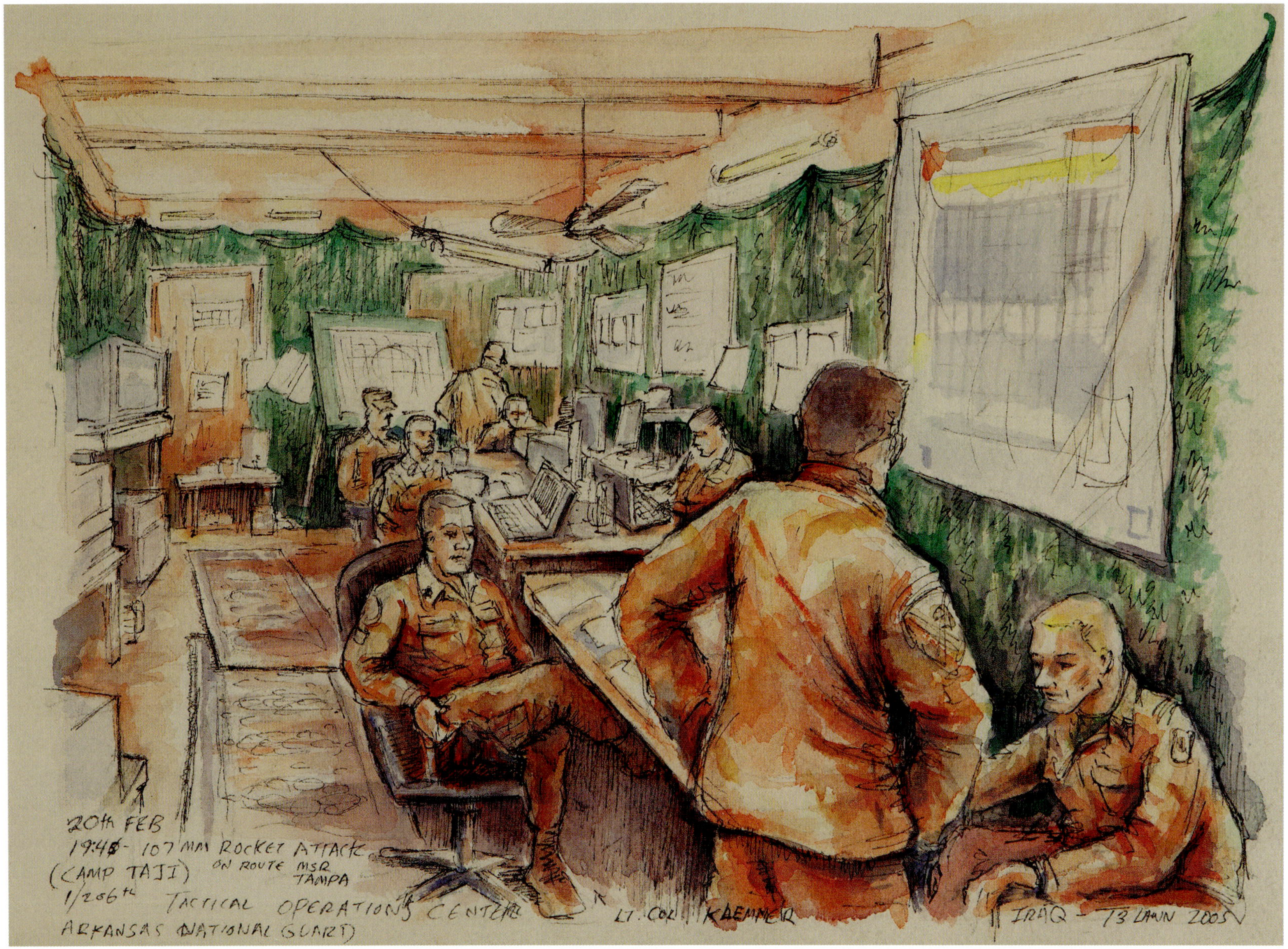

I/206TH TACTICAL OPERATIONS CENTER
Camp Cooke, At Taji, Iraq, Operation Iraqi Freedom
Timothy B. Lawn, 2005
Ink and watercolor on paper, 9" x 12"

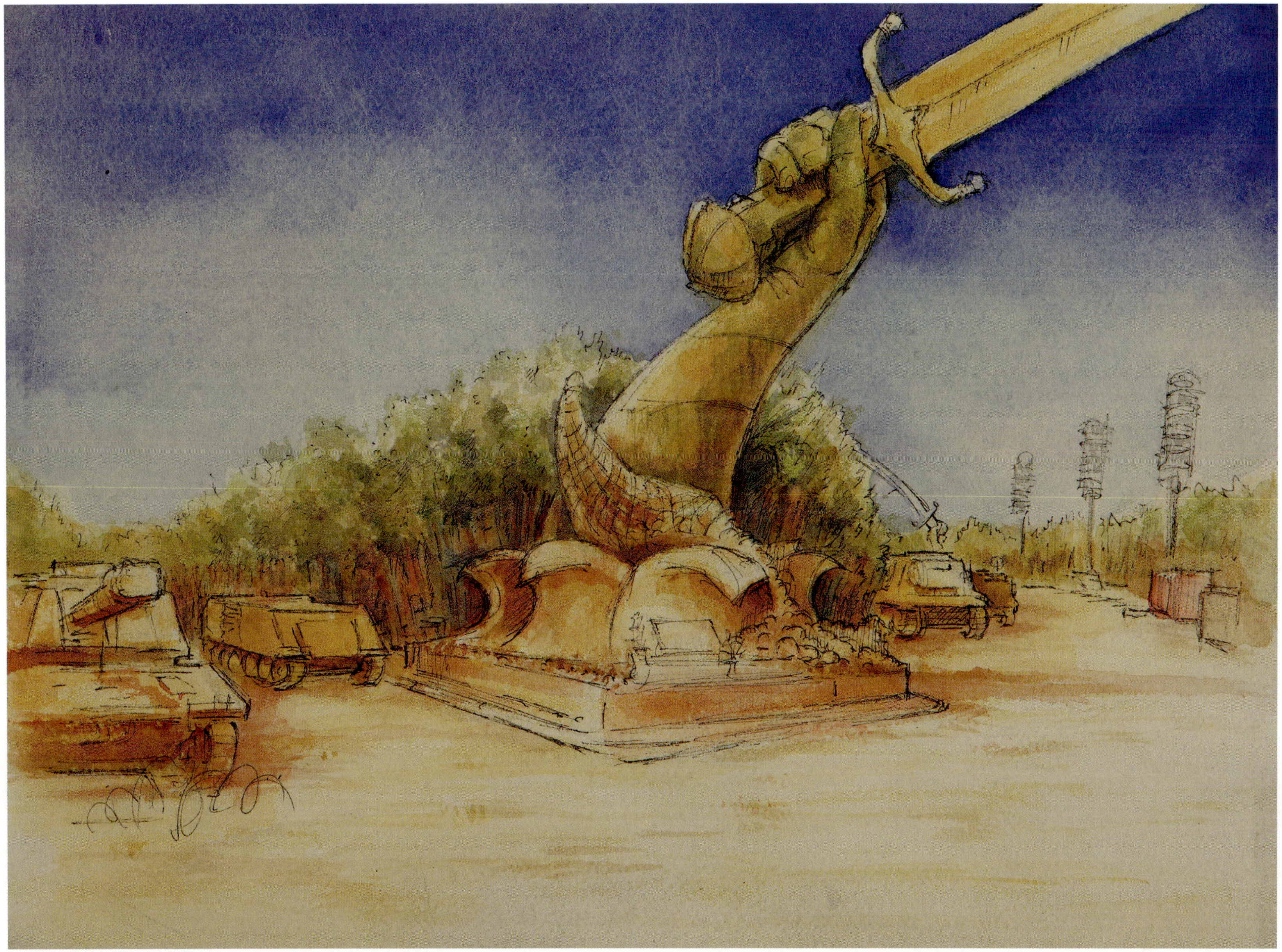

SADDAM'S SWORD
Baghdad, Iraq, Operation Iraqi Freedom
Timothy B. Lawn, 2005
Ink and watercolor on paper, 9" x 12"

Along the bottom edge, the artist wrote a quote by George Orwell: "People sleep peaceably in their beds at night only because rough men stand ready to do violence on their behalf."

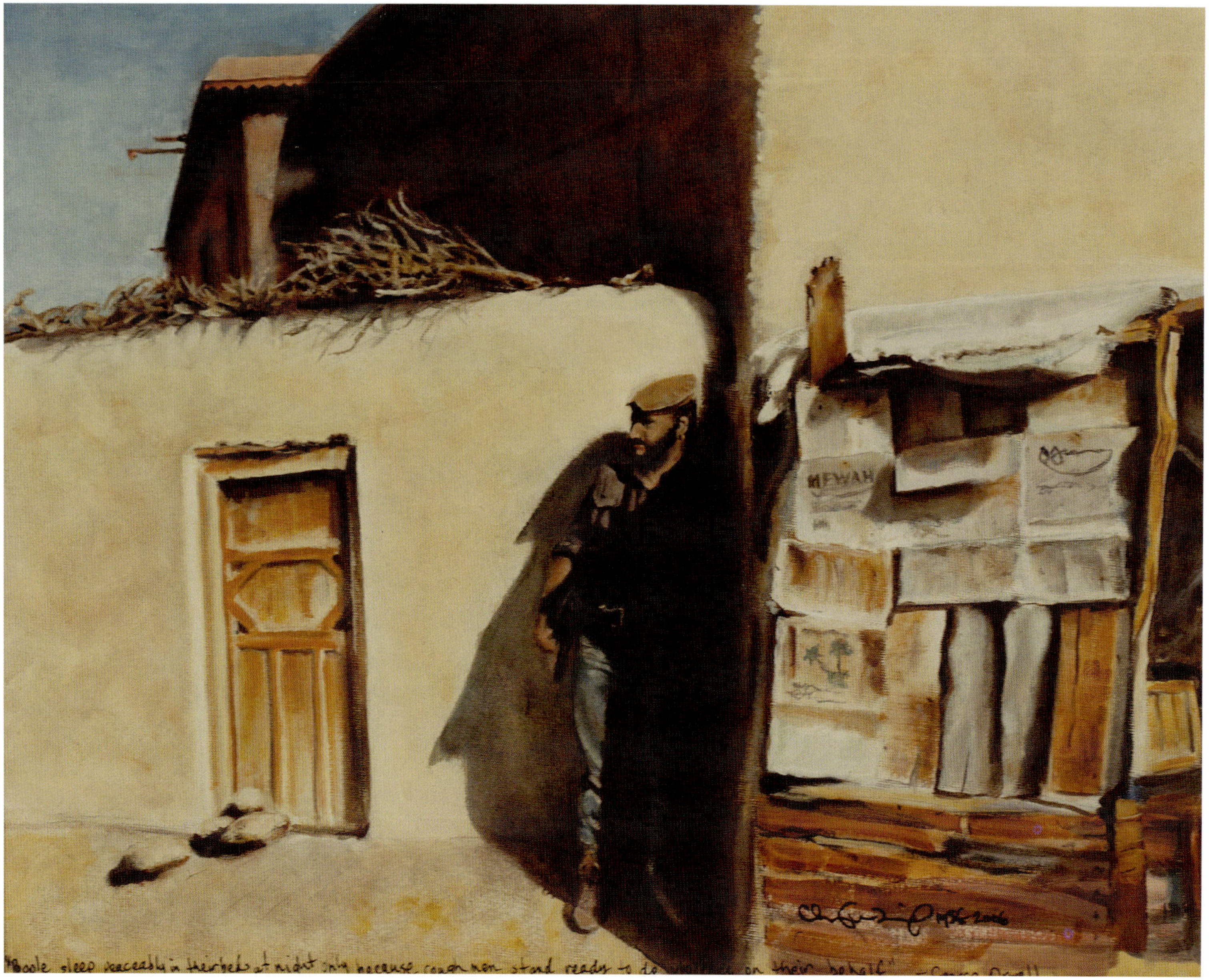

MASTERS OF CHAOS—ODA563, 5SFG(A), DECEMBER 2001
Afghanistan, Operation ENDURING FREEDOM
Christopher W. Thiel, 2006
Oil on canvas, 16" x 20"

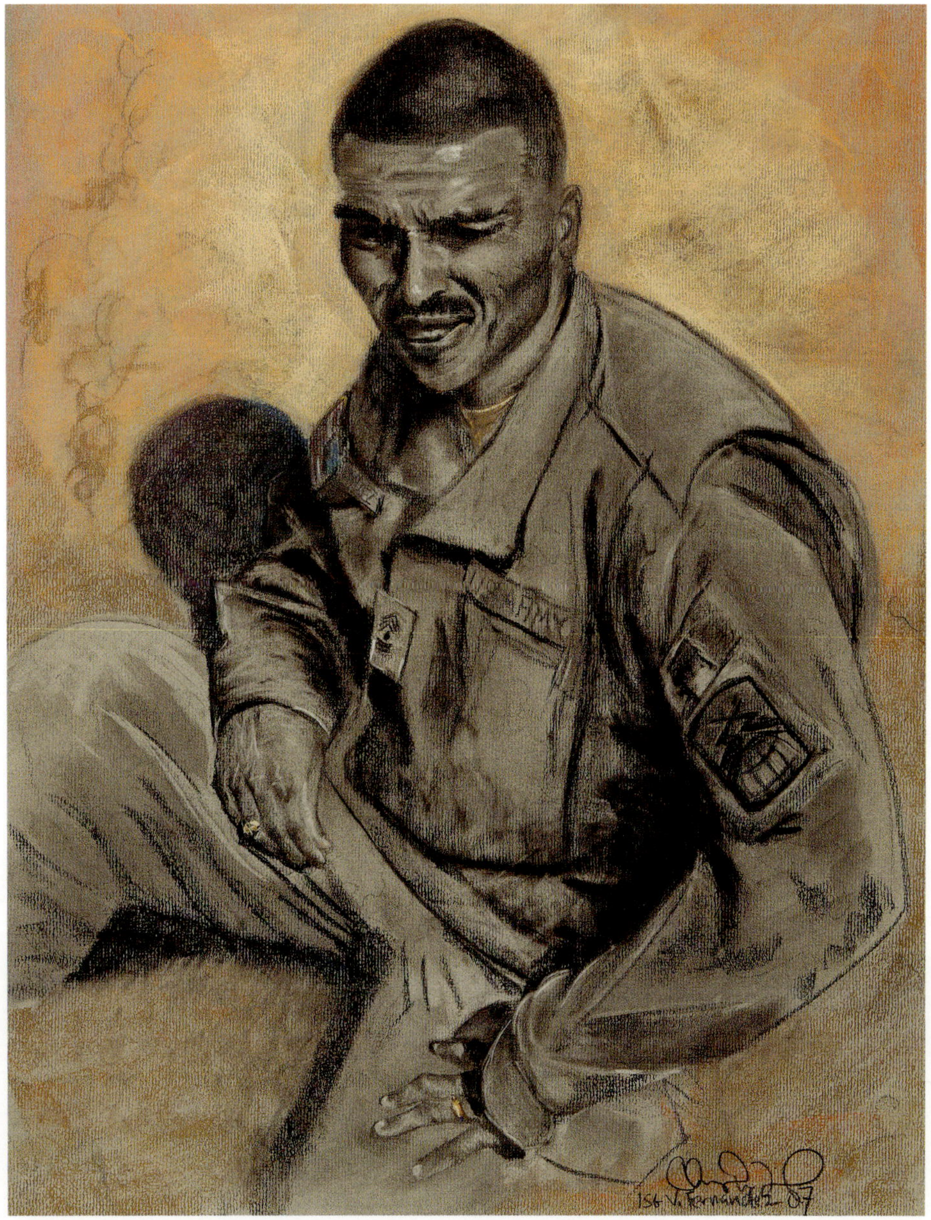

1SG VIC FERNANDEZ
Ali Al Salem, Kuwait, Operation Iraqi Freedom
Christopher W. Thiel, 2007
Charcoal and pastel on paper, 24" x 18"

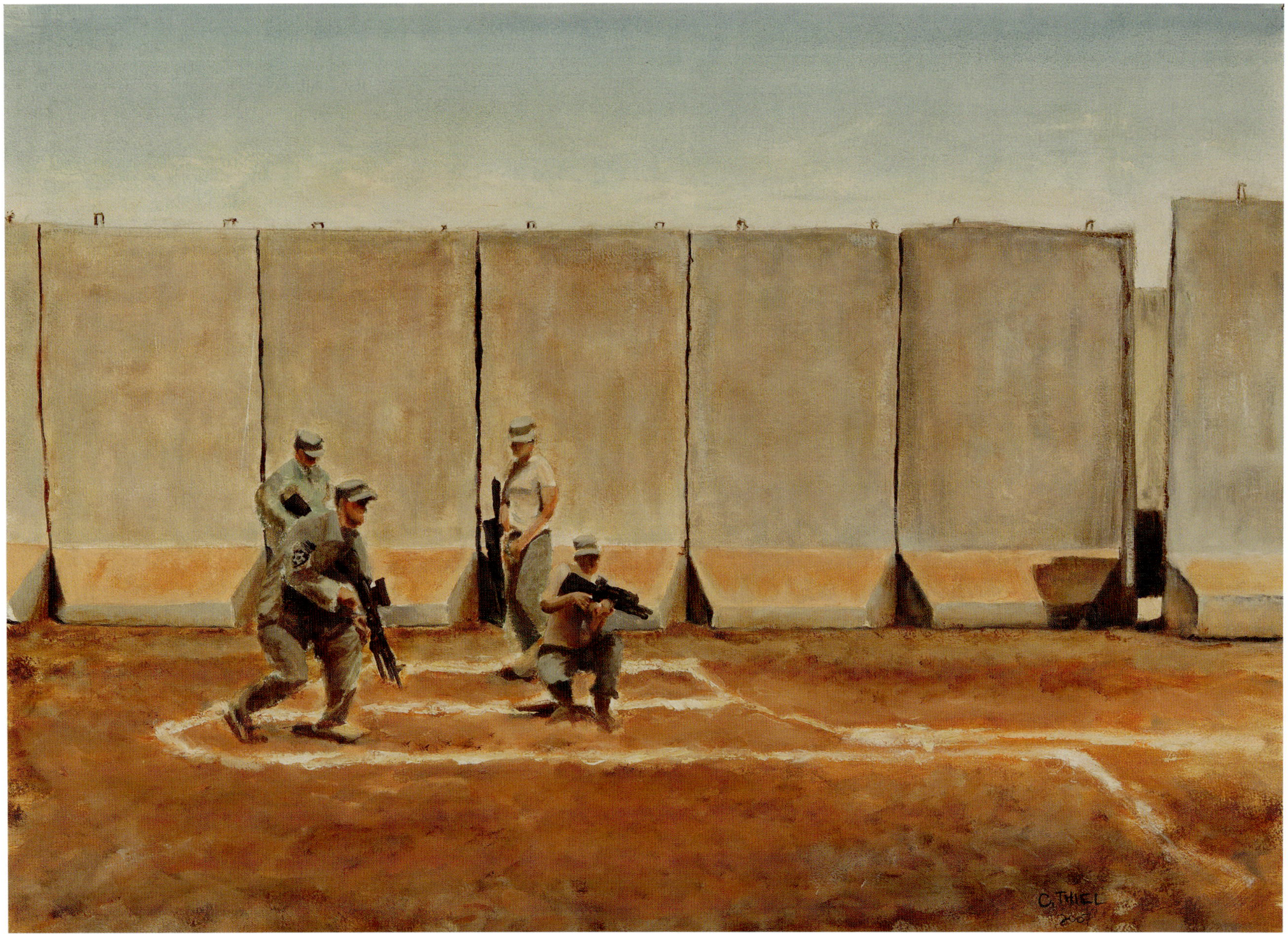

TOMAHAWK—GLASS HOUSE DRILL
Camp Liberty, Baghdad, Iraq, Operation Iraqi Freedom
Christopher W. Thiel, 2007
Oil on paper, 22¾" x 30½"

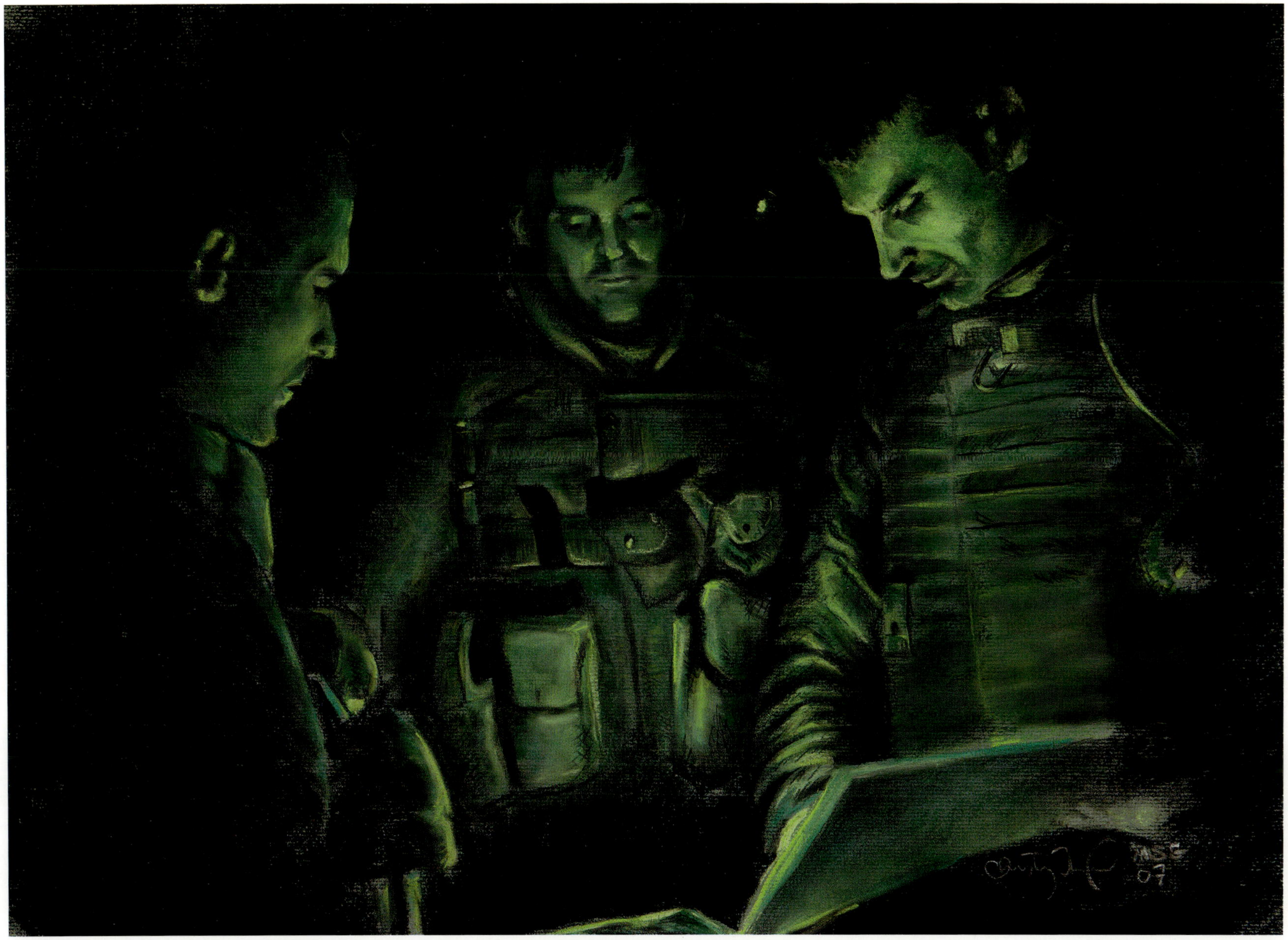

BAILEY'S PRE-COMBAT CHECKS
Tall Afar, Iraq, Operation Iraqi Freedom
Christopher W. Thiel, 2007
Conte crayon on paper, 18" x 24"

"My goal during this process is to capture what soldiers experience and how it affects them and the community. It's a thrill for me to have a soldier see a piece of art and say that's who I was and that's what I did."
—Martin J. Cervantez

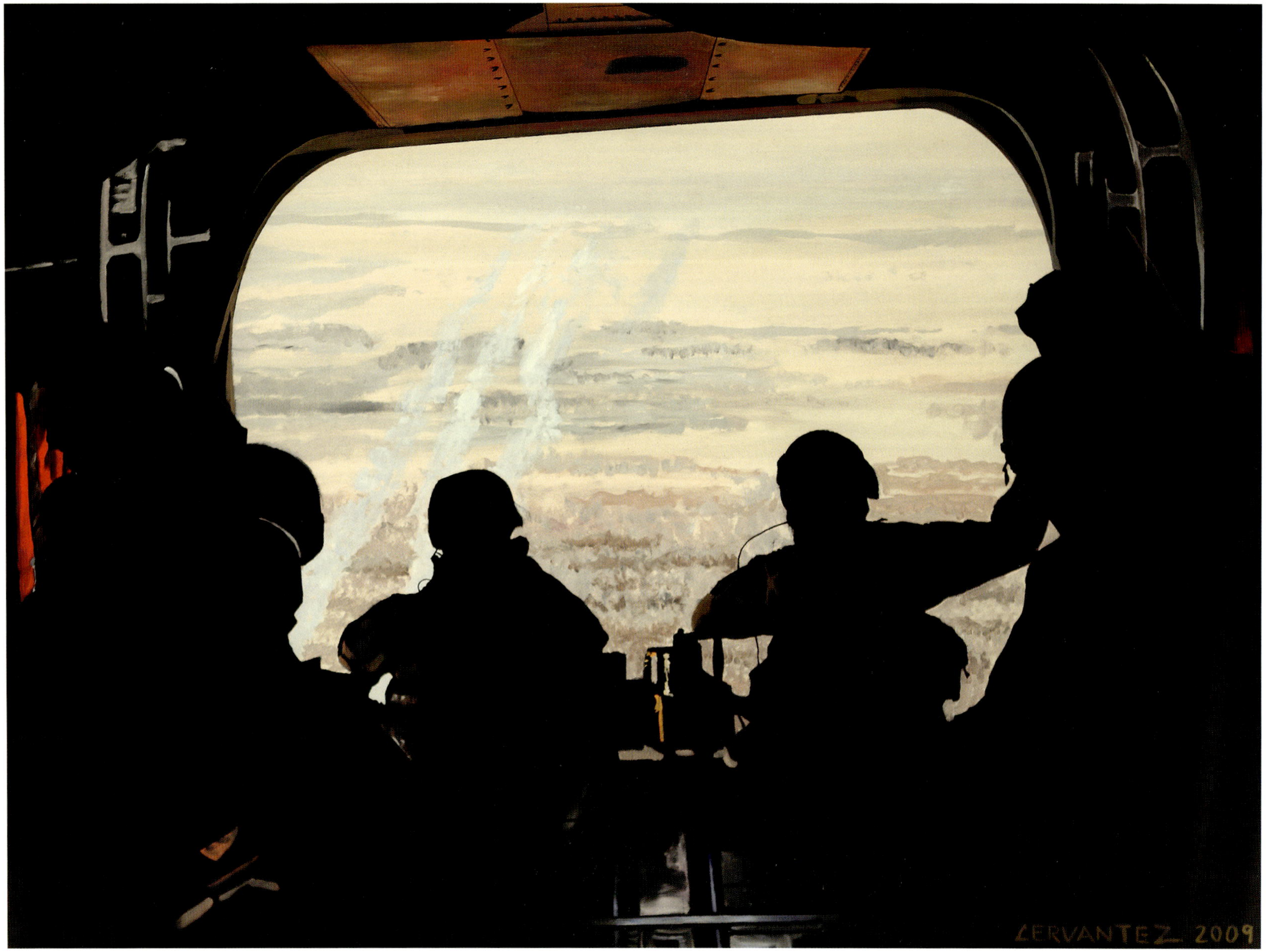

TAILGATING OVER THE VALLEY
Afghanistan, Operation Enduring Freedom
Martin J. Cervantez, 2009
Oil on canvas, 36" x 48"

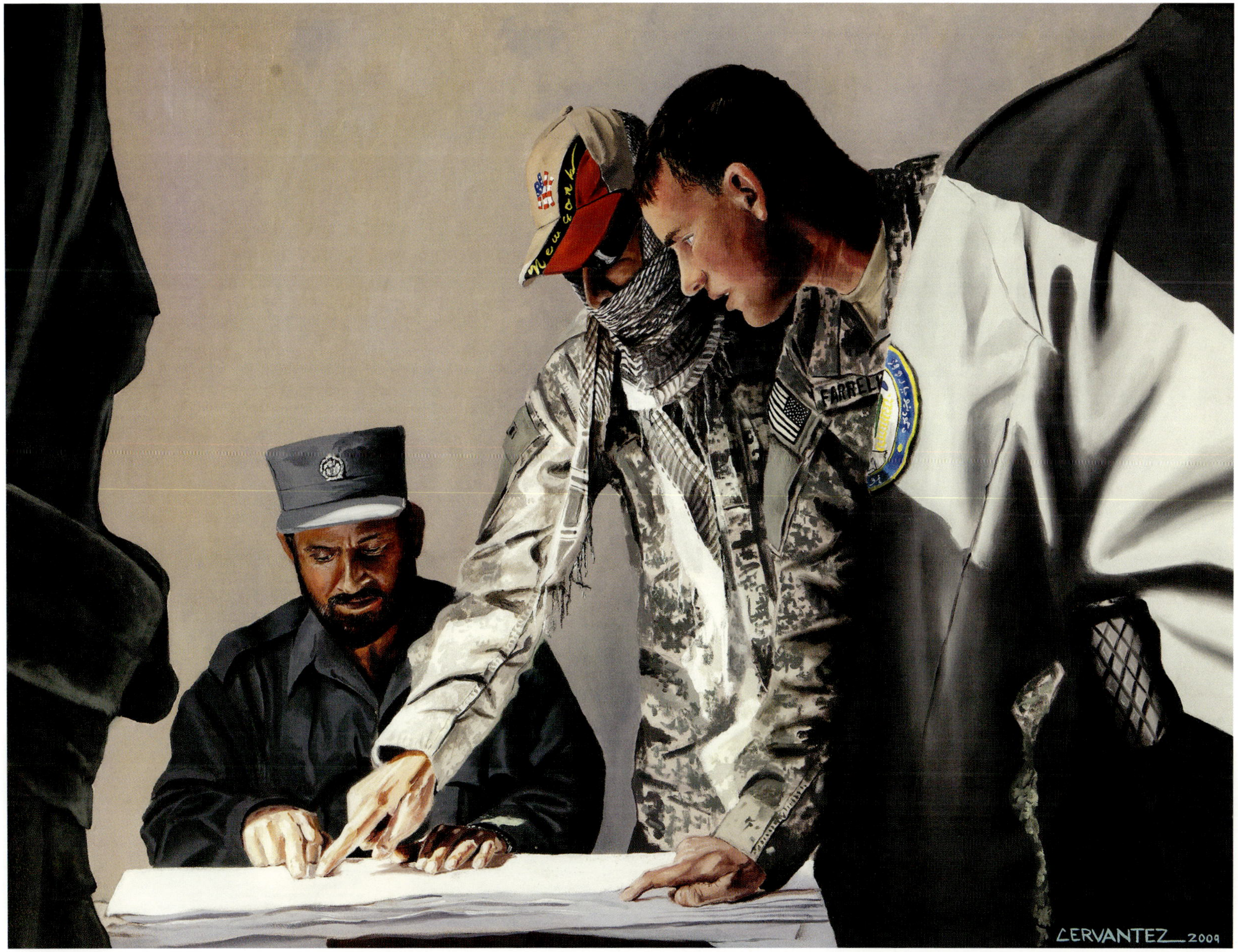

A HUGE RESPONSIBILITY
Khost, Afghanistan, Operation Enduring Freedom
Martin J. Cervantez, 2009
Oil on canvas, 30" x 40"

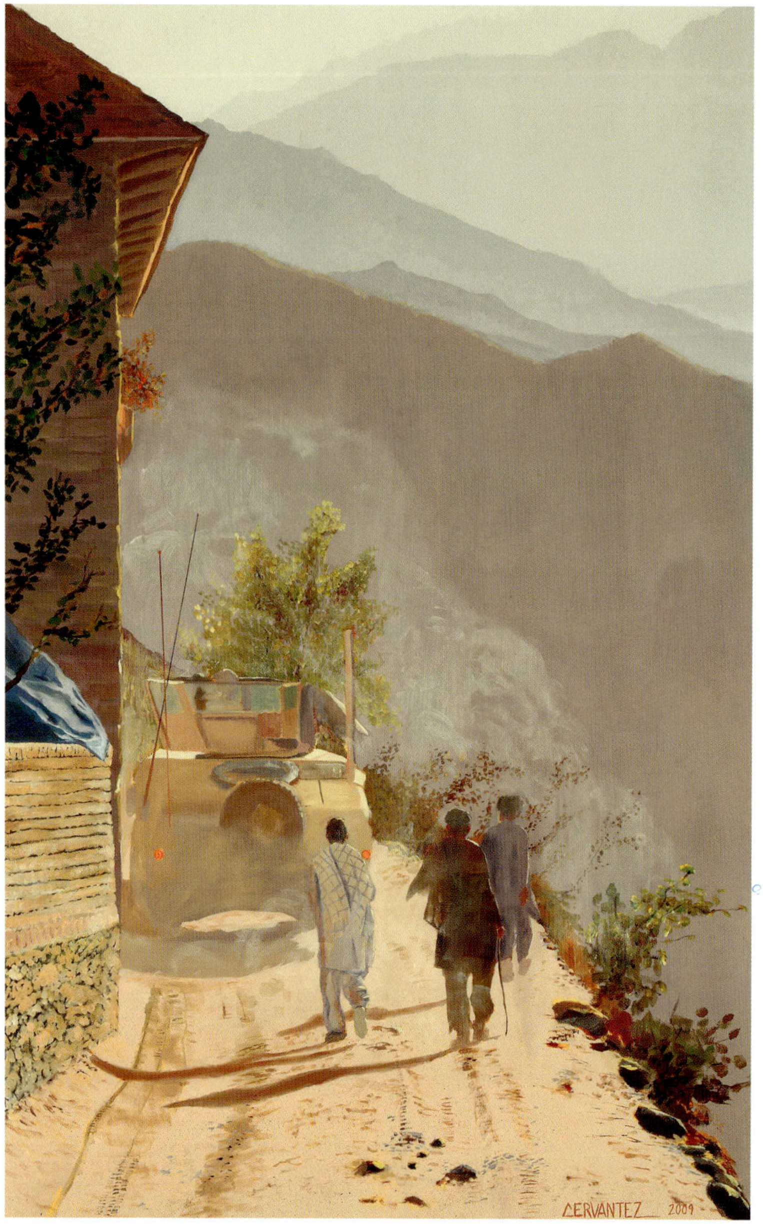

TREACHEROUS CORNER
Afghanistan, Operation Enduring Freedom
Martin J. Cervantez, 2009
Oil on canvas, 48" x 30"

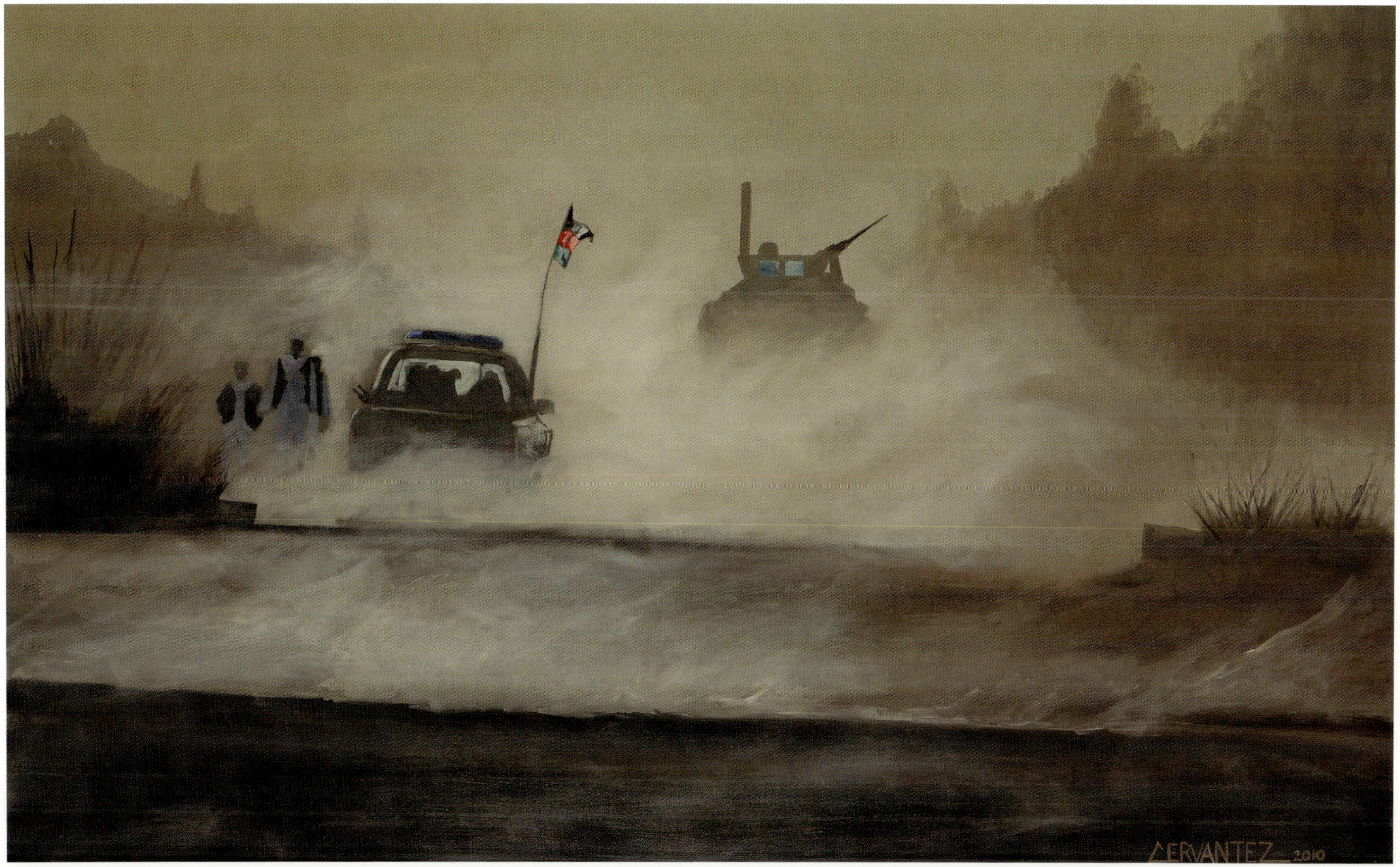

DUST DEVILS
Road near Forward Operating Base Salerno, Afghanistan, Operation Enduring Freedom
Martin J. Cervantez, 2010
Oil on canvas, 30¼" x 48"

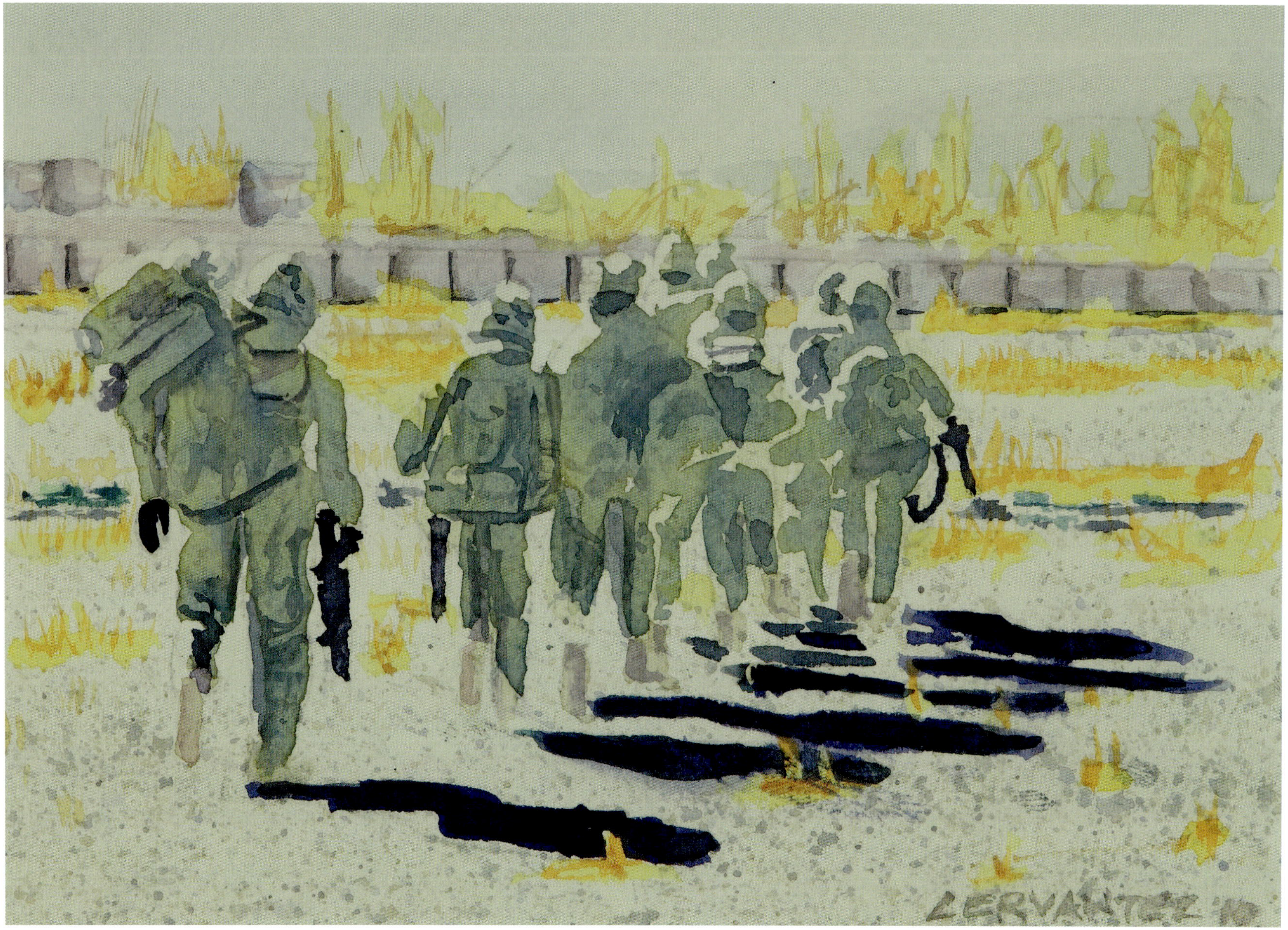

HEADING OUT
Afghanistan, Operation Enduring Freedom
Martin J. Cervantez, 2010
Watercolor on paper, 5" x 7"

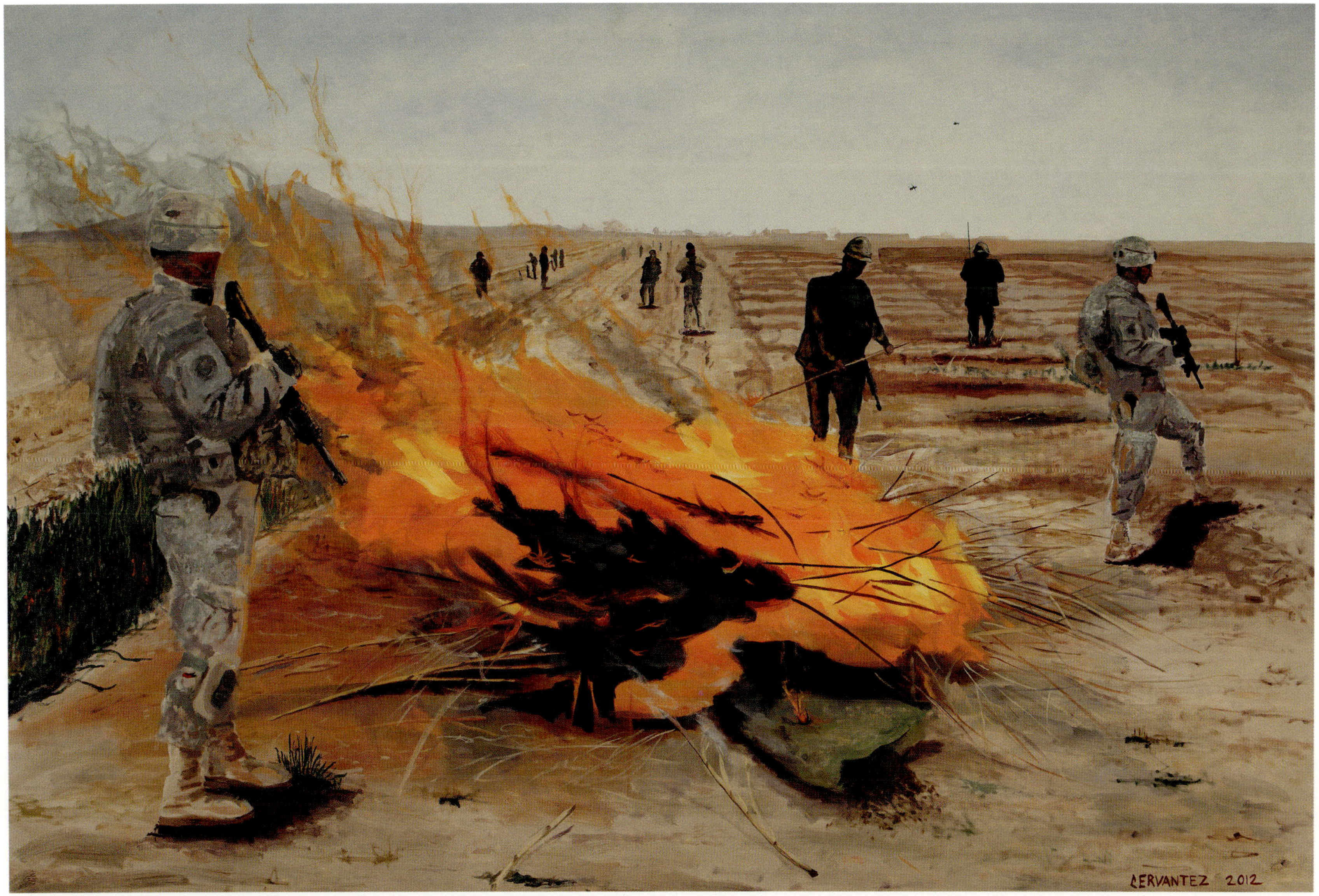

BURNING CONTRABAND
Afghanistan, Operation Enduring Freedom
Martin J. Cervantez, 2012
Oil on canvas, 48" x 60"

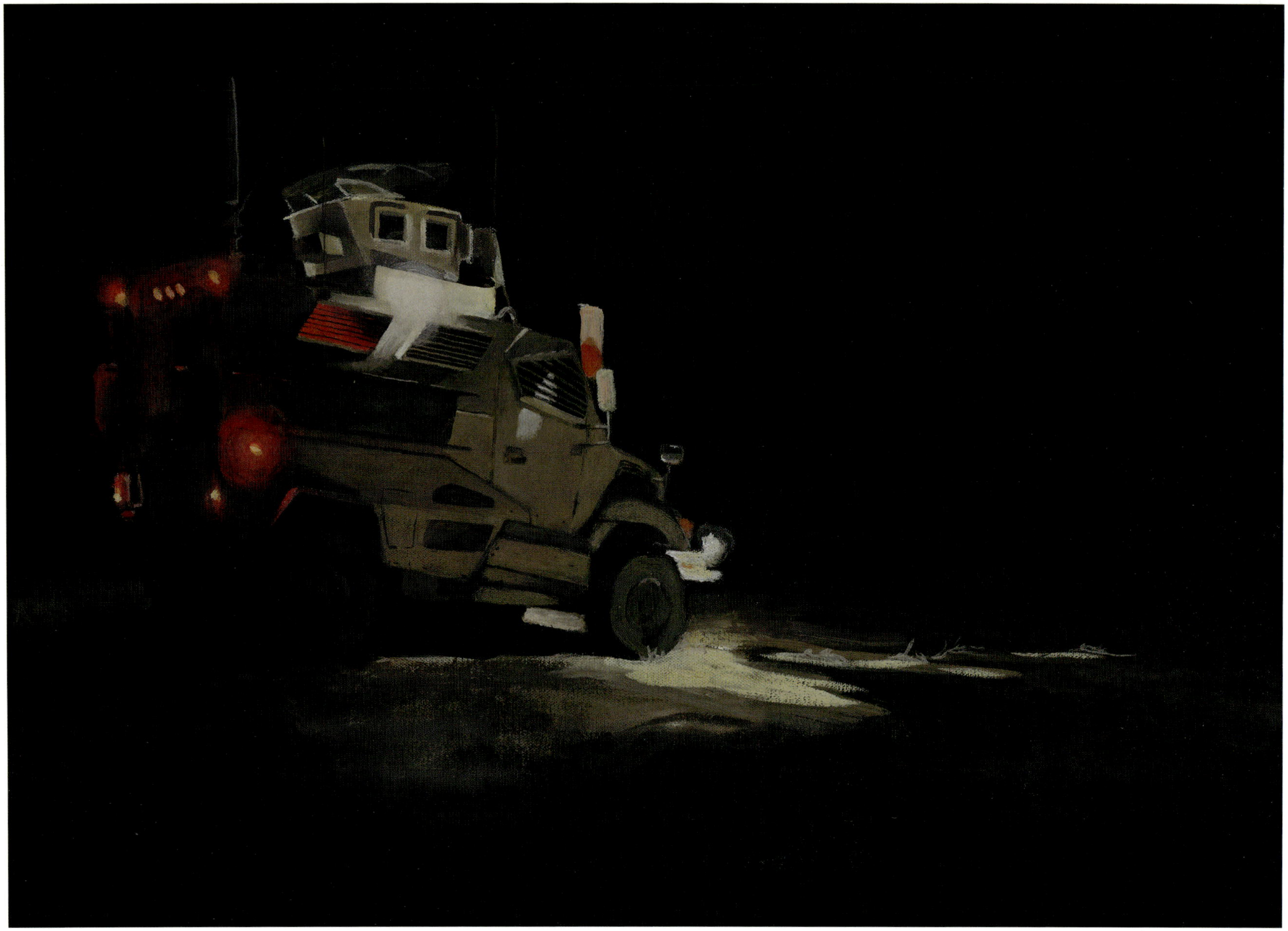

NO MAN'S LAND
Afghanistan
Amy Brown, 2014
Acrylic and pastel on canvas, 20" x 24"

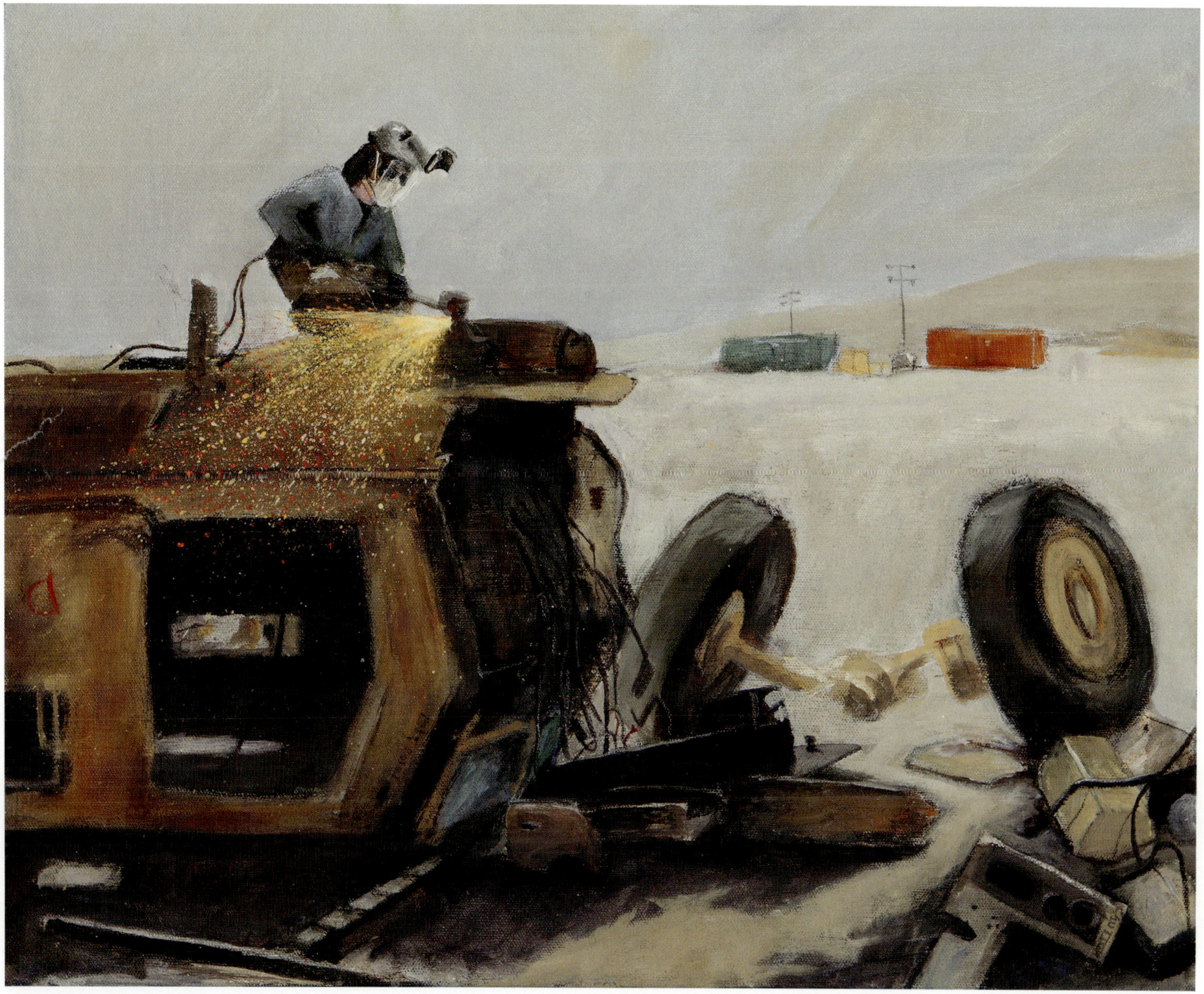

"I entered the . . . yard to sketch contracted workers dismantling damaged mine-resistant, ambush-protected (MRAP) vehicles for scrap metal. I attempted to make conversation with the workers about their mission and it became quite obvious that our communications would be limited due to a language barrier. The one thing I heard repeatedly as I walked through the yard or sketched was the repeated phrase, 'too much chop chop.'"

—Amy Brown

TOO MUCH CHOP CHOP
Afghanistan
Amy Brown, 2014
Acrylic and pastel on canvas, 16" x 20"

"When you deploy, you feel like you have prepared yourself for every scenario. I took a few tubes of watercolors, a couple of jars of neutral colored acrylics and, of course, a lot of pens and pencils. Unfortunately, I did not return with many pens. Most just could not hold up to the 100 degree plus heat and sitting in the sun for extended periods of time on the flight line. Almost every pen I brought over exploded from the heat."
—Amy Brown

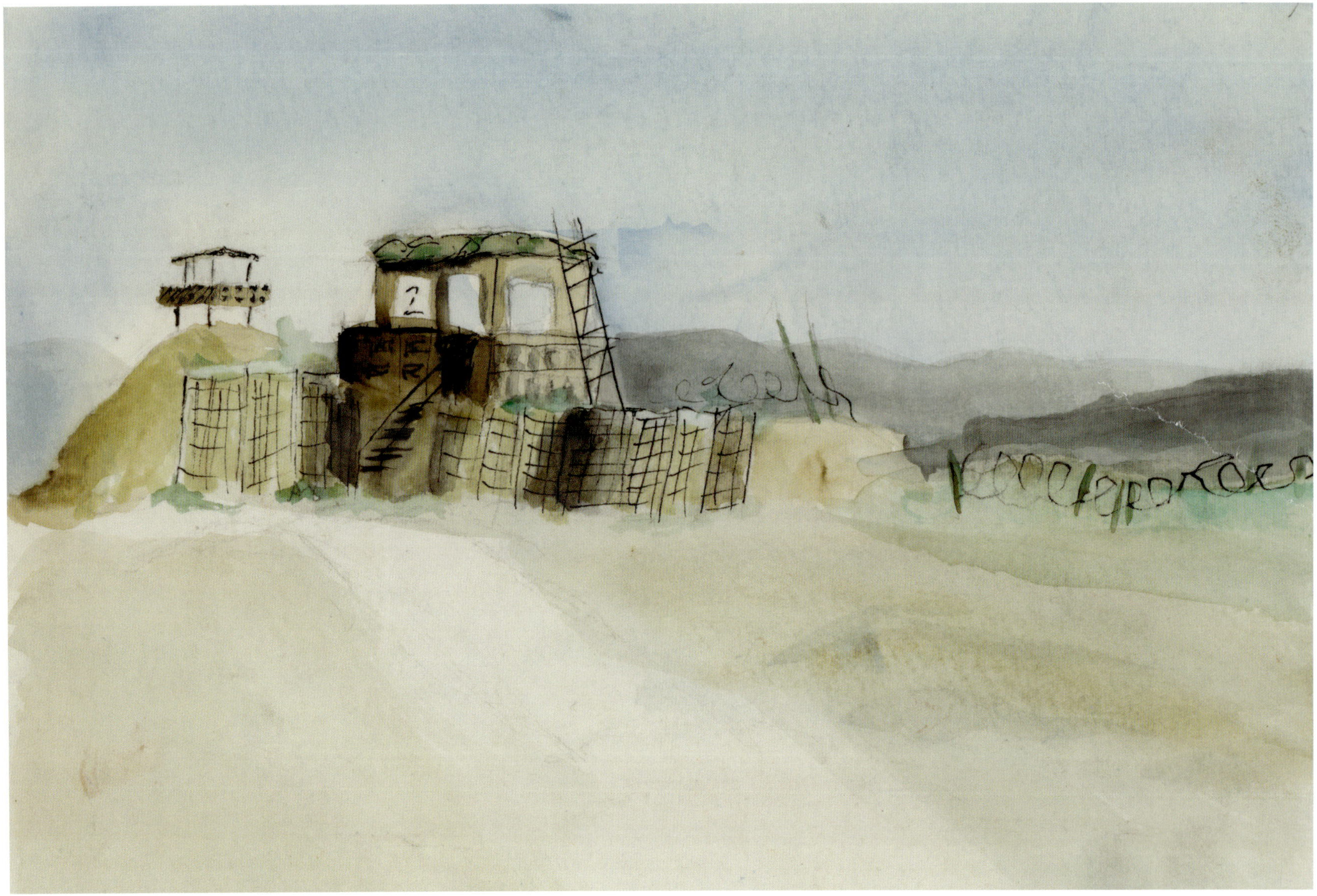

FOB WOLVERINE: GUARD TOWER #2
Afghanistan
Amy Brown, 2013
Pen and watercolor on paper, 7" x 9"

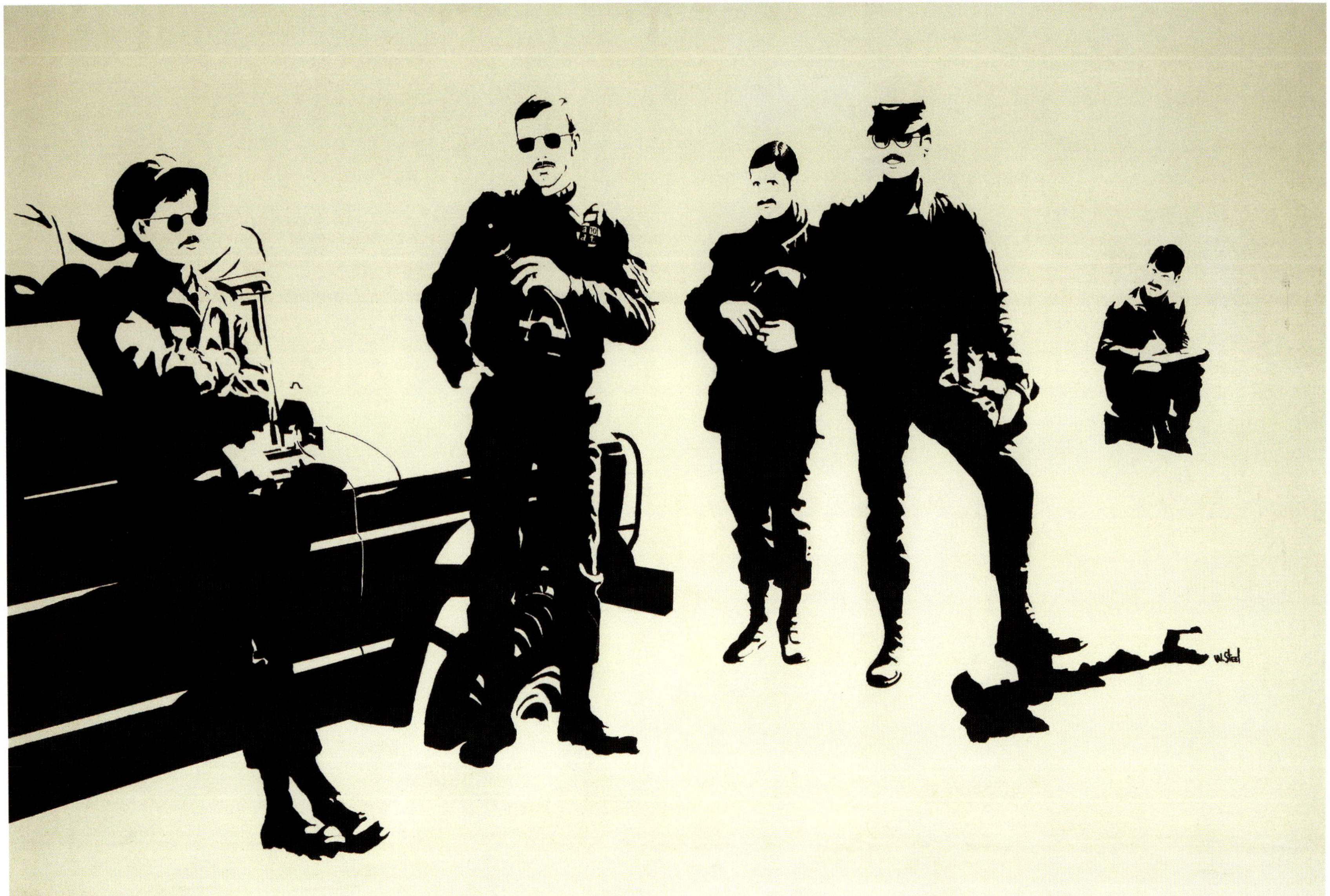

ARMY ARTIST TEAM X
Korea
William H. Steel, 1970
Acrylic on canvas, 32" x 48"

The Artists

Theodore J. Abraham was born in 1923. He documented Army activities in Vietnam in 1967 and in Alaska in 1970 as part of the Volunteer Civilian Artist Program. **(24, 25)**

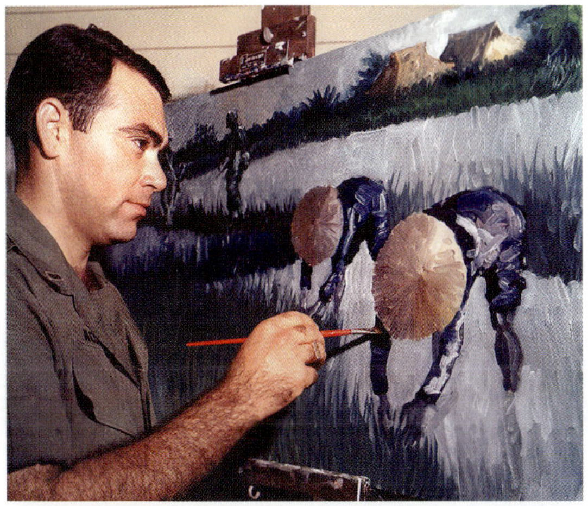

2d Lt. Augustine G. Acuña

Capt. Mario H. Acevedo was born in El Paso, Texas, in 1955 and studied at New Mexico State University. He was a captain in the U.S. Army Reserve when called to active duty and served six months as a combat artist in Operation DESERT STORM. **(107, 108, 109, 110, 111)**

2d Lt. Augustine G. Acuña, born in 1941, served as team leader of Combat Artist Team II in Vietnam from October 1966 to February 1967. He studied architecture at the University of Arizona. **(3, 4)**

Pfc. Samuel E. Alexander was born in 1943. He received his bachelor of fine arts from the Ringling School of Art in Sarasota, Florida, in 1966. At the time of his selection to Combat Artist Team IV, he was assigned to the Information Office, Headquarters Company, U.S. Army, Vietnam. He served as an artist in Vietnam from August to December 1967. **(26, 27)**

S. Sgt. Emilio E. Arias was selected as a member of Army Artist Team XXIX, recording Army activities in Korea from February to May 1992. He also served as an instructor at the Defense Information School at Fort Meade, Maryland. **(116, 117)**

S. Sgt. George Banagis served in the U.S. Navy from 1968 to 1972, and during that time one of his duties was art director on the USS *Midway*. He joined the U.S. Army in 1981 as an illustrator. He was selected for Army Artist Team XXX to document Army activities in Panama and Honduras from September to December 1992. **(121, 122)**

Numbers in parentheses indicate pages where an artist's work appears.

George Biddle was born in 1885 in Philadelphia, Pennsylvania. He studied law at Harvard University and art at the Pennsylvania Academy of Fine Arts and the Académie Julian in Paris. He was chairman of the War Art Advisory Committee, which helped the Army select the artists who would paint a record of the war. He reached North Africa just as the Tunisian Campaign was coming to an end. About ten days after the first attack, he went to Sicily and covered the Sicilian Campaign. He was then assigned to an American cruiser that landed English commando groups for the attack on Italy. **(x)**

Sp4c. Roger Blum, born in 1941, was a member of Combat Artist Team I in Vietnam from August to December 1966. After his service, he earned a master's degree in illustration from San Jose State University. **(xi, 5, 6, 7, 8)**

Sfc. Amy Brown was an art teacher in the North Carolina public school system when she witnessed the events of 11 September 2001 and enlisted in the Army as a multimedia illustrator. At the time of this publication, she is assigned as Artist-in-Residence. She deployed to New York and New Jersey in the fall of 2012 to document relief efforts after Hurricane Sandy and to record the Army in Afghanistan in the summer of 2013. Her work won first and second places in the illustration category and third place in the fine arts category in the 2013 Military Graphic Artist of the Year competition. **(184, 185, 186)**

2d Lt. Mary Colleen Cain was assigned to Fort Story, Virginia, when she was selected for Army Artist Team XXIV to document the training of the Wisconsin Army National Guard in 1989. She has a fine arts degree from St. Mary's College, Notre Dame. **(89)**

M. Sgt. Martin J. Cervantez is a self-taught artist from Michigan. He was assigned as Artist-in-Residence shortly after graduating from the Sergeants Major Academy. Between June 2008 and June 2012, he deployed twice to Afghanistan in support of Operation ENDURING FREEDOM and once to Haiti to document relief efforts after the 2010 earthquake. In addition to his work as Artist-in-Residence, he has participated in the Combat Paper Project and the Joe Bonham Project to record wounded warriors. **(178, 179, 180, 181, 182, 183)**

Sp5c. Brian H. Clark entered the Army in 1966 as an illustrator and was assigned to psychological operations at Fort Bragg, North Carolina. He was a member of Combat Artist Team VII in Vietnam from August to December 1968. He earned a bachelor of fine arts in 1965 from Wayne State University in Detroit, Michigan. He subsequently studied at the Académie de la Grande Chaumière in Paris. **(44)**

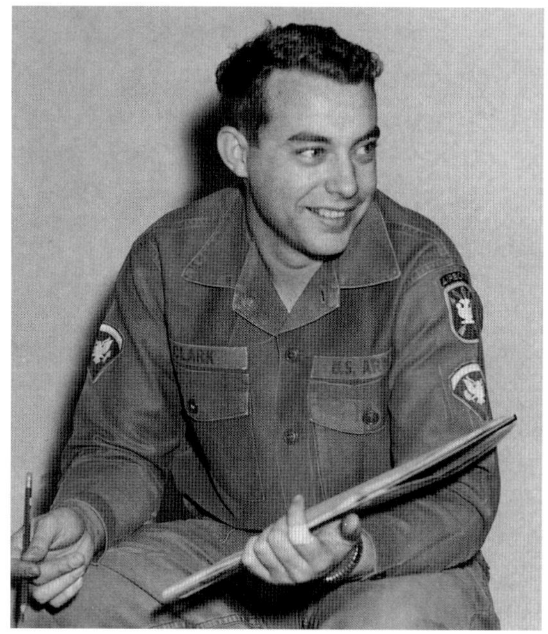

Sp5c. Brian H. Clark

Pfc. Robert T. Coleman attended college at the Kendall School of Design in Grand Rapids, Michigan. He volunteered for the draft and traveled to Vietnam as part of Combat Artist Team VI from February to March 1968. **(43)**

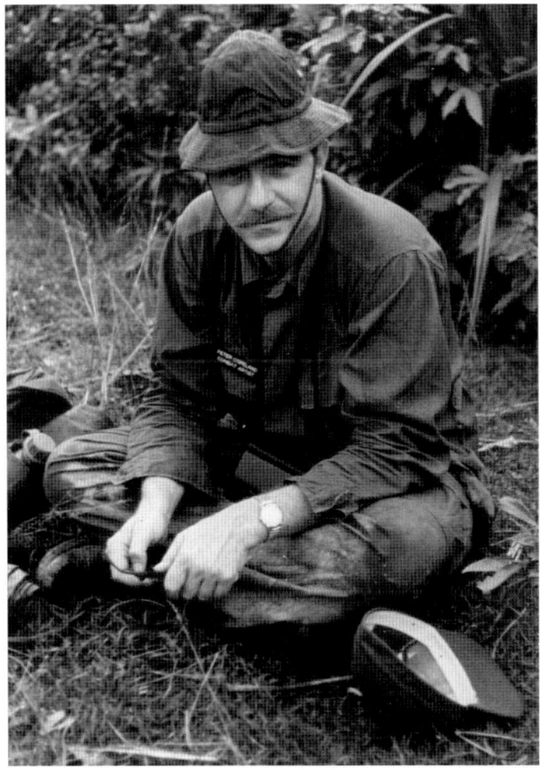

Peter F. Copeland

Peter F. Copeland was born in New York. He joined the Merchant Marines in World War II, serving on an aviation fuel tanker in the Pacific. After the war, he spent twelve more years as a Merchant Marine. He volunteered as a civilian artist for the Combat Artist Program in 1967. **(30)**

Sp4c. Michael R. Crook was born in 1941 in Lincoln, Nebraska, and was raised in Sierra Madre, California. He received his art education at Pasadena City College, Chouinard Art Institute, University of California–Los Angeles, and California State University–Los Angeles. Crook was a member of Combat Artist Team III in Vietnam. **(28, 29)**

2d Lt. James R. Drake studied painting at Leone Cooper School of Art in Webster Groves, Missouri. He received his bachelor of fine arts from the University of Cincinnati in Ohio. He was a member of Combat Artist Team VIII in Vietnam from February to June 1969. **(53, 54, 55)**

Sp4c. Theodore E. Drendel is a native of Naperville, Illinois, and majored in art at Northern Illinois University. In 1965, he enlisted in the Army. As a member of Combat Artist Team II, he documented the Vietnam War from October 1966 to February 1967. **(9)**

Capt. Heather C. Englehart was born in 1978, attended North Dakota State University, and graduated with a degree in architecture. She deployed with the Louisiana Army National Guard from 2003 to 2004 in support of Operation Iraqi Freedom. She donated her completed paintings and drawings to the Army Art Collection. **(164, 165, 166, 167, 168, 169)**

S. Sgt. Brian D. Fairchild was born in 1961 and grew up in Freeland, Pennsylvania. He enlisted in the Army as an illustrator out of high school. As a staff sergeant, he became an illustration instructor at the Graphics Documentation Course, Lowry Air Force Base, Colorado. He was selected as a member of Army Artist Team XXIX to record Army activities in Korea from February to May 1992. **(123, 124, 125)**

Sp4c. David N. Fairrington studied at Texas Technological College in Lubbock, Texas. After five years, he received a bachelor of advertising art and design. He was drafted during the Vietnam War and served two years before being selected as a member of Combat Artist Team VI. **(45)**

Sgt. Patrick Farrell graduated from Rhode Island School of Design in 1984 with a bachelor of fine arts. A year later, he entered the Army as a motion picture specialist, and from 1985 to 1988 he produced videotapes for the Training Support Center, Fort Lewis, Washington. From July to October 1988, he served as a member of Army Artist Team XXII, documenting training activities at the National Training Center in Fort Irwin, California. **(83, 84, 85)**

Sp5c. Janet Fitzgerald was born in 1949 in Portland, Maine. She has a bachelor of fine arts from Columbus College of Art and Design in Ohio. She enlisted as an illustrator in 1976 and was selected for Army Artist Team XVIII in 1979. Traveling to various locations on the East Coast, she documented women's training and Corps of Engineers activities. Following her career as a soldier, she was a civilian graphic designer for the Corps of Engineers before retiring in 2008. **(81, 82)**

Pfc. William E. Flaherty Jr. studied art at the Columbus College of Art and Design in Ohio. He freelanced as an illustrator in Columbus and Louisville, Kentucky, before joining the Army. He was a member of the 25th Infantry Division combat art program and the 18th Military History Detachment. He was selected to participate on Combat Artist Team VII in Vietnam from August to December 1968. **(46)**

Sgt. Adam Glenday received a bachelor of fine arts in commercial design from Jacksonville State University in 1995. He entered the Army in 1984 and at Fort McPherson became his unit's artist. He was assigned to the 570th Military Police Company when he was chosen to participate on Army Artist Team XXV. As a team member from June to September 1990, he recorded training experiences of the 1st Armored Division in Ansbach, Germany. **(91, 92, 93)**

Sfc. Elzie Golden was born in 1953 and entered active military service in 1984. He was assigned to Training Support Activities, Eighth Army, East, Korea, where he was an instructor in the computer graphics section. He was a sergeant when chosen in 1990 to be on Army Artist Team XXV, which documented Reserve Officers' Training Corps training at Fort Lewis, Washington. He was a senior graphics instructor at the Defense Information School at Fort Meade, Maryland. He later served as the Artist-in-Residence from 2002 to 2005. His artwork has won numerous awards, including first place in the Military Graphic Artist of the Year competition in 2000 and 2002. He studied at the School of Visual Arts in New York City and the University of Arizona. **(157, 158, 159, 160, 161, 162, 163)**

S. Sgt. Roberta Lynn Goschke was born in Newcastle, Pennsylvania, in 1954. She joined the Army in 1975 and took painting and drawing classes while stationed in Heidelberg, Germany. She graduated from Northern Virginia Community College in 1991 with a major in advertising design. She was selected for the Army Artist Program to document Army activities in Panama from September to November 1992. **(126)**

George Harding studied illustration with Howard Pyle. He was one of eight official artists sent overseas with the American Expeditionary Forces during World War I. He subsequently taught illustration at the Pennsylvania Academy of the Fine Arts. **(x)**

Pvt. James S. Hardy was born in 1944. He studied at the École des Beaux-Arts in Paris and at the Sorbonne. He received his bachelor of arts with an emphasis in graphics media from San Diego State College. In 1974, he completed graduate school at San Francisco Art Institute. He documented the Vietnam War as a member of Combat Artist Team IX. **(56)**

William C. Harrington received a bachelor of fine arts in sculpture from the University of Illinois and a master of fine arts in sculpture from the University of Hartford. He was a member of Combat Artist Team VII in Vietnam in 1968. **(47)**

Sfc. Sieger Hartgers was born in Apeldoorn, Netherlands, in 1949. He has a fine arts degree in drawing and printmaking from the Akademie Voor Beeldende Kunsten in Arnhem, Netherlands; a bachelor of fine arts in painting from the University of Colorado in Denver; and a master's degree in painting and drawing from the University of Northern Colorado. He joined the Army in 1972 as an illustrator assigned to the Sergeants Major Academy. In 1979, he was selected for an Army Artist Team to record Army jungle and desert training. He was also a graphics instructor and supervisor at the Graphics Documentation School at Lowry Air Force Base, Colorado. In 1990, he documented Operation Desert Shield. Later, he was selected as the Artist-in-Residence to chronicle training at Fort Polk, Louisiana, and the Combat Maneuver Training Center in Hohenfels, Germany, in the mid-1990s. **(xiii, 99, 100, 101, 102, 141, 142, 143)**

Horatio A. Hawks studied at the Wentworth Institute, Boston, and the Worcester Junior College in Massachusetts. He served in the U.S. Marine Corps from 1951 to 1952. He traveled to Vietnam in February 1969 as part of the Volunteer Civilian Artist Program. **(57)**

Sp4c. Joseph S. Hindley was born in 1949 and attended the Newark School of Fine and Industrial Art in New Jersey, Kansas State University, and the University of Kansas. He was assigned as a photographer to the 121st Signal Battalion at Fort Riley, Kansas, when he was selected as a team member for Army Artist Team XIII to document the REFORGER exercise in 1972. **(68, 69, 70)**

Sp4c. Steven M. Hyatt was a member of Army Artist Team XV and traveled to Panama in April 1974 to document Army activities there. He observed the Jungle Operations Training Center, the Tropical Test Center, and the Panama Canal Company. **(76, 77, 78)**

William G. Jacobson Jr. studied art at Indiana University, though the biggest influence on his artistic development was his father, a professional illustrator. He was a member of Army Artist Team XIII and traveled to Europe in 1972 to observe and document Army activities. **(71, 72, 73)**

CWO2 Chester Jezierski studied at the Rhode Island School of Design. He was a member of Army Artist Team X and documented the Army in Korea from February to June 1970. **(63)**

Pfc. Phillip W. Jones was born in 1944. He studied at the Ringling School of Art in Sarasota, Florida. He was a private at the time of his acceptance to Combat Artist Team V, which served in Vietnam from November 1967 to March 1968. **(31)**

Sp4c. Robert Knight is from Newark, New Jersey. He graduated from the Newark School of Fine and Industrial Art with a major in pictorial illustration. Shortly after graduation, he was drafted and served in 1966 in Vietnam as a member of Combat Artist Team I. **(10, 11)**

Pvt. John D. Kurtz is from Wilmington, Delaware. He studied at the Pennsylvania Academy of Fine Arts and the Accademia di Belle Arti in Florence, Italy. As a member of Combat Artist Team VI, he traveled to Vietnam in 1968 to document the war. **(48)**

Sp5c. David M. Lavender is from Daytona Beach, Florida. He graduated from a three-year commercial art course at the Ringling School of Art in Sarasota, Florida. In October 1966, he was sent to Vietnam as a member of Combat Artist Team II. **(12, 13, 14)**

M. Sgt. Timothy B. Lawn deployed to Iraq as part of Operation Iraqi Freedom from 2004 to 2005 as an Army photographer, print journalist, and combat illustrator and donated his watercolors and drawings to the Army Art Collection. After deployment, he served in the U.S. Army Reserve with the Special Operations Command in Tampa, Florida. When this book went to press, he was deployed to Afghanistan, serving as the public affairs noncommissioned officer in charge with the 1st Theater Sustainment Command at New Kabul Compound, Afghanistan. He has contributed art related to this latest deployment to the Army Art Collection. He was due to rotate home in early summer of 2014. **(170, 171, 172, 173)**

Pfc. Gary Lewis was born in Goldsboro, North Carolina, in 1967. He studied at Southern Wayne Senior High, Goldsboro, North Carolina. He received the Army Achievement Medal for graphic arts at Fort Gordon, Georgia, and the Commander's Award for mural painting at Fort Sam Houston, Texas. He held the rank of private first class when he served as a member of an Army Artist Team attached to the 116th Military History Detachment for the REFORGER exercise in 1987. **(82)**

Sp4c. Daniel T. Lopez is from Fresno, California. He studied art at Fresno City College and Fresno State College before he was drafted. He served as a member of the Combat Artist Team IV in Vietnam from August to October 1967. **(32, 33)**

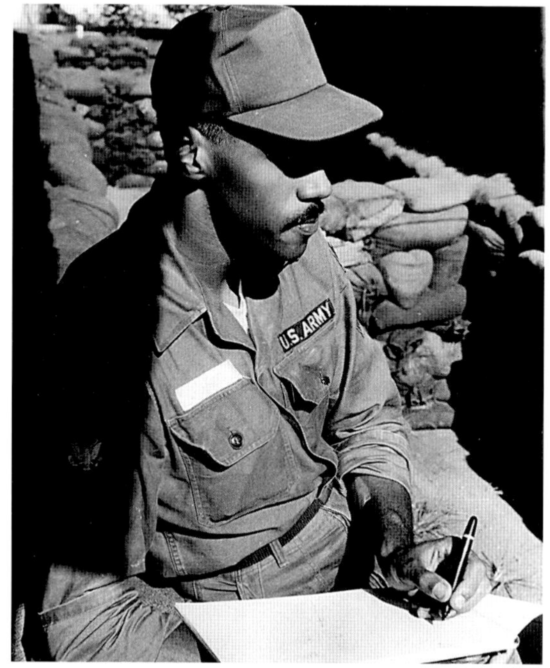

Sp4c. Robert Knight

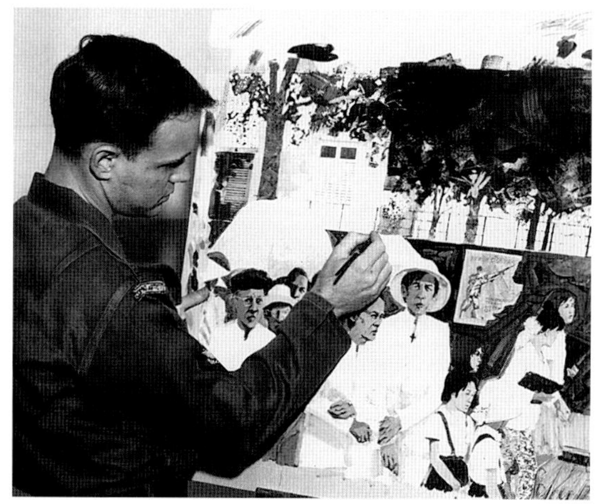

Sp5c. David M. Lavender

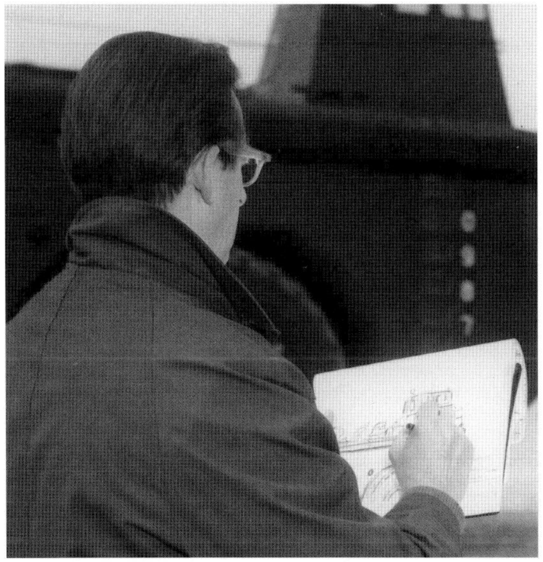

Thomas C. O'Hara

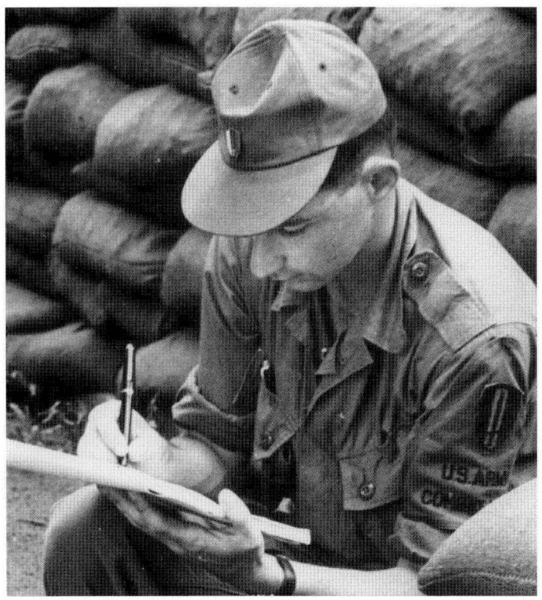

Pfc. Ronald E. Pepin

Sgt. Jeffrey T. Manuszak was born in Hammond, Indiana, in 1968. He studied at the Ray College of Design in Chicago before enlisting in the Army as an administrative specialist in 1992. He was stationed at the Pentagon in the U.S. Army Chief of Staff's office when selected by artist Peter Varisano to be part of the first full-time Army Art Program. His assignments included documenting Army activities in Haiti, training at Fort Polk, and activities of the air assault school, Fort Campbell, Kentucky. After separating from the Army, he earned his graduate degree in educational design. **(114, 133, 134, 135, 136)**

Stephen J. Matthias was born in Washington, D.C., in 1935. He studied at Brown University, the Rhode Island School of Design, American University, and the École des Beaux-Arts in Paris. He served two years in the U.S. Navy during which time he was the personnel director for the Light Photographic Squadron on the USS *Randolph* in the Mediterranean. He received a bachelor of arts in architecture from the University of Virginia and was working at an architecture firm in Baltimore when he traveled to Vietnam as part of the Volunteer Civilian Artist Program in December 1967. **(49)**

Pfc. Dennis O. McGee was born in 1940 in Chicago. He received a bachelor of arts in commercial art from San Jose State University in 1964. He was serving in Vietnam with a tank battalion near Long Binh when he was selected to be a member of Combat Artist Team III. **(34)**

Sp5c. Burdell Moody is from Mesa, Arizona, and attended Arizona State University. In 1966, he was working toward his bachelor's degree in secondary art education when he was drafted. He was a member of Army Artist Team IV in Vietnam from August to October 1967. **(36, 37)**

Sp5c. Robert T. Myers studied at the Pennsylvania Academy of Fine Arts from 1958 to 1962. In 1961, he received the William Emlen Cresson European travel scholarship to study fine art abroad. He served in Vietnam as a member of Army Artist Team III from February to June 1967. **(35)**

Thomas C. O'Hara graduated from Cambridge School of Art and received a bachelor of fine arts from Rhode Island School of Design in 1947. He spent August 1968 in Vietnam as a participant in the Volunteer Civilian Artist Program. **(64)**

Sp5c. Michael P. Pala is from Bridgeport, Connecticut, and studied art at the Warren Harding School. Prior to the Vietnam War, he was an apprentice industrial engraver in Bridgeport. As a member of Combat Artist Team VI, he documented the Army in Vietnam from February to March 1968. **(50)**

Pfc. Ronald E. Pepin, born in 1943, was a member of Combat Artist Team I in Vietnam. Prior to entering the Army, he was an electronic draftsman, a technical illustrator, and a freelance artist. **(15, 16)**

Carlton Plummer was born in Brunswick, Maine. He graduated from the Vesper School of Art and later received a bachelor's degree from the Massachusetts College of Art and a master of fine arts from Boston University of Art. In 1969, he traveled to Bangkok, Thailand, as a participant in the Volunteer Civilian Artist Program. **(58)**

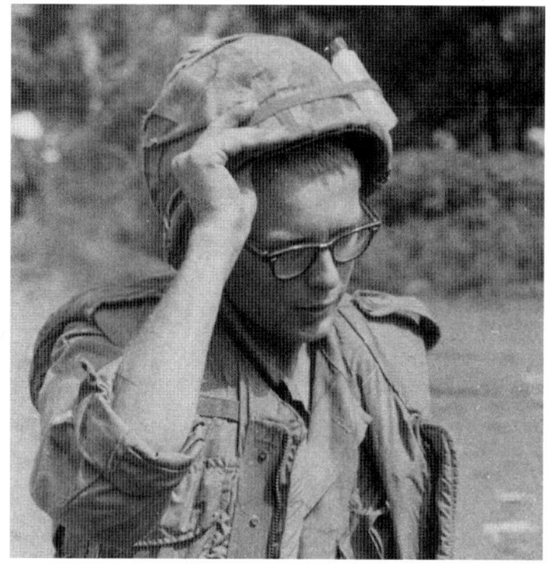

Sp4c. James Pollock

Sp4c. James Pollock was born in South Dakota in 1943. He graduated from South Dakota State University in 1965 with a major in art. He served on Combat Artist Team IV from 1966 to 1967. **(38, 39)**

Pfc. Gary W. Porter studied at San Diego State College. He was a member of Combat Artist Team II, which spent October 1966 to February 1967 in Vietnam. **(17)**

S. Sgt. Roger W. Price graduated from high school in Reidsville, North Carolina, and immediately joined the Army and worked in design, layout, illustration, and photography. As a member of Army Artist Team XXV, he documented 1st Armored Division training in Germany and the fiftieth anniversary of World War II from June to September 1990. **(94, 95)**

Sp4c. Stephen H. Randall earned a bachelor's degree in applied art from Iowa State University. From August to December 1968, he was a member of Combat Artist Team VII in Vietnam. **(51)**

Sgt. Raymond R. Reuter was born in North Wildwood, New Jersey. He earned a bachelor's degree in art education from Marshall University, Huntington, West Virginia. He was serving with the 101st Airborne Division when selected in 1975 for Army Artist Team XVI. The team's focus was documenting the resettlement activities at the Vietnamese refugee camp at Fort Chaffee, Arkansas, and the U.S. Army in Japan. **(79, 80)**

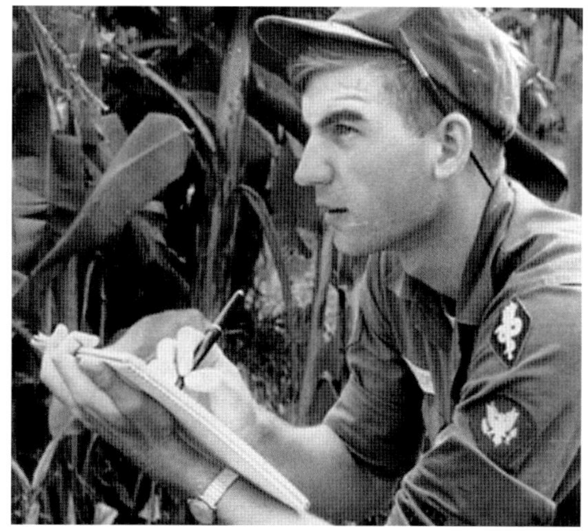

Sp4c. Paul Rickert

Sp4c. Paul Rickert was born in Philadelphia in 1947. He studied art with his father, William Rickert, and with the artist Nelson Shanks. He also attended the Los Angeles Art Center College of Design. Rickert was a member of Combat Artist Team I in Vietnam from August to December 1966. **(18)**

Col. Robert B. Rigg received his art training from his mother, Irene Rigg, as well as some limited training from the Art Institute of Chicago. He painted a number of scenes based on his time in Vietnam in 1963, which he donated to the Army Art Collection. His work is the only eyewitness art in the collection from Vietnam that predates the start of the Combat Artist Program in 1966. **(40)**

Sp5c. Felix R. Sanchez studied commercial art at the Chicago School of Fine Arts and life drawing at the Chicago Art Institute. He enlisted in October 1964 and served in Vietnam as a member of Combat Artist Team I. **(19)**

Maj. Robert Sankner was born in Carteret, New Jersey, in 1957. He attended Newark State College and graduated in 1973 with a bachelor of fine arts. Sankner enlisted in October 1975 and later completed officer candidate school and was commissioned in 1976. As a major in the U.S. Army Reserve, he was selected for Army Artist Team XXX to document the Army in the Canal Zone, Panama, from August to November 1992. **(127, 128, 129, 130)**

Sp6c. Kenneth J. Scowcroft studied at the Famous Artist Commercial School and the Graphic Arts Facility at Fort Huachuca, Arizona, and received a degree from the University of Arizona. He was a member of Combat Artist Team III, which served in Vietnam from February to June 1967. **(41)**

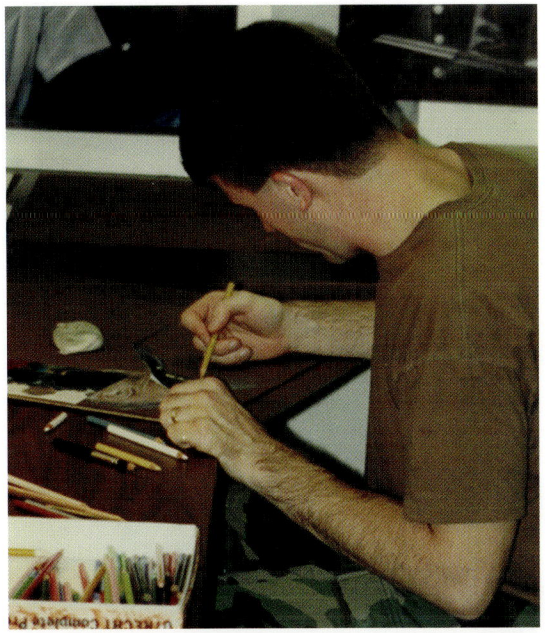

Sgt. Carl E. "Gene" Snyder

M. Sgt. Henrietta M. Snowden joined the Army in 1982. In 1991, she was one of several artists who volunteered to chronicle homecoming parades for veterans of the conflict in the Persian Gulf and reunions of those soldiers with their families. She was Artist-in-Residence from 1999 to 2002 and deployed to Kosovo to record day-to-day activities and documented the damaged Pentagon following the 11 September 2001 attacks. **(xv, 118, 148, 149, 150, 151, 155, 156)**

Sgt. Carl E. "Gene" Snyder was born in Baltimore, Maryland, in 1968. He enlisted in the Army as an infantryman in 1989 and reenlisted as a graphic illustrator in 1991. He was a corporal, managing the graphics office of The Old Guard in Washington, D.C., when selected by artist Peter Varisano to become a resident artist for the Army Art Program from 1994 to 1997. He was later a graphics instructor for the Defense Information School at Fort Meade, Maryland, from 1997 to 2000. He has a bachelor's degree in studio art and art history from Vermont College–Norwich University. **(xiv, 115, 137, 138, 139, 140)**

Al Sprague was born in 1938 in Colon, Panama. He received his bachelor's degree from American University in Washington, D.C., in 1962 and his master's degree in 1965. He has taught art to high school and college students in the United States and Panama. As a volunteer civilian artist, he spent twelve days in Panama documenting Operation Just Cause in 1989. **(xiii, 86, 87, 88)**

Pfc. William H. Steel graduated from Redwood High School, Larkspur, California, in 1964 and the California School of Arts and Design in Oakland, California, in 1968. He was a member of Combat Artist Team X from February to June 1970 to document U.S. Army activities in Korea. **(65, 66, 67, 188)**

2d Lt. Craig L. Stewart has a bachelor's degree in agriculture from the University of Washington. He was a member of Combat Artist Team IX in Vietnam from August to December 1969. **(59)**

M. Sgt. Christopher W. Thiel is a native of Smithtown, New York. He was Artist-in-Residence from 1996 to 1999 and documented Army activities in Bosnia-Herzegovina and Army training throughout the United States. He returned as Artist-in-Residence in 2007 after graduating from the Sergeants Major Academy and deployed to Iraq to record Operation IRAQI FREEDOM. He retired from the Army in 2008. **(144, 145, 146, 147, 174, 175, 176, 177)**

Lt. Col. Frank M. Thomas was born in February 1939 in La Habra, California. He studied art and education at Brigham Young University and the University of South Carolina. He was in the U.S. Army Reserve when called to six months' active duty to document DESERT STORM. **(112, 113)**

Sfc. Peter G. Varisano was born in 1956 in Kane, Pennsylvania, and entered the Army in 1974. He was a graphics instructor at the basic and advanced courses at the Noncommissioned Officer Illustrators School, Lowry Air Force Base, Colorado, from 1984 to 1991. He participated in two art teams: the first in 1989 documenting training with the Wisconsin Army National Guard, and the second in 1990 covering Operation DESERT SHIELD in the Persian Gulf. In 1991, he was selected by Army Chief of Staff General Gordon R. Sullivan to be the first full-time Army Artist-in-Residence. During this assignment, he documented Army activities in Florida following Hurricane Andrew and operations in Somalia in 1993 and in Rwanda in 1995. He earned a bachelor's degree in studio art from Vermont College–Norwich University in 1994 and a master's degree from the Vermont College of Fine Arts in 1997. After retiring from the Army, he became director of graphics instruction at the Defense Information School at Fort Meade, Maryland. In 2001, he was the only artist allowed into the security perimeter of the World Trade Center in New York after the 11 September 2001 attacks to paint search-and-rescue efforts. **(90, 103, 104, 105, 106, 131, 132)**

Sp4c. William F. Voiland was born in 1948 and educated at Washington State University, Pullman, Washington. While still in college, he started his own design business, creating silkscreen posters, commissioned artwork, and portraits. He was drafted in 1971. As a member of Army Artist Team XIII, he documented Army activities in Germany from October to December 1972. **(74, 75)**

Sp4c. Victory Von Reynolds was born in 1942 and received a bachelor's degree and a master of fine arts from the University of Idaho. He later taught at the University of Nevada, the University of Alabama, and Wayne State College in Nebraska. He enlisted in the Idaho Army National Guard in 1966, and, while serving in Vietnam in the 116th Engineer Battalion, he was selected to be part of Combat Artist Team VIII from February to June 1969. **(60)**

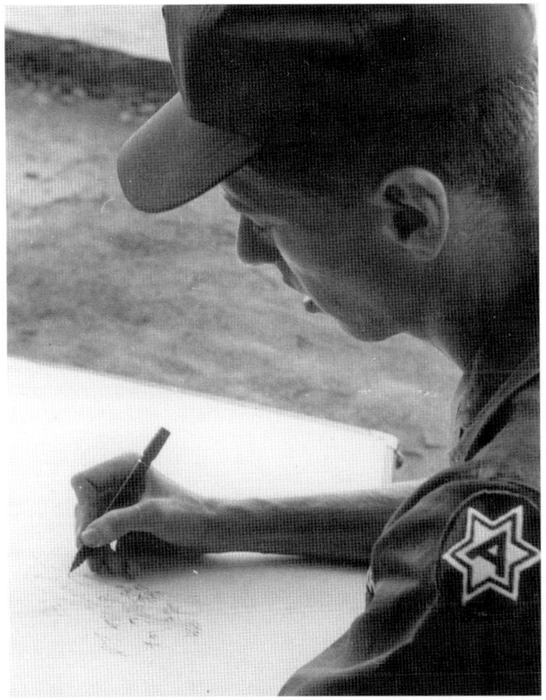

1st Lt. John O. Wehrle

James Walker was born in 1818 in England. He was largely a self-taught artist. He was living in Mexico City in 1846 when the Mexican War began. He offered his services as interpreter to the staff of Brig. Gen. William J. Worth and was present at the battles of Contreras, Churubusco, and Chapultepec and the capture of Mexico City. In 1864, he spent several months with the Army of the Cumberland and ultimately was commissioned by Maj. Gen. Joseph Hooker to paint the *Battle of Lookout Mountain.* **(ix)**

1st Lt. John O. Wehrle was born in 1941 in San Antonio, Texas. He received his bachelor's degree at Texas Technological College in Lubbock, Texas, in 1964 and a master of fine arts in painting and printmaking from the Pratt Institute, Brooklyn, New York, in 1969. He was the officer in charge of Combat Artist Team I in Vietnam from August to December 1966. **(20, 21, 22, 23)**

Lawrence A. Westby received a degree in advertising design from the Colorado Institute of Art. He was a member of Army Artist Team X and documented the Army in Korea in 1970. **(67)**

John Potter Wheat was born in 1920 and studied at the Art Students League at the Yale School of Fine Arts and graduated with a master of fine arts. His works are in the New Britain Museum of American Art in New Britain, Connecticut; the Springfield Museum of Fine Arts in Springfield, Massachusetts; and the Butler Institute of American Art in Youngstown, Ohio. During World War II, he was an artist with the Office of Strategic Services. He was a civilian artist for the Army in Vietnam from February to March 1968 through the Copley Society of Boston. **(52)**

Sp4c. Ronald A. Wilson was born in Los Angeles, California, in 1941. He studied commercial art at Brigham Young University. He was a member of Combat Artist Team IV in Vietnam from August to December 1967. **(42, 43)**

PIN : 104685-000